Circular Economy and Sustainability

This book series aims at exploring the rising field of Circular Economy (CE) which is rapidly gaining interest and merit from scholars, decision makers and practitioners as the global economic model to decouple economic growth and development from the consumption of finite natural resources. This field suggests that global sustainability can be achieved by adopting a set of CE principles and strategies such as design out waste, systems thinking, adoption of nature-based approaches, shift to renewable energy and materials, reclaim, retain, and restore the health of ecosystems, return recovered biological resources to the biosphere, remanufacture products or components, among others.

However, the increasing complexity of sustainability challenges has made traditional engineering, business models, economics and existing social approaches unable to successfully adopt such principles and strategies. In fact, the CE field is often viewed as a simple evolution of the concept of sustainability or as a revisiting of an old discussion on recycling and reuse of waste materials. However, a modern perception of CE at different levels (micro, meso, and macro) indicates that CE is rather a systemic tool to achieve sustainability and a new eco-effective approach of returning and maintaining waste in the production processes by closing the loop of materials. In this frame, CE and sustainability can be seen as a multi-dimensional concept based on a variety of scientific disciplines (e.g., engineering, economics, environmental sciences, social sciences). Nevertheless, the interconnections and synergies among the scientific disciplines have been rarely investigated in depth.

One significant goal of the book series is to study and highlight the growing theoretical links of CE and sustainability at different scales and levels, to investigate the synergies between the two concepts and to analyze and present its realization through strategies, policies, business models, entrepreneurship, financial instruments and technologies. Thus, the book series provides a new platform for CE and sustainability research and case studies and relevant scientific discussion towards new system-wide solutions.

Specific topics that fall within the scope of the series include, but are not limited to, studies that investigate the systemic, integrated approach of CE and sustainability across different levels and its expression and realization in different disciplines and fields such as business models, economics, consumer services and behaviour, the Internet of Things, product design, sustainable consumption & production, bio-economy, environmental accounting, industrial ecology, industrial symbiosis, resource recovery, ecosystem services, circular water economy, circular cities, nature-based solutions, waste management, renewable energy, circular materials, life cycle assessment, strong sustainability, and environmental education, among others.

Harshita Jain · Osheen Martolia ·
Anamika Shrivastava
Editors

The Circular Fallacy: Closing the Loop in a Consumerist World

Springer

Editors
Harshita Jain
Amity Institute of Environmental Sciences (AIES)
Amity University
Noida, Uttar Pradesh, India

Osheen Martolia
University of Southampton
Southampton, UK

Anamika Shrivastava
Amity Institute of Environmental Sciences (AIES)
Amity University
Noida, Uttar Pradesh, India

ISSN 2731-5509 ISSN 2731-5517 (electronic)
Circular Economy and Sustainability
ISBN 978-3-032-22686-0 ISBN 978-3-032-22687-7 (eBook)
https://doi.org/10.1007/978-3-032-22687-7

This Springer imprint is published by the registered company Springer Nature Switzerland AG
The registered company address is: Gewerbestrasse 11, 6330 Cham, Switzerland

To those who dare to question the norm,
Who believe sustainability is not a choice but a responsibility,
And who work tirelessly—through research, policy, innovation, and awareness—To shape a world where progress does not come at the cost of the planet.

We also dedicate this book to our families, mentors, and students—For their unwavering support, inspiration, and belief in our vision. Your encouragement fuels our purpose.

—Harshita Jain and Anamika Shrivastava

Foreword

In an era defined by escalating environmental crises, rapid technological advancement, and relentless consumerism, the idea of a circular economy has emerged as a promising solution to global waste and sustainability challenges. The notion of "closing the loop" offers a compelling vision, one where waste is minimized, resources are reused, and economic growth harmonizes with environmental responsibility. However, as the contributors of this book insightfully argue, the circular economy is not without its limitations, contradictions, and blind spots.

The Circular Fallacy: Closing the Loop in a Consumerist World presents a timely and critical examination of the real-world applicability of circularity. This book challenges the utopian perception of a seamless loop by interrogating the technological, social, psychological, and policy-based constraints that hinder its true implementation. It uniquely goes beyond the surface-level narratives of sustainability to delve into underexplored dimensions such as algorithmic waste, product obsolescence, behavioral economics, and biomimicry.

The interdisciplinary approach adopted by the editors, Dr. Harshita Jain, Dr. Anamika Shrivastava, and Ms. Osheen Martolia, brings together diverse perspectives rooted in environmental science, public policy, and practical implementation. Their effort to integrate global trends with grounded realities in developing nations provides a much-needed balance between theory and practice.

This book is not merely a critique, it is a call to action. It urges scholars, practitioners, policymakers, and consumers to rethink prevailing paradigms, ask uncomfortable questions, and explore innovative pathways beyond conventional circularity. By fostering such critical discourse, this volume contributes meaningfully to the global conversation on sustainability.

I commend the editors and contributors for their bold, reflective, and well-researched work. This book will undoubtedly serve as an essential resource for academics, professionals, and anyone concerned with building a more sustainable and just future.

Dr. Renu Dhupper
Director and Professor
Amity Institute of Environmental
Sciences (AIES)
Amity University
Noida, Uttar Pradesh, India

Preface

The concept of a circular economy has gained significant traction over the past decade, celebrated as a promising antidote to the environmental consequences of linear consumption and waste generation. Governments, industries, and academic institutions have embraced the idea of "closing the loop," a vision where materials are reused, recycled, and reintegrated seamlessly into production cycles. However, this idealized model often overlooks the socio-economic, psychological, and infrastructural challenges that complicate its practical realization.

This book, *The Circular Fallacy: Closing the Loop in a Consumerist World*, emerged from a collective urge to critically assess the true potential and limitations of circularity in the real world. Rather than accepting the model at face value, we seek to explore its unspoken assumptions, technological constraints, and the broader cultural and behavioral factors that influence waste and consumption.

The chapters included in this volume offer a multidisciplinary perspective, spanning environmental sciences, public policy, industrial ecology, consumer behavior, and sustainability studies. Our goal is not to discredit the circular economy but to enrich the conversation by addressing its oversights and complexities. We have also chosen to spotlight often-ignored areas such as e-waste, construction and demolition waste, waste from data centers, and even the possibility of biomimicry as a regenerative model that nature has already mastered.

As editors, we bring together voices from academia and practice, from the Global North and South, and from diverse disciplinary backgrounds. We believe this book will be valuable for researchers, practitioners, policymakers, and students who are seeking to understand not just the "what" of circularity but also the "why" and "how" of its application in a consumer-driven world.

We are grateful to all the contributing authors for their commitment and scholarly insights. We also extend our appreciation to Springer for recognizing the importance of this work and providing us with a platform to publish it.

We hope this book sparks meaningful dialogue, provokes critical thought, and encourages practical, inclusive, and future-oriented strategies for a truly sustainable future.

Noida, India Dr. Harshita Jain
Noida, India Dr. Anamika Shrivastava
Southampton, UK Ms. Osheen Martolia

Acknowledgements

We extend our heartfelt gratitude to all those who contributed directly or indirectly to the development of this book. This collaborative work would not have been possible without the support, guidance, and encouragement of several individuals and institutions. First and foremost, we thank Amity University, Noida and the University of Southampton for providing an enabling academic environment, access to research resources, and continuous encouragement throughout the conceptualization and development of this book. Their commitment to interdisciplinary research and sustainability has deeply influenced the spirit of this volume.

We express our sincere appreciation to our colleagues, research scholars, and students, whose insightful discussions, questions, and feedback helped shape several chapters of this manuscript. Their active engagement has been an invaluable asset. We also acknowledge the reviewers and editorial team at Springer for their interest in our work, and for providing a platform to bring this critical discourse on circularity and sustainable consumption to a broader audience. Lastly, we are indebted to our families for their unwavering support, patience, and understanding during the writing and compilation of this volume. Their constant encouragement enabled us to persist through the complexities of the publishing process.

This book is dedicated to all those working towards a sustainable and equitable future—questioning norms, challenging fallacies, and striving for meaningful change.

Dr. Harshita Jain
Assistant Professor, Amity Institute of
Environmental Sciences (AIES)
Amity University
Noida, Uttar Pradesh, India
hjain@amity.edu

Dr. Anamika Shrivastava
Assistant Professor, Amity Institute of
Environmental Sciences (AIES)
Amity University
Noida, Uttar Pradesh, India
ashrivastava1@amity.edu

Ms. Osheen Martolia
University of Southampton
Southampton, UK
o.martolia@soton.ac.uk

Contents

Editors and Contributors

About the Editors

Dr. Harshita Jain an Assistant Professor at the Amity Institute of Environmental Sciences, Amity University Noida, specializes in sustainable waste management. She holds a Ph.D. in Environmental Sciences, and has an M.Sc. in Environmental Sciences and a B.Sc. Hons. in Botany. Dr. Jain has made significant contributions through her extensive publications in SCI/SCIE journals and international conferences, covering groundwater remediation, sustainable waste management, climate change impact assessment, and machine learning in disaster management. She has authored several books, including *Sustainable Solid Waste Management* and other environmental science texts. Her research achievements also include patent filings. Dr. Jain actively supervises Ph.D. scholars, M.E./M.Tech dissertations, reviews international journal papers, and engages in community service and continuous learning. She is trained in environmental management systems and disaster resilience and participates in faculty development programs to promote innovative teaching and sustainable development. Dr. Jain is a dedicated advocate for sustainable waste management and environmental sustainability, committed to ensuring a healthy environment for future generations.

Ms. Osheen Martolia is an experienced Environmental Consultant with a strong background in environmental project management. With over 5 years of expertise in the field, Osheen has demonstrated a remarkable ability to foster collaboration among project stakeholders, technical specialists, subcontractors, and clients to ensure effective communication and coordination within multidisciplinary teams. Notable achievements include leading the development and implementation of a strategic greening and plantation action plan that resulted in the successful plantation of 30 million trees across Delhi NCR in 2023. Osheen is known for a proactive approach to ensuring successful project outcomes and has a proven track record in air quality compliance, leading to significant reductions in violations and public complaints. With robust interpersonal skills and a dedication to sustainable practices, Osheen continues to make a positive impact in the environmental sector.

Dr. Anamika Shrivastava is an accomplished professional with over 6 years of expertise in Environment Analysis, Climate Change, ESG, Sustainable Development, and Sustainability. She has demonstrated a result-oriented approach throughout her career and possesses strong communication skills with international clients. Dr. Shrivastava has played a key role in providing solutions across projects, displaying proficiency in report writing, teaching, research, and publication. Her skills extend to cutting-edge instrumentation and laboratory techniques, particularly in sample preparation and analysis of soil, water, and plants. Dr. Shrivastava is recognized for her creative thinking and possesses honed interpersonal, problem-solving, and analytical abilities.

Contributors

Tajudeen Alaburo Abdulsalam Department of Management and Marketing, KFUPM Business School, Dhahran, Saudi Arabia

Ghaith Al-Abdallah University of Kurdistan Hewlêr, Erbil, Iraq, and Catholic University in Erbil, Erbil, Iraq

Meenam Bhatia Department of Botany, Daulat Ram College, University of Delhi, Delhi, India

Niketa Bhati Amity Institute of Environmental Sciences, Amity University Uttar Pradesh, Noida, India

Digambar Chavan CII - ITC Center of Excellence for Sustainable Development, New Delhi, India

Renu Dhupper Amity Institute of Environmental Sciences, Amity University Uttar Pradesh, Noida, India

Sakshi Garg Amity Institute of Environmental Sciences, Amity University, Noida, Uttar Pradesh, India

Harshita Jain Amity Institute of Environmental Sciences, Amity University Uttar Pradesh, Noida, India

Shikhar Jain CII - ITC Center of Excellence for Sustainable Development, New Delhi, India

Karina Bhatia Kakkar School of Management, GD Goenka University, Sohna, Haryana, India

Dominic R. Karuthedath Christ University, Bengaluru, Karnataka, India

Yanshik Khatri School of Management, GD Goenka University, Sohna, Haryana, India

J. Kishore Kumar Christ University, Bengaluru, Karnataka, India

Navya Vikraman Nair School of Environment, Enterprise, and Development (SEED), University of Waterloo, Waterloo, ON, Canada;
Department of Civil Engineering, SCMS School of Engineering and Technology, Karukutty, Ernakulam, Kerala, India

Jyoti Saini Integrated Centre for Adaptation to Climate Change, Disaster Risk Reduction and Sustainability (ICARS), Indian Institute of Technology Roorkee (IIT Roorkee), Greater Noida, Uttar Pradesh, India

Salman Shooshtarian School of Property, Construction and Project Management, RMIT University, Melbourne, Australia

Anamika Shrivastava Amity Institute of Environmental Sciences, Amity University Uttar Pradesh, Noida, India

Lovepreet Singh Department of Chemical Engineering, Thapar Institute of Engineering and Technology, Patiala, India

Pooja Singh CII - ITC Center of Excellence for Sustainable Development, New Delhi, India

Shanu Singh Christ University, Bengaluru, Karnataka, India

Sanju Sreedharan Department of Civil Engineering, SCMS School of Engineering and Technology, Karukutty, Ernakulam, Kerala, India

Surbhi Srivastava Christ University, Bengaluru, Karnataka, India

Monika Tiwari GSL Associates Pvt Ltd, Saidulajab, New Delhi, India

Chapter 1
Global Waste Problems, Rise of Consumerism, Waste Management Models and Their Limitation

Pooja Singh, Digambar Chavan, and Shikhar Jain

1.1 Introduction

The solid waste generation and its management is one of the major global issues. The rapid urbanisation, industrialisation, financial abundance and population growth are major factors contributing to the generation of solid waste (Zhang et al. 2024). The quantity of the waste generated is positively correlated with the economic growth of the nation, population explosion, urbanization and industrialization (Rana et al. 2015). The waste generated is no more of economic value mostly ends up into landfills. It is estimated that world generates about 2 billion tonnes of municipal solid waste annually and it is expected to increase by 70% by 2050 (IFC 2024). The urbanization as one of the major causes of rise in waste and the urban population is expected to constitute nearly 70% of the total population by 2050 (Kurniawan et al. 2022). Climate change is global crisis mainly due to the higher emissions of greenhouse gases (GHGs). Waste is responsible for 20% of the world's human-related methane emissions (IFC 2024) contributes to climate change. On a global scale a major portion of municipal solid waste (MSW) accounts for 44% followed by recycling waste 38 percent and other waste 18 percent (Yatoo et al. 2024). Unscientific disposal of solid waste leads to many environmental and health hazards. The improper solid waste management cause air, water, land pollution and responsible for GHG emissions and climate change (Abubakar et al. 2022). The main limitation

P. Singh (✉) · D. Chavan · S. Jain
CII - ITC Center of Excellence for Sustainable Development, 3rd Floor, Andhra Association, 24-25 Lodi Institutional Area, Lodi Road, New Delhi 110003, India
e-mail: pooja@cii.in

D. Chavan
e-mail: digambar.chavan@cii.in

S. Jain
e-mail: shikhar.jain@cii.in

H. Jain et al. (eds.), *The Circular Fallacy: Closing the Loop in a Consumerist World*, Circular Economy and Sustainability, https://doi.org/10.1007/978-3-032-22687-7_1

in optimising sustainable waste management practices includes many considerations as socio economic aspects, technological & financial limitations, cultural considerations and policy implications (Zhang et al. 2024). In India the per capita average waste generation is 123. 45 gms/day, the highest waste generation varies according to cities highest is recorded for the state of Delhi (526.5 gms/capita/day) and lowest for Meghalaya (21.0 gms/capita/day) (CPCB 2023). The population increase combined with increased economic power is responsible for generation of large quantities of MSW (Bolton and Rousta 2019. In European countries per habitant waste generation was reported to be highest for Austria (Table 1.1) (Statista 2022).

The rapid urbanisation and industrialisation in developing countries causes a major challenge in management of the solid waste. Apart from rapid urbanisation financial constraints, in efficient policy regulation, lack of advanced technology is considered as major factors for poor solid waste management. Environmental degradation is one of the major negative impacts due to mismanagement of solid waste (Muheirwe et al. 2022) and cause various form of pollutions (air, water, & land). The municipal solid waste consists of major portion of organic waste which is mainly responsible for GHG emissions. The improper management of solid waste may lead to contamination of river, oceans, and occurrence of floods and vector borne diseases (Hoang and Fogarassy 2020). The improper solid waste management impacts human includes various health issues such waterborne infections such as intestinal worms, diarrhoea, dengue fever, hepatitis A, leptospirosis, and Zika virus (Penteado and de Castro 2021). The open dumping of waste is considered as an inexpensive way to dispose of waste or

Table 1.1 The per capita municipal solid waste generation in European countries (Statista 2022)

S. no	Global North countries	Average municipal solid waste generation in Kg in Europe /per habitant/per year
1	Austria	835
2	Denmark	787
3	Norway	768
4	Luxembourg	721
5	Belgium	677
6	Switzerland	677
7	Cyprus	673
8	Iceland	659
9	Finland	630
10	Malta	618
11	Germany	593
12	France	539
13	Greece	509
14	Italy	495
15	Coratio	478
16	Slovakia	478

into landfills creating potential risk to the nearby inhabitants (Mahanta and Rafizul 2019). The open landfills and incinerator near residential areas cause health risks to the nearby residents. The effective solid waste management is only a solution to mitigate the impact of solid waste on environment and human health. The mismanagement of waste including improper sorting, transportation and dumping leading to many consequences as shown in Fig. 1.1 (Zhang et al. 2024). A more balanced waste management approach focusses on practices such as decreased production, waste categories, reuse, recycling, and energy recovery over the common practices of landfilling, open dumps, and open incineration (Kabera and Nishimwe 2019). Despite the extensive efforts in numerous WM frameworks, existing WM models sometimes operates in silos, which restricts their capacity to provide comprehensive solutions for the diverse and complicated nature of modern waste streams. Advanced WM models focuses on traditional techniques which focuses on accounting environmental impact though life-cycle assessment (LCA), cost optimization using linear programming optimization (LPO) and policy driven waste diversion programs like zero waste to landfills (ZWL). It has been observed that implementing individual waste management (WM) models may lack comprehensive, data-driven insights needed for system-wide optimization, resulting in fragmented outcomes and limited scalability. Therefore, this chapter provides a core framework for incorporating optimization, environmental assessment, and policy-oriented certification into a unified waste management strategy by defining model complementarities. The future view calls for a hybrid paradigm that combines the analytical rigor of LPO, the sustainability measures of LCA, and the implementation strength of ZWL. This integrated strategy can considerably improve the performance of trash diversion initiatives, assist evidence-based policy formulation, and help industry and municipalities meet long-term sustainability goals. The concepts presented in this chapter aim to spark

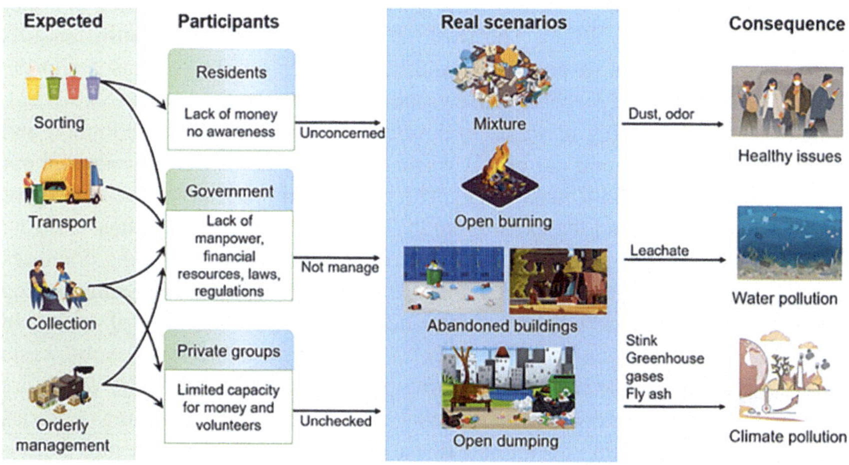

Fig. 1.1 Environmental challenges from inappropriate treatment of MSW (Zhang et al. 2024)

interdisciplinary collaboration and drive future research and practice toward more resilient, adaptable, and circular waste management systems.

1.1.1 Global Waste Problems

The nature of solid waste arising from developed and developing countries is variable therefore generalised waste management practice cannot be implemented. In developed countries only 2 percent of the waste is dumped while in developing or low-income countries approximately 93% of the waste is dumped (Yatoo et al. 2024). Similarly, it was estimated that low-income countries only collect 10% of the waste generated in sub urban areas that results in various environment consequences (Abubakar et al. 2022). The municipalities in low-income countries spend about 20% of their municipal budget in managing solid waste, middle income and high-income countries accounts for 10% and 4% respectively (Kaza et al. 2018). The municipal budget is governed by local authorities with limited funds and partial capacity for planning, narrow management, and operational monitoring (Kaza et al. 2018). The majority of MSW and sewage are collected, reused, and recycled in industrialised nations, but contrarily most developing nations do not yet have access to collection and proper disposal, much alone well-designed municipal treatment facilities (Stafford 2020). Waste generation for developing or low-income countries would be about more than three times of the high-income countries (Kaza et al. 2018). The waste generated in high-income countries relative more of dry type of waste that can be recycled while low- and middle-income countries generates more of food waste and green waste and less amount of dry waste that can be recycled.

In developing countries unscientific method of waste disposal is often practiced as an open dumping, burning, throwing garbage near communities and in water bodies (Abubakar et al. 2022). The waste is generally dumped off into Landfills as the easiest method of disposal, but landfills generate lot of toxic gases into the atmosphere and released leachate cause groundwater and land pollution (Swati et al. 2019). The open dumping of waste in developing countries is quite common and scavengers make livelihood out of this waste by sorting and collecting material out of this waste (Bolton and Rousta 2019). The land constraint restricts the buildup of new landfills in cities and the existing landfills have already reached their carrying capacities. The wealth gap within the different population in developing countries influence the waste generation pattern. The economic status is directly influencing the disposal method pattern than impoverished populations (Kalina et al. 2024). In most of the developing countries due to the lack of infrastructure for the solid waste management, most of the solid waste is dumped in landfills or dumpsite which are about to reach carrying capacities (Abubakar et al. 2022). The developing countries for e.g. South Africa, exhibit economic disparity and prominent wealth gaps defines the efficient behaviour of affluent population in managing solid waste, while poor faces many challenges of improper solid waste management (Kalina et al. 2024).

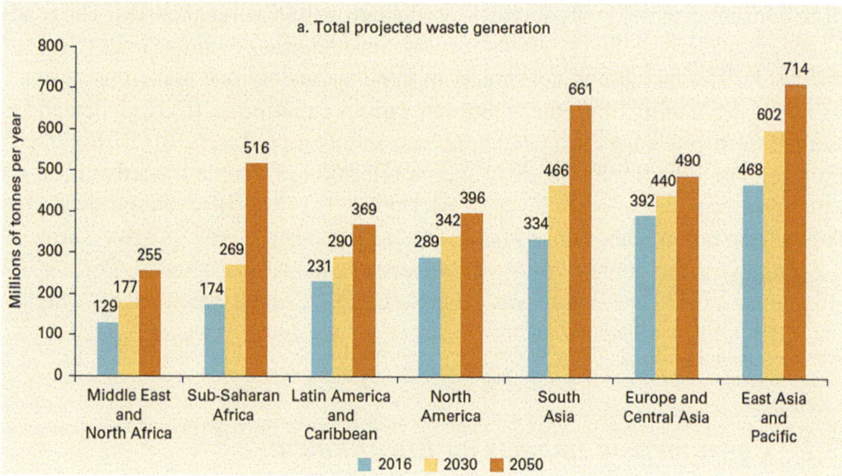

Fig. 1.2 Total Projected waste generation trend for different Regions (Kaza et al. 2018)

The developed countries have adopted better mechanism to manage solid waste. The amount of waste generated in developed countries is much higher than developing countries. Figure 1.2 depicts total waste generation for different regions by 2050 (Kaza et al. 2018). The municipal solid waste was only 2% of the total waste generated in Sweden in 2014, and remaining 83% waste generated comes from mining industries (SMED 2014). The financial and management infrastructure is highly developed and only 2% of the waste is dumped in landfills (Agarwal et al. 2025).

1.2 Consumerism

Consumerism trends greatly influence the waste generation, the greater the demand for products and services that ultimately turned into waste generation. Fast fashion, e -platform shopping, packed food items, over consumption and ease of availability of resources leads to the unexpected generation of waste. It is estimated that per capita waste generation in developed countries is expected to increase by 19% by 2050 while in low- & middle-income countries the waste is expected to increase by 40% or more by 2050 (Kaza et al. 2018). The consumerism trend varies between the developing and developed countries so is the generation of solid waste. In this industrialized era the pattern of consumption is so intensive that extraction of material to produce products is continuously expanding (Lopez 2017). It is considered that highly developed countries will generates less amount of waste generation growth by 2030 as they have achieved a point of economic development where less products will be consumed as per gross domestic product growth (Kaza et al. 2018). The products once considered as luxury goods are now considered as common goods (Crocker et al. 2013). The increase in consumerism trends is mainly due numerous reasons. (a) Ease of availability of products through various e platforms. (b) High demand is observed form consumers. (c) Competitiveness among producers. (d) Rising Population. (e) The urbanisation leads to the mobilisation of people towards cities. (f) E- platform provides another layer of packaging. (g) short lifecycle products. (h) E-Waste generation. Due to rise in consumerism trends the type of waste varies as per economy of the country, most of the waste remains unaccounted. For e.g. 62 million tonnes (MT) of e- waste was generated in 2022 which was about 82% higher from 2010 (Global waste 2024).

1.2.1 Consumerism Impacts on Environment

The consumerism demands new products; the new product are manufactured from natural resources as a result creates huge scarcity for natural resources. Landfills are over filled with heterogeneous wastes dues to rise in over consumerism. Single-use plastic products (SUPPs) may be the pinnacle of convenience, but they pose a serious risk to human health and the environment during production, distribution, and littering (UNEP 2021a). SUPPs products designed to be both lightweight strong and long lasting to maximize reusability to decrease the impact on the Environment. The damage to environment can only be prevented through eliminating single use products. Elimination of SUPPs is challenging due geographical, political and technological constraints. Alternatively, SUPPS can be manufactured utilising recyclable materials and renewable energy sources to create sustainable production (UNEP 2021b).

The rise in electronic products consumerism leads to increase in E-waste generation, any rejected product with a plug or battery into the open environment leads to health and environmental hazards, containing toxic additives or hazardous substances such as mercury, which can enter the food chain damage the human brain and coordination system (Global waste 2024). A record 62 million tonnes (Mt) of e-waste was produced in 2022, increased about 82% from 2010; and expected to rise about another 32% to 82 million tonnes, in 2030 (UNEP 2021b). The fast fashion trends lead to over generation of textile waste for e.g. in America alone, an approximate 11.3 million tons of textile waste—equivalent i.e. 85% of all textiles—end up in landfills on a yearly basis (Fast Fashion 2024). The textile and apparel industries are among the most polluting; their high energy, water, and chemical consumption, as well as the production of textile waste and the release of microfibres into the environment after laundry, all contribute to their negative ecological impact (Niinimäki et al. 2020). Globally, about 87% of disposed of textiles ended up in landfill, of which more than 90% are reusable and recyclable (Moazzem et al. 2021). The fast fashion and short product life cycle, post consumerism textile waste that end up in Landfills or incinerated release considerable amount of GHG s and toxic emissions (Leal et al. 2022).

1.3 Linear Programming Model for Waste Reduction and Optimization

The planning of effective municipal solid waste (MSW) management system for long-term for a specified geographical area requires complex decision making along with consideration of several factors. Due to the highly hetrogenous nature of waste, it gets difficult to evaluate each waste item separately for exploring their effective management options. To solve this problem, the integration of waste management models based on linear programming helps to tackle the multiple decision layers issues which arises simultaneously for tackling waste collection, storage, processing, transport and recycling etc. (Batur et al. 2020). For ideal waste management system, it's important to consider and evaluate sixfold structure including waste collection, transportation, process selection, allocation of waste to its treatment process, selection of ideal location for the treatment and processing and capacity assessment (Ghiani et al. 2014). One of the most pressing challenges in municipal solid waste management (MSWM) lies in the substantial financial burden posed by transportation logistics. Study conducted by Belien et al. (2011) highlights that in some cases, transportation standalone can consume a staggering proportion of the total operational budget of up to 80%. Therefore, finalizing waste processing method before considering the facility location and transportation planning from broader system design often leads to insignificant outcome. Effective waste management system requires integrated, or holistic approach selection, routing, and operational workflows are woven together from the outset.

Neglecting the interconnected nature of waste management system design can lead to inefficient results and questions the planning that may seem theoretically sound on paper but prove impractical in real-world applications. System engineering models (SEMs) serve as the backbone of modern waste management planning, integrating a suite of analytical methods to address complex challenges. These models encompass predictive simulations, optimization frameworks, and forecasting techniques designed to balance efficiency and feasibility. They also incorporate tools like cost–benefit evaluations to weigh financial trade-offs and multi-dimensional decision frameworks that account for social, environmental, and economic priorities. These are complemented by system evaluation tools (SETs), which provide critical lenses for assessing sustainability. For instance, life cycle assessments (LCAs) evaluate environmental footprints across a product's lifespan, while material flow tracking maps resource distribution and waste generation patterns. Risk assessment protocols further identify vulnerabilities, ensuring systems are resilient to operational or external disruptions. Together, SEMs and SETs create a holistic toolkit for designing waste management strategies that are both data-driven and adaptable to real-world uncertainties (Cobo et al. 2018; Batur et al. 2020). SEMs are decision-making mechanisms used to create a system from scratch. In contrast, SETs are evaluation tools for measuring a system's performance. This section of this chapter provides case studies of linear programming model in the context of waste management and limitations for execution as well as scaling up for better waste management.

1.3.1 Role of Linear Programming Optimization in Waste Reduction

Many industries and companies are forced by the global economic crises to focus more on cutting needless costs and getting rid of any waste that may be present in the production lines (Kumar and Authier 2012). As a result, majority of the businesses are focusing on lean manufacturing processes and the central notion is quite easy to implement, i.e. continuously striving for elimination or prevention of waste from the core manufacturing process to lower down material cost and improve product profits (Dillion and Shingo 1985). The lean production method seeks to continuously improve operations and cut costs by reducing waste. Waste, or Muda in Japanese, refers to any activity that does not add value to the customer's experience. It is divided into seven categories: overproduction, inventory, waiting, faults, overprocessing, motion, and transportation Fig. 1.3. Lean production uses a variety of tools

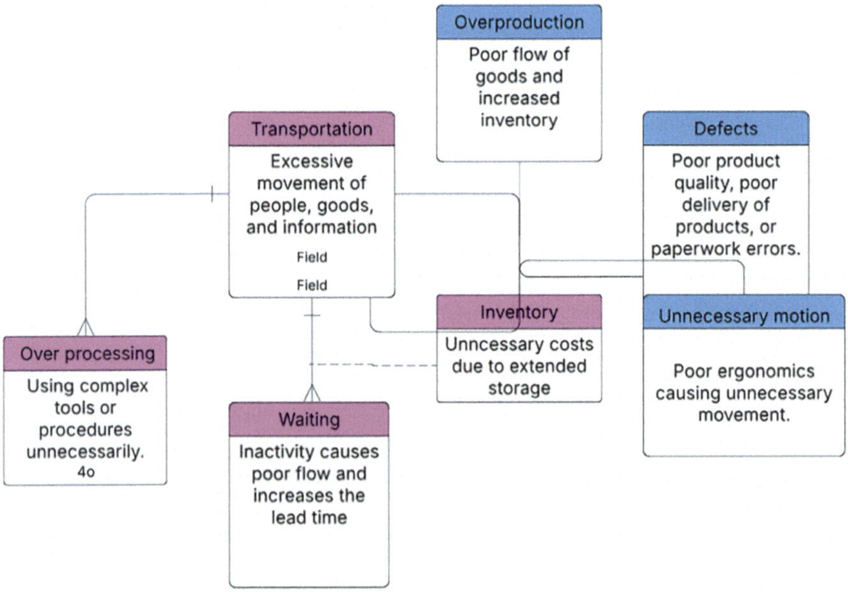

Fig. 1.3 Sources of seven types of waste in the lean manufacturing process (Modified and adopted from Eldein and Sobhi 2019)

and techniques, including SMED, 5S, visual control, and standard work, to reduce inefficiencies and increase efficiency (Eldein and Sobhi 2019).

1.3.2 Case Studies

1.3.2.1 Linear Programming Optimization for Waste Reduction in Production Line

The study was conducted by Eldein and Sobhi (2019) to reduce the waste generation from a conceptual production line of a typical manufacturing industry. To reduce the overall waste quantum from a production line Single Minute Exchange of Die methodology (SMED) was executed to lower down the changeover time between various units of the production line. The achieved results from analysing all types of waste showed that, two types of wastes are significant such that, material waste during cutting the sheet metal and the waiting time due changeover procedures in press station of production line. This study findings showed that linear models could be effective tools to reduce the waste quantum from 15.1 to below 6.6%. This estimate showed that implementing linear models in decision making can reduce the operational cost of waste management in shorter as well as longer periods for manufacturing industry. The use of mathematical models alongside digitization may help to enhance the overall circularity of material intensive manufacturing industries. Following are the vital steps designed for the reduction of waste from the production line of manufacturing process.

a. *Defining the problem*

Significant material waste is frequently the result of concentrating just on generating one product during the production stage without considering possible by-products or secondary outputs. Finding ways to reduce total waste generation and make better use of garbage is made easier using a systems thinking or holistic approach. To produce a range of goods from a single sheet, the author in this instance presented multiple techniques for altering the cutting patterns of the same sheet type. According to estimates, this modification to the pre-processing step—where several products are planned and combined—significantly decreased the overall quantity of trash, underscoring the need of effective and integrated production planning.

b. *Model Development*

This is an important step in applying a linear programming model for waste minimization. It defines the problem mathematically by establishing goals, constraints, and relationships. In essence, it's like determining the most optimal way to cut big rolls of material into smaller ones to meet demand and minimize excess scrap. The aim is to reduce the excess scraps by identifying the most efficient cutting strategy.

Model development step essentially considers all the possibilities and scenarios for reducing waste form a manufacturing process and helps to incorporating uncertainties in waste prevention which may arise due to the local and global materials supply chain issues.

c. *Data Preparation*

The Sheet metals are normally offered in standard lengths and are used to cut a variety of pieces varying in size and quantity to meet the individual needs of each project. In this case, the optimization model was run on a set of ten given sheets. Ten of the 27 sheets utilized in the project were particularly examined for efficiency. The cutting patterns generated by the model allowed more than two different items to be cut from each of these 10 sheets, demonstrating the model's capacity to maximize material use and reduce waste across many product kinds.

d. *Running the model*

The LINGO optimization technique was used to identify the most economical cutting patterns for decreasing waste across various sheet kinds. The model's output gave the best alternatives for minimizing material loss while achieving production requirements. To validate its effectiveness, the results were implemented over a two-month period, during which the organization tracked both the quantity of waste generated and the number of sheets needed for manufacturing. This real implementation demonstrated that the model effectively reduced waste and optimized resource utilization.

e. *Output Data Evaluation for Waste Reduction*

Using a linear programming (LP) approach, the organization can reduce material waste across diverse sheet types by optimizing how goods are cut from the sheets. The model output includes key information such as the number and kind of sheets needed, the overall waste generated, the waste connected with each individual sheet, and the specific goods cut from each sheet, as well as their quantities. This data-driven strategy facilitates more efficient planning and resource consumption, allowing the business to detect and decrease wasteful waste. As a result of deploying the LP solution, waste per sheet decreased dramatically across all sheet types, confirming the model's success in improving operational efficiency and sustainability Fig. 1.4.

In typical linear production schemes, resources are extracted, manufactured into products, and ultimately discarded. But due to growing concern for the environment and regulatory pressure, industries are looking forward with anticipation to measures to reduce the generation of wastes, maximize use of resources, and increase sustainability without fundamentally modifying their current systems of production. This strategy is referred to as linear waste minimization and revolves around waste reduction, resource optimization, closed-loop recycling, and industrial symbiosis to maximize material lifecycle and minimize reliance on virgin materials. Many of the leading players in industries such as automobiles, electronics, FMCG, textiles,

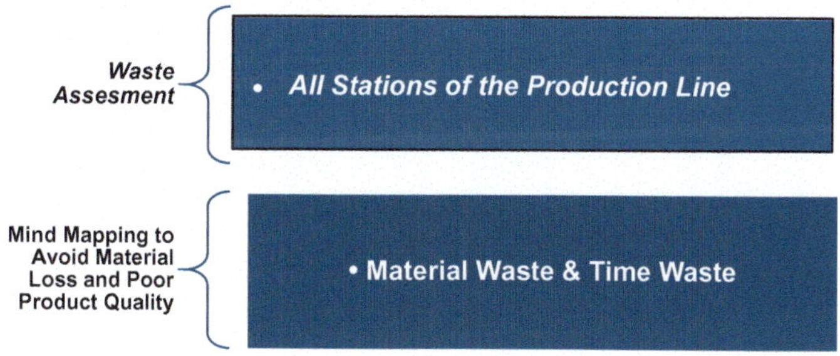

Fig. 1.4 Fundamental approach to initiate waste reduction for a production line

and metallurgy have managed to incorporate best-in-class waste reduction strategies. The strategies involve lean manufacturing, green packaging, closed-loop recycling, energy-saving processes, and take-back initiatives that all culminate into reduced landfill waste, lower carbon footprint, and higher circularity. The following table outlines 12 actual case studies of how businesses have implemented waste reduction models in their business, their implications for sustainability, and the wider implications for industry change. Each case is backed by an actual reference, providing insights into tried and tested best practice for businesses wishing to move towards resource-efficient and waste-reduced operations. Table 1.2 represents a compilation of real-world case studies from diverse industries demonstrating effective strategies for waste reduction and circular economy implementation. Each example highlights the type of waste addressed, the reduction or recycling approach adopted, and the measurable impact achieved. These cases reflect how industries are integrating sustainability into their core operations to minimize environmental footprint and enhance resource efficiency.

1.3.3 Limitations of Linear Programming Models

a. *Handling Uncertainty*

Traditional linear programming (LP) models often use deterministic parameters, with fixed values for variables like waste generation rates, operational expenses, and regulatory conditions. However, real-world waste management systems are fundamentally variable and dynamic, driven by shifting trash amounts, market volatility, and changing regulatory landscapes. To address these complexities and improve system resilience, stochastic or hybrid modelling approaches such as combining LP with machine learning or probabilistic analysis are becoming more common in modern waste planning and decision-making frameworks (Ochoa-Barragán et al. 2024).

Table 1.2 Industry case studies demonstrating circular economy and waste reduction strategies

Sl. no	Case study (industry)	Waste type	Reduction strategy	Outcome/impact	References
1	Interface (Carpet manufacturing)	Post-production carpet waste	Closed-loop recycling of carpet tiles using bio-based and recycled materials	Reduced virgin material usage by 50%, promoting circular flooring solutions	Interface (2023). *2023 ESG impact report*
2	Unilever (FMCG—sustainable packaging)	Plastic packaging waste	Developed refillable packaging and recycled-content packaging	Eliminated 100,000 tons of plastic use by 2025	*Unilever's sustainable packaging strategy* (2024)
3	Starbucks (Reusable cup initiative—food & beverage)	Single-use plastics	Launched Bring Your Own Cup and sustainable cup design initiatives	Avoided 150 million single-use cups annually	*Starbucks' sustainability commitment* (2024)
4	Tata Steel (Zero-waste steel plant—metallurgy)	Slag and industrial by-products	Implemented 100% slag utilization for cement and road construction	Diverted over 10 million tons of waste from landfills	*Tata steel sustainability report* (2020)
5	Dell Technologies (Closed-loop recycling—electronics)	E-waste	Developed closed-loop plastics recycling for computer parts	30% of Dell laptops now made with recycled plastics	*Dell's circular economy practices* (2024)
6	Adidas (Parley ocean plastic initiative—apparel)	Ocean plastic waste	Developed sportswear from 100% recycled ocean plastic	Produced over 10 million pairs of sustainable shoes	*Adidas* (2015)
7	Nestlé (Food waste reduction—food & beverage)	Expired and surplus food waste	Optimized food redistribution and upcycled food by-products	Achieved 50% food waste reduction across supply chains	*Nestlé* (2023)
8	Siemens (Energy-efficient manufacturing—industrial)	Process waste & energy loss	Integrated Industry 4.0 smart manufacturing technologies	Reduced energy waste by 20%, lowering emissions	*Siemens' sustainability initiatives* (2021)

(continued)

Table 1.2 (continued)

Sl. no	Case study (industry)	Waste type	Reduction strategy	Outcome/impact	References
9	H&M (Garment recycling program—fashion)	Textile waste	Launched take-back and upcycling programs for used garments	Diverted 40,000 tons of textile waste per year	*H&M's garment collecting program* (2013)
10	IKEA (Furniture & home goods—retail)	Wood and packaging waste	Transitioned to circular models with sustainable sourcing and material reuse	98% of products now designed for circularity	*IKEA's circular economy initiatives* (2024)

b. *Complexity in Multi-Objective Optimization*

Balancing conflicting objectives, such as maximizing profit while limiting environmental effect, necessitates advanced multi-objective optimization methods. One such strategy is the epsilon constraint approach, which was successfully used in a recent assessment of New York City's pandemic waste management. This work combined linear programming (LP) and machine learning to achieve both economic and environmental goals, demonstrating a novel technique to optimizing complicated urban garbage systems during a crisis (Batur et al. 2020).

c. *Scalability Challenges*

Large-scale waste management systems, such as municipal or city-wide networks, involve a vast number of interacting variables and constraints, making them computationally intensive to model and optimize. While the ALOMWASTE model developed by Batur et al. (2020) addressed some of these challenges by improving solution efficiency, real-time application and decision-making remain constrained by the high computational demands of such complex systems.

d. *Data Dependency*

Accurate outcomes from linear programming (LP) models depend heavily on the availability and reliability of high-quality input data, including well-defined constraints and precise coefficient values. For instance, in the case of Medan's waste bank, successful optimization required comprehensive records of waste inflows, categorization, and established buyer partnerships. However, in less-organized or informal waste management systems, such granular data is often lacking, posing significant limitations to the effective application of LP-based approaches (Auliani et al. 2022; Eldein and Sobhi 2019).

e. *Linear Assumptions*

The accuracy of linear programming (LP) results is highly dependent on the availability of high-quality data to define model constraints and coefficients. For example, the effective implementation of LP in Medan's waste bank relied on detailed records of waste inflows, segregation data, and established buyer partnerships. Such structured and consistent datasets are often unavailable in less organized or informal systems, limiting the practical applicability and reliability of LP models in those contexts as shown in Fig. 1.5. (Hernández-Romero et al. 2024; Ochoa-Barragán et al. 2024).

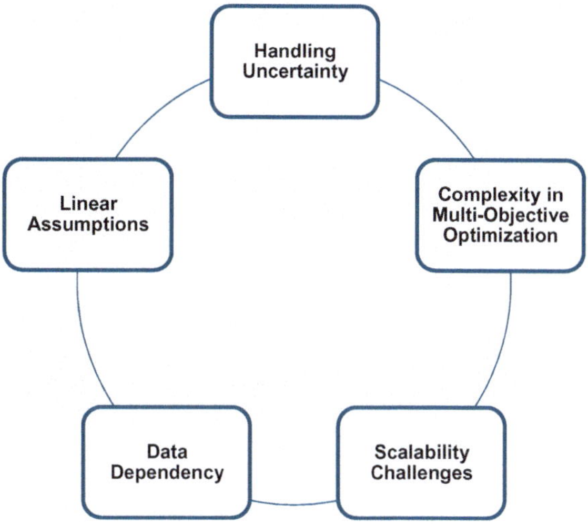

Fig. 1.5 Limitation of the linear waste reduction models

1.4 Zero Waste to Landfill (ZWL) Model

The waste generated is mainly heterogenous in nature, which makes difficult to streamline sustainable management of the waste. The complex nature of waste leaves no options, except discarding into unscientifically into landfills (Zaman 2015). The disposal of waste into landfills is although a unscientifically managed but considered as the easiest way of waste disposal, but landfills are limited and has certain carrying capacity. To reduce the amount of waste that ends up in landfills a model of "Zero Waste to Landfill (ZWL)" is practised across many cities of the world. The main objective of ZWL model is to divert the waste upto 99% from landfills (Zero waste 2024). ZWL model explore the opportunities for producers to realise the importance of waste. The ZWL model for a city requires lot of inputs from research fields and disciplines, including smart cities, urban resilience, green/smart mobility, the circular economy and complimentary, enabling undertakings such as the fab lab and makers movements (Awasthi et al., 2021). Zaman (2015) stated based on available literature review ZWL can be implemented into four phases Fig. 1.6. Phase one: the phase one emphasizes on waste prevention, through use of available resources instead of using virgin materials and non-renewable sources. Phase two: underlines the importance of sustainable consumption enable individual attitudes and decision-making during consumption activities, results in considerable reduction of waste. Phase three: This phase stresses upon 'downcycling' follows various steps such as waste sorting, collection, recycling and maximum resource recovery with minimal environment impact. Phase four: Last phase is policy implementation that focus on recycling activities by creating waste recycling jobs and hence promotes circular economic growth.

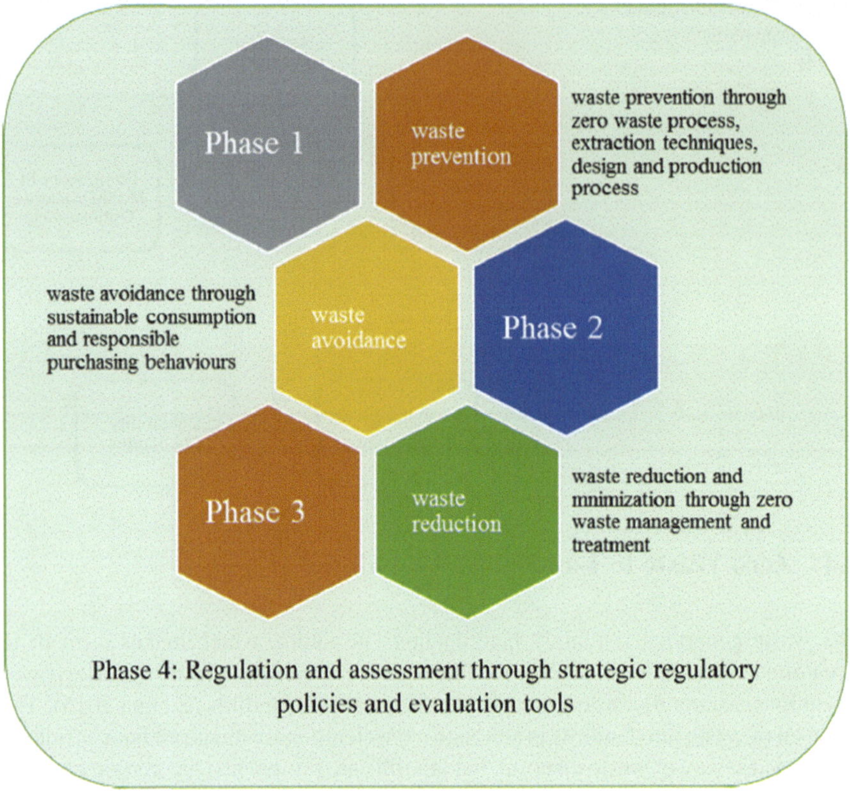

Fig. 1.6 Phases in implementing zero waste guidelines (Zaman 2015)

1.4.1 Case Studies on Zero Waste to Landfill (ZWL)

1.4.1.1 Swedish ZWL Model

The Sweden is known for its zero-waste strategy and considered as a world leader in sustainable waste management. Waste is considered as a useful commodity in Sweden and used to produce energy and multiple useful products (Bolton and Rousta 2019). A small city Known as Boras in Sweden is segregating waste into three waste streams as shown in Fig. 1.7.

The Swedish model is considered as one of best example of ZWL strategy as this model has successfully diverted more than 99% of the waste. The first stream deals food and residual waste collected in black and white bags respectively. The black and white bags are distributed free of cost. The residual waste generally consists of diapers, paper, plastic are meant to be stored in white bags. The food waste collected in black bags is treated biologically and waste collected in white bags is sent to combined heat and power plant. The second waste stream deals with mainly metal,

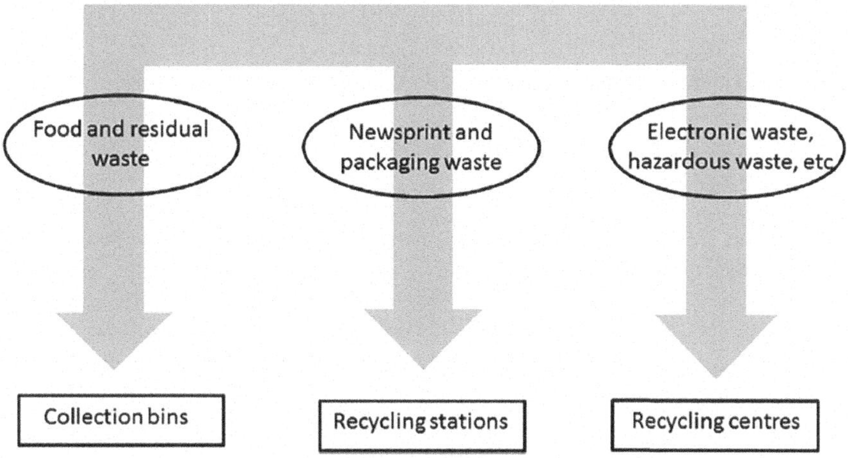

Fig. 1.7 Three municipal waste stream in (Boras) Sweden (Bolton and Rousta 2019)

plastic, coloured glass, transparent glass, and paper packaging. These second waste stream fractions are collected at the recycling stations. The producers and importers of packaging are responsible arranging the recycling stations and gathering and recycling the dispose of packaging. Third waste stream is consists of, hazardous waste, electronic waste, and garden waste, discarded smart phones, TVs, etc.; municipality is solely responsible for providing waste collection bins and manage them accordingly.

1.4.1.2 Jakarta ZWL Model

Jakarta is one of the most populus country in Southeast Asia (Rustiadi et al. 2021). A study conducted by (Suryawan and Lee 2024), on Municipal solid waste management preferences for Jakarta's slum and non-slum dwellers. The solid waste is dumped in open is the most pressing issues as landfills have limited carrying capacity raises a huge land consumption challenge. There is an urgent need for ZWL approach to overcome waste issue in Jakarta. They have selected 12 indicators across five adaptive capacities and corelated them with importance performances (I-P) preferences. The indicators were (a) waste treatment infrastructure (b) Circular Economy (c) Smart MSWM (d) Straightforward Waste Separation (e) Enforcement of Green Consumption (f) Entrepreneurship in Waste Recycling Policy (g) Skilled Staff (h) Collaborative Stakeholders (i) Portals for Effective Response (j) Involvement and Monitoring (k) Project Platform. Long-Term Public Attention: Among the 12 indicator there was a strong emphasizes on a circular economy model and robust waste segregation, considered as for optimal land use and to accomplish Jakarta's ambitions for a sustainable urban environment. Further this study highlighted the differences among perceptions of slum and non-slum dwellers. The non-slum dwellers prioritised waste treatment, while slum dwellers not favoured waste treatment due to the cost factors. To achieve ZWL for Zakarta it can be concluded that they must prioritise segregation, circular economy and waste treatment as priority measures, while slum dweller can be engaged in segregation.

1.4.1.3 San Francisco, ZWL Model

San Francisco a city in United States visioned a ZWL model in 2002 to be achieved by 2020. ZWL model in San Francisco is a successful case study achieved nearly 80% diversion rate considered to be highest in any major city of the United States (EPA 2017). The success of ZWL model in San Francisco was achieved mainly due to implementation of strong public private partnership, strict policy execution, education, financial penalties and incentives (Kaza et al. 2018). To help inhabitants more evidently know their waste disposal practices and financial impact, each house or building was provided with a detailed bill of fees for waste management. To strictly introduce the practice of segregation among residents' payments were reduced if there was a shift in waste from mixed waste bins to ones designated for recycling or composting. The waste management authorities provided prime importance to recycling toward efficient system, the size of the provided mixed waste bins were halved, and the size of recycling containers were doubled. Waste bins were frequently reviewed, and households that fail to conform with policies were warned and penalised (Kaza et al. 2018).

1.4.1.4 Limitation of Zero Waste to Landfill (ZWL) Models

Swedish ZWL was a prosperous model to manage MSW but as MSW is very complex in nature it requires lot of effort to segregate at source. The ZWL needs to be modified as per the needs of the households and consumption patterns (Bolton and Rousta 2019). The Swedish ZWL model can be successfully implemented in developed countries to the ease available financial and technological resources. Jakarta ZWL model was another alternative to manage solid waste, in a populated country. But implementation of ZWL strategy is often considered constraint as due to financial and technological constraints. San Francico ZWL model can be considered as an achievement as it had diverted 80% of the waste from Landfills. The main limitation is the implementation of the strategy that requires extensive financial, technological and human resources as hundred percent diversion is a difficult task to achieve.

1.5 Life Cycle Assessment Models in Waste Management

The Life Cycle Assessment (LCA) is a systematic methodology for analysing the environmental implications of products or processes throughout their entire life cycle, from raw material extraction to end-of-life disposal (Curran 2016). LCA is a tool which provides important technique for assessing the environmental implications of waste management systems which further allows policymakers and practitioners to evaluate environmental scenarios and identify sustainable solutions. In industrial waste management, LCA models evaluate and analyse pathways of waste treatment, processing and disposal methods (such as recycling, incineration and landfilling). It helps assisting industries in identifying methods to reduce the environmental impact of their waste streams. LCA provides a comprehensive framework for quantifying emissions, resource consumption, and energy recovery possibilities across a wide range of waste streams by assessing processes from its waste generation to the final disposal. There are several studies which has demonstrated usefulness of LCA in optimizing municipal solid waste (MSW) management, industrial waste treatment, and circular economy activities LCA is becoming increasingly popular and widely accepted methods for crucial decision making in SWM systems. The LCA helps to enhance waste management by assessing existing processes of WM, secondary products and residues (Allesch and Burner 2017: Thushasri et al. 2020). LCA assesses the benefits and stress of the environment in relation to waste management systems. It helps to analyse system performance, allow comparisons of alternatives, and identify potential system improvements.

The environmental assessments of SWM should be based on the flows defined by quantified approach using LCA, as such assessments give useful and transparent information for decision making.

1.5.1 Applications of LCA for Waste Management

The LCA aids in scenario analysis MSW management by enabling decision-makers to compare and evaluate the environmental implications of various waste treatment and disposal techniques throughout the waste's life cycle. LCA enable in determining trade-offs between various waste management activities such as landfilling, recycling, composting, and energy recovery by quantifying emissions, resource usage, and pollution. LCA also influences strategic planning and policy by simulating baseline and future situations to prioritize low-impact, resource-efficient solutions. LCA has power to enable smooth and greener transition to a circular economy by highlighting approaches that increase material recovery while reducing environmental impacts. Furthermore, LCA can be combined with techno-economic analysis (TEA) to help guide decisions that are both environmentally friendly and financially effective (Laurent et al. 2014). Figure 1.8 represents the different steps involved in life cycle assessment.

a. *Scenario Analysis*

LCA is a powerful technique for systematically comparing alternative municipal solid waste (MSW) management options by modelling and analysing their environmental impacts throughout the life cycle. It advocates for the comparison of various scenarios

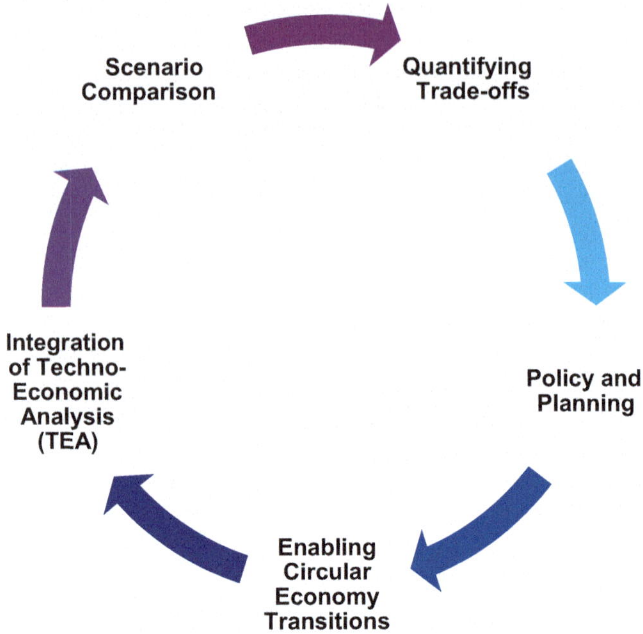

Fig. 1.8 Different steps involved in life cycle assessment

and decision making such as landfilling versus incineration, source-segregated recycling systems versus mixed waste disposal, and organic waste treatment using composting versus anaerobic digestion. Each scenario of waste management can be examined based on a range of impact categories, including global warming potential (expressed in CO_2 equivalents), human toxicity, resource depletion, eutrophication, and acidification. This comparison method provides vital insights for identifying ecologically preferred strategies, as well as guiding evidence-based waste management policy and planning (Gentil et al. 2010).

b. *Trade-offs*

LCA is crucial for assessing the environmental trade-offs associated with various municipal solid waste (MSW) management systems. By examining the entire waste life cycle from creation to collection, treatment, and final disposal. LCA allows for the discovery of both the advantages and unforeseen environmental burdens of each alternative. For example, while incineration with energy recovery can greatly reduce trash volume and fossil fuel use, it can also result in greater emissions of particulates and hazardous substances like dioxins and furans (Schott and Andersson 2015). By means of such comparative assessment, LCA clarifies the environmental trade-offs natural in every situation, hence helping decision-makers to choose solutions that maximize net environmental performance while considering system-wide consequences.

c. *Information & Policy Planning*

LCA also provides a science-based framework for influencing waste management policy and infrastructure development by simulating baseline and future intervention scenarios. LCA can be used by policymakers and planners to assess the environmental impact of transitioning from landfill-dominated systems to more integrated solutions such as recycling, composting, and waste-to-energy (WtE). These assessments may be region-specific, reflecting differences in waste composition, treatment capacity, and regulatory frameworks. By analysing the life cycle consequences of various management approaches, LCA assists in prioritizing interventions that reduce greenhouse gas emissions, conserve natural resources, and comply with national sustainability goals. Furthermore, LCA promotes the incorporation of circular economy ideas into municipal solid waste (MSW) plans by emphasizing opportunities for material recovery and waste prevention, ultimately directing evidence-based decision-making for long-term system optimization (Khoo 2009).

d. *Enables Circular Transition*

By identifying waste treatment alternatives that put material recovery and resource efficiency first, life cycle assessment (LCA) is a useful analytical tool in supporting the shift toward circular economy models in MSW management. LCA assists in identifying procedures that closes material loops and lessen reliance on virgin resources by examining environmental costs associated with various waste valorization methods, including recycling, composting, anaerobic digestion, and the manufacture of refuse-derived fuel (RDF). By identifying environmental hotspots, the technique helps stakeholders create interventions that keep materials' inherent value in the market intact while simultaneously diverting waste from landfills (Bhardwaj et al. 2025; Agarwal & Jain 2024; Jain 2025). By doing this, LCA encourages the creation of circular waste systems that reduce emissions, prolong product life, and improve resource circularity all of which are in line with decarbonization objectives and global sustainability frameworks (Huysman et al. 2017).

e. *Integrates with Techno-Economic Analysis (TEA)*

In addition to environmental analysis, LCA can be combined with TEA to offer a more comprehensive assessment of MSW management tactics that considers both economic viability and sustainability. Stakeholders may evaluate the ecological effects, capital and operating expenses, possible material recovery income, and overall cost–benefit trade-offs of various waste treatment scenarios thanks to this unified methodology. A waste-to-energy facility, for instance, might do well in LCA for lowering GHG emissions, but TEA might highlight issues with feedstock availability or costly initial investment. Decision-makers can choose solutions that are socially acceptable, economically feasible, and environmentally sound by combining the two approaches. This multidimensional assessment is especially useful in developing countries and rising cities, where resource allocation must be improved without jeopardizing environmental goals (Corominas et al. 2013).

1.5.2 Case Study: Life Cycle Assessment of a 1 kg PET Bottle

The polyethylene terephthalate (PET) bottles are frequently used in packaging, particularly for beverages, because they are lightweight, resilient, and inexpensive. However, their environmental impact is significant over the whole life cycle, from raw material extraction to end-of-life treatment. This section provides a simplified LCA of a 1 kg PET bottle, comparing important environmental consequences across several disposal pathways, including mechanical recycling, landfilling, and burning with energy recovery (Yu et al. 2015). The manufacturing of virgin PET plastic is highly resource-intensive, relying mostly on fossil-based raw materials such as oil and natural gas, with an associated energy consumption of 83.8 MJ/kg and significant emissions including 2330 g CO_2/kg, 25 g SO_2/kg, and 20.2 g NOx/kg. Though used in small numbers, additives in PET manufacturing frequently have a higher environmental toxicity than the polymer itself, emphasizing the need to reduce reliance on virgin PET through circular processes. While the use phase of PET bottles has no environmental impact, transit across the manufacture, consumption, and disposal stages adds about 2 MJ/kg to the overall energy footprint, assuming a 200 km journey distance. At the end of life, three disposal scenarios have diverse environmental profiles. Form the results of this LCA analysis carried out by Yu et al. (2015) revealed that mechanical recycling was the most sustainable method, using only 27.07 MJ/kg of energy and producing 163 g CO_2/kg with low levels of NOx and VOCs. While landfilling emits just 94.6 g/kg of CO_2 it produces 2080.6 g/kg of hydrocarbons, posing major ecological challenges and climate dangers. Incineration with energy recovery has a moderate energy requirement (32.5 MJ/kg) but emits large levels of CO_2 (2016 g/kg) and NOx (2.436 g/Kg) generating concerns about air pollution and harmful residues. These study findings revealed that environmental trade-offs associated with different stages of PET bottle life cycle as well as the importance of prioritizing recycling in integrated waste management systems (Yu et al. 2015).

1.5.3 Limitations and Challenges in LCA Implementation

LCA has transformed waste management by allowing data-driven comparisons of treatment options and emphasizing the trade-offs between energy recovery, emissions, and resource conservation. Its use in municipal, industrial, and construction waste systems demonstrates its adaptability in facilitating circular economy transitions. However, restrictions arising from data gaps, modelling assumptions, and boundary definitions need caution in interpretation. The developments in real-time data gathering, dynamic LCA techniques, and stakeholder interaction are required to improve LCA's precision and relevance, cementing its position as a pillar of sustainable waste governance. LCA in waste management has significant limitations that affect the accuracy and relevance of its findings. One significant limitation is that existing LCA models may not adequately capture the actual efforts and environmental burdens involved with resource recovery and recycling operations, frequently underrepresenting the complexity of recycling systems and their downstream benefits (Rigamonti et al. 2009a, b; Shen et al. 2010). The availability and quality of inventory data continue to be critical challenges, as robust, region-specific, and up-to-date data on waste technologies, emissions, and outputs are frequently unavailable, limiting comprehensive modelling of waste management systems and alternative treatment routes (Vadenbo et al. 2017).

Furthermore, despite their significant impact on environmental outcomes, the selection and depiction of technological configurations within system limits is frequently simple or inconsistent, and this is commonly disregarded in trash LCAs (Astrup et al. 2015). Several studies used a zero-burden assumption that ignores upstream impacts from production and use phases, it is particularly challenging to quantify waste prevention measures like reuse and source reduction. This limits the evaluation of prevention's full environmental benefits (Dolci et al. 2016). The use of various software tools, impact assessment techniques, and occasionally out-of-date or inconsistent data sources also presents methodological issues, which can result in disparities and lower comparability among research (Ortego et al. 2018; Astrup et al. 2015). Lastly, site-specific insights that are essential for evaluating local environmental consequences like toxicity or noise pollution are frequently absent from LCA's normal focus on global or regional impacts (Astrup et al. 2015). These drawbacks show that to increase LCA's dependability and policy relevance in waste management, better data quality, methodological harmonization, and larger system perspectives are required. As the global MSW generation is rising due to population increase, life-style changes, consumerism therefore it becomes priority to develop effective and sustainable waste management (WM) systems (Kaza et al. 2018). LCA is widely acknowledged and accepted tool for assessing environmental performance of waste management systems due to is feasibility for systematic and quantitative evaluation of consequences across full waste life cycle (Laurent et al. 2014). However, the use of LCA in waste management has limitations. High data uncertainty, irregular system boundary definitions, and difficulty in assigning environmental responsibilities in multi-output systems (e.g., recycling or energy recovery) frequently jeopardize result accuracy and comparability Fig. 1.9. Furthermore, LCA overlooks socioeconomic, behavioural, and regulatory elements that are critical to real-world waste management (Koch and Salou 2015). The tool's application in developing countries is further limited by its omission of spatial and temporal variability, particularly when estimating emissions from landfills or informal waste sectors. These limitations highlight the need to supplement LCA with other approaches such as Techno-Economic Analysis (TEA), Social Life Cycle Assessment (S-LCA), or Multi-Criteria Decision Analysis (MCDA) to enable more comprehensive, context-sensitive evaluations of waste systems in the face of escalating global waste issues.

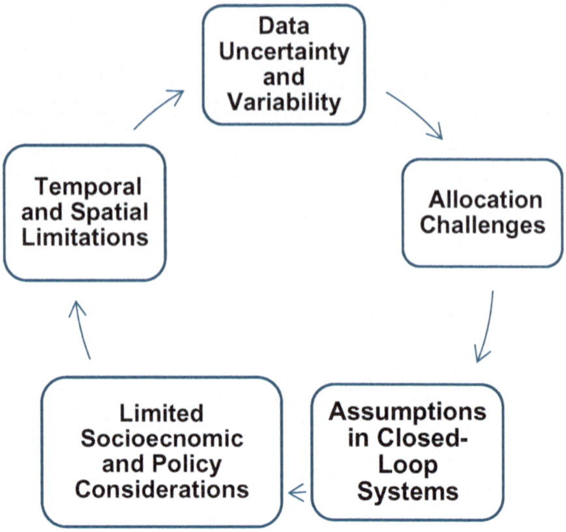

Fig. 1.9 Limitation of LCA for application of SWM

1.6 Conclusion

Solid waste management is a critical global issue, which need to be addressed instantly. The improper solid waste management creates end number of problems, but sound solid waste management strategy leads to aesthetics, preserve resources, mitigates adverse health impacts and generates employment and financial resources. The policy regulations, governance can be dominant player in managing effective solid waste management, but above all the public awareness and responsible solid waste segregation at source is of prime importance. The accelerated urbanization, consumerism, and increasing resource use have become solid waste management a critical global issue. Resolving this issue necessitates a transition from disjointed methods to more cohesive and data-informed tactics. This chapter critically analyzed three significant waste management models: Linear Programming Optimization (LPO), Life Cycle Assessment (LCA), and Zero Waste to Landfill (ZWL), emphasizing their respective strengths, limits, and possibilities for integration. The findings suggest that improved planning, performance assessment, and policy synchronization for urban and industrial waste systems can be facilitated by a unified, data-centric approach that integrates many models. To provide flexible, scalable, and sustainable solutions in the face of growing waste generation and environmental concerns, future waste management strategies must place a high priority on system-wide optimization, stakeholder participation, and the concepts of the circular economy.

References

Abubakar IR, Maniruzzaman KM, Dano UL, AlShihri FS, AlShammari MS, Ahmed SMS, Al-Gehlani WAG, Alrawaf TI (2022) Environmental sustainability impacts of solid waste management practices in the global South. Int J Environ Res Public Health 19(19):12717

Adidas (n.d.) Adidas x Parley for the Oceans. https://www.adidas.com

Agarwal S, Jain H (2024) Investigating the potential of floral waste as a vermicompost and dual-functional biosorbent for sustainable environmental management. Water Air Soil Pollut 235:322. https://doi.org/10.1007/s11270-024-07144-y

Agarwal S, Jain H, Dhupper R, Mathur A (2025) Optimizing methyl orange dye removal from aqueous solutions using white chrysanthemum floral waste-derived bioadsorbent: a study of kinetics, thermodynamics, isotherms, and RSM optimization. Int J Environ Res 19(3):104

Allesch A, Brunner PH (2017) Material flow analysis as a tool to improve waste management systems: the case of Austria. Environ Sci Technol 51(1):540–551

Astrup TF, Tonini D, Turconi R, Boldrin A (2015) Life cycle assessment of thermal waste-to-energy technologies: review and recommendations. Waste Manage 37:104–115

Auliani R, Wibowo R, Zuska F (2022) Optimization of waste bank management using linear programming: Case study in Medan, North Sumatera. In: Proceedings of the International Conference on Natural Resources and Sustainable Development (ICNRSD 2018), pp 361–366

Awasthi AK, Cheela VS, D'Adamo I, Iacovidou E, Islam MR, Johnson M, Miller TR, Parajuly K, Parchomenko A, Radhakrishan L, Zhao M (2021) Zero waste approach towards a sustainable waste management. Resour, Environ Sustain 3:100014

Batur ME, Cihan A, Korucu MK, Bektaş N, Keskinler B (2020) A mixed integer linear programming model for long-term planning of municipal solid waste management systems: Against restricted mass balances. Waste Manage 105:211–222

Beliën J, De Boeck L, Van Ackere J (2014) Municipal solid waste collection and management problems: A literature review. Transp Sci 48(1):78–102

Bhardwaj LK et al (2025) Resilience in the digital age: Emerging technologies for climate adaptation. In: Choudhury M, Dixit G, Majumdar S (eds) Sustainable synergy: Harnessing ecosystems for climate resilience. Climate change management. Springer, Cham. https://doi.org/10.1007/978-3-031-77957-2_11

Bolton K, Rousta K (2019) Solid waste management toward zero landfill: A Swedish model. In: Sustainable resource recovery and zero waste approaches. Elsevier, pp 53–63

Cobo S, Dominguez-Ramos A, Irabien A (2018) From linear to circular integrated waste management systems: a review of methodological approaches. Resour Conserv Recycl 135:279–295

Corominas L, Foley J, Guest JS, Hospido A, Larsen HF, Morera S, Shaw A (2013) Life cycle assessment applied to wastewater treatment: State of the art. Water Res 47(15):5480–5492. https://doi.org/10.1016/j.watres.2013.06.049

CPCB (2023) Annual report solid waste management. https://cpcb.nic.in/uploads/MSW/MSW_AnnualReport_2021-22.pdf. Accessed on 4 April 2025

Crocker R (2013) From access to excess: consumerism, 'compulsory' consumption and behaviour change. In: Lehmann S, Crocker R (eds) Motivating change: Sustainable design and behaviour in the built environment. Earthscan Publication, London

Curran MA (2016) Overview of goal and scope definition in life cycle assessment. In: Goal and scope definition in life cycle assessment. Springer Netherlands, Dordrecht, pp 1–62. https://doi.org/10.1007/978-94-024-0855-3_1

Dell Technologies (2022) Circular economy in action: Leading the fight against e-waste. https://www.dell.com. Accessed on 1st April 2025

Dillion AP, Shingo S (1985) A revolution in manufacturing: The SMED system. CRC Press

Dolci F, Hischier R, Zampori L (2016) Challenges in the application of life cycle assessment to waste prevention strategies. Int J Life Cycle Assess 21(6):861–871. https://doi.org/10.1007/s11367-016-1063-3

EPA (United States Environmental Protection Agency) (2017) "Zero waste case study: San Francisco." Managing and transforming waste streams—a tool for communities. https://www.epa.gov/transforming-waste-tool/zero-waste-case-study-san-francisco

Eldein SAS, Sobhi N (2019) Waste reduction by linear programing optimizing. In: IOP conference series: Materials science and engineering, vol 610, no 1. IOP Publishing, p 012079

Fast Fashion (2024) Concerning fast fashion waste statistics. https://earth.org/statistics-about-fast-fashion-waste/#:~:text=1.,the%20end%20of%20the%20decade. Accessed on 28th April 2025

Gentil EC, Damgaard A, Hauschild M, Finnveden G, Eriksson O, Thorneloe S, Kaplan PO, Barlaz M, Muller O, Matsui Y, Ii R (2010) Models for waste life cycle assessment: Review of technical assumptions. Waste Manage 30(12):2636–2648

Ghiani G, Laganà D, Manni E, Musmanno R, Vigo D (2014) Operations research in solid waste management: A survey of strategic and tactical issues. Comput Oper Res 44:22–32

Global E-Waste (2024) Global e-waste monitor 2024: Electronic waste rising five times faster than documented e-waste recycling. https://unitar.org/about/news-stories/press/global-e-waste-monitor-2024-electronic-waste-rising-five-times-faster-documented-e-waste-recycling. Accessed on 15th April 2025

Hernández-Romero IM, Niño-Caballero JC, González LT, Pérez-Rodríguez M, Flores-Tlacuahuac A, Montesinos-Castellanos A (2024) Waste management optimization with NLP modeling and waste-to-energy in a circular economy. Sci Rep 14(1):19859

H&M (n.d.) Garment collecting—close the loop. https://www2.hm.com. Accessed on 1st April 2025

Hoang NH, Fogarassy C (2020) Sustainability evaluation of municipal solid waste management system for Hanoi (Vietnam)—Why to choose the 'Waste-to-Energy' concept. Sustainability 12(3):1085

Huysman S, De Schaepmeester J, Ragaert K, Dewulf J, De Meester S (2017) Performance indicators for a circular economy: a case study on post-industrial plastic waste. Resour Conserv Recycl 120:46–54

IFC (2024) The world has a waste problem. Here's how to fix it. Accessed 27 February 2024. https://www.ifc.org/en/blogs/2024/the-world-has-a-waste-problem

IKEA (n.d.) Our circular agenda. https://www.ikea.com. Accessed on 1st April 2025

Interface (2020) Mission zero progress report. https://www.interface.com. Accessed on 1st April 2025

Jain H (2025) Synergizing carbon reduction technologies and nature-based solutions for a climate-neutral future. Mitig Adapt Strateg Glob Change 30:42. https://doi.org/10.1007/s11027-025-10237-6

Kabera T, Nishimwe H (2019) Systems analysis of municipal solid waste management and recycling system in east Africa: benchmarking performance in Kigali city, Rwanda. In: E3S web of conferences, vol 80, p 03004. EDP Sciences

Kalina M, Makwetu N, Tilley E (2024) "The rich will always be able to dispose of their waste": a view from the frontlines of municipal failure in Makhanda, South Africa. Environ Dev Sustain 26(7):17759–17782

Kaza S, Yao L, Bhada-Tata P, Van Woerden F (2018) What a waste 2.0: A global snapshot of solid waste management to 2050. World Bank. https://doi.org/10.1596/978-1-4648-1329-0

Khoo HH (2009) Life cycle impact assessment of various waste conversion technologies. Waste Manage 29(6):1892–1900

Koch P, Salou T (2015) Agri-footprint 2.0—part 2: Description of data—feed products. Blonk Consultants

Kumar BS, Abuthakeer SS (2012) Productivity enhancement by implementing lean tools and techniques in an automotive industry. Ann Fac Eng Hunedoara 10(1):167

Kurniawan TA, Maiurova A, Kustikova M, Bykovskaia E, Othman MHD, Goh HH (2022) Accelerating sustainability transition in St. Petersburg (Russia) through digitalization-based circular economy in waste recycling industry: a strategy to promote carbon neutrality in era of Industry 4.0. J Clean Prod 363:132452

Laurent A, Bakas I, Clavreul J, Bernstad A, Niero M, Gentil E, Hauschild MZ, Christensen TH (2014) Review of LCA studies of solid waste management systems–part I: lessons learned and perspectives. Waste Manage 34(3):573–588. https://doi.org/10.1016/j.wasman.2013.10.045
Leal Filho W, Perry P, Heim H, Dinis MAP, Moda H, Ebhuoma E, Paço A (2022) An overview of the contribution of the textiles sector to climate change. Front Environ Sci 10:973102
Lopez R (2017) The environment as a factor of production: The effects of economic growth and trade liberalization 1. In: International trade and the environment. Routledge, pp 239–260
Mahanta PK, Rafizul IM (2019) Human health risk assessment due to the presence of heavy metalsin soil of waste disposal site at Khulna in Bangladesh. J Eng 10(1):01–12
Moazzem S, Wang L, Daver F, Crossin E (2021) Environmental impact of discarded apparel landfilling and recycling. Resour Conserv Recycl 166:105338
Muheirwe F, Kombe W, Kihila JM (2022) The paradox of solid waste management: a regulatory discourse from Sub-Saharan Africa. Habitat Int 119:102491
Nestlé (n.d.) Food loss and waste. https://www.nestle.com. Accessed on 1st April 2025
Niinimäki K, Peters G, Dahlbo H, Perry P, Rissanen T, Gwilt A (2020) The environmental price of fast fashion. Nat Rev Earth Environ 1(4):189–200
Ochoa-Barragán R, Munguía-López ADC, Ponce-Ortega JM (2024) A hybrid machine learning-mathematical programming optimization approach for municipal solid waste management during the pandemic. Environ Dev Sustain 26(7):17653–17672
Ortego MI, Rieradevall J, Gabarrell X (2018) Influence of software and data sources on life cycle assessment results: comparison of IWM-2 and SimaPro. Environ Impact Assess Rev 68:1–10. https://doi.org/10.1016/j.eiar.2017.11.002
Penteado CSG, de Castro MAS (2021) Covid-19 effects on municipal solid waste management: what can effectively be done in the Brazilian scenario? Resour Conserv Recycl 164:105152
Rana R, Ganguly R, Gupta AK (2015) An assessment of solid waste management system in Chandigarh city, India. Electron J Geotech Eng 20(6):1547–1572
Redwood Materials (n.d.) Recycle with us. https://www.redwoodmaterials.com. Accessed on 1st April 2025
Rigamonti L, Grosso M, Møller J (2009a) Life cycle assessment of recycling: influence of assumptions and methodological choices. Waste Manage 29(2):986–994. https://doi.org/10.1016/j.wasman.2008.07.005
Rigamonti L, Grosso M, Møller J (2009b) Environmental benefits of recycling: a review of life cycle assessment studies. Waste Manage 29(2):986–994. https://doi.org/10.1016/j.wasman.2008.07.005
Rustiadi E, Pravitasari AE, Setiawan Y, Mulya SP, Pribadi DO, Tsutsumida N (2021) Impact of continuous Jakarta megacity urban expansion on the formation of the Jakarta-Bandung conurbation over the rice farm regions. Cities 111:103000
Schott ABS, Andersson T (2015) Food waste minimization from a life-cycle perspective. J Environ Manage 147:219–226
Shen L, Worrell E, Patel MK (2010) Open-loop recycling: a LCA case study of PET bottle-to-fibre recycling. Resour Conserv Recycl 55(1):34–52
Siemens USA (2021) Sustainability initiatives across U.S. Manufacturing. Retrieved from https://www.siemens.com. Accessed on 1st April 2025
SMED (2014) Avfall i sverige 2014 [waste in Sweden 2014]. Swedish Environmental Protection Agency, Stockholm, Sweden (in Swedish)
Stafford WH (2020) WtE best practices and perspectives in Africa. In: Municipal solid waste energy conversion in developing countries. Elsevier, pp 185–217
Starbucks (2024) Reusable cup program expansion. https://about.starbucks.com. Accessed on 1st April 2025
Statista (2022) Per capita municipal waste generation in Europe in 2022. https://www.statista.com/statistics/789638/production-waste-tons-by-inhabitant-union-european/. Accessed on 3 Mar 2025

Tata Steel (2020) Sustainability report 2019–20. https://www.tatasteel.com. Last Accessed 1st April 2025

Suryawan IWK, Lee CH (2024) Importance-performance dynamics and willingness to pay in coastal areas for climate-adaptive marine debris management. Regnal Stud Mar Sci 77:103596

Swati, Thakur, IS, Vijay VK, Ghosh P (2019) Scenario of landfilling in India: Problems, challenges, and recommendations. In: Hussain C (eds) Handbook of environmental materials management. Springer, Cham. https://doi.org/10.1007/978-3-319-73645-7_167

Thushari I, Vicheanteab J, Janjaroen D (2020) Material flow analysis and life cycle assessment of solid waste management in urban green areas, Thailand. Sustain Environ Res 30:1–17

Toyota Motor Corporation (n.d.) Toyota production system. https://global.toyota/en/company/vision-and-philosophy/production-system/. Accessed on April 2025

UNEP (2021a) How to reduce the impacts of single-use plastic products. https://www.unep.org/news-and-stories/story/how-reduce-impacts-single-use-plastic-products. Accessed on 2nd May 2025

UNEP (2021b) The global e-waste monitor 2024. https://ewastemonitor.info/the-global-e-waste-monitor-2024/. Accessed on 2nd May 2025

Unilever (2020) Unilever makes progress on its sustainable packaging goals. https://www.unilever.com. Accessed on 1st April 2025

Vadenbo C, Hellweg S, Astrup T (2017) Life cycle assessment of waste management systems: influence of data quality and methodological choices. Waste Manage 61:593–607. https://doi.org/10.1016/j.wasman.2016.11.034

Yatoo AM, Hamid B, Sheikh TA, Ali S, Bhat SA, Ramola S, Ali MN, Baba ZA, Kumar S (2024) Global perspective of municipal solid waste and landfill leachate: generation, composition, eco-toxicity, and sustainable management strategies. Environ Sci Pollut Res 31(16):23363–23392

Yu V, Borodin A, Kulakov K, Sereda V (2015) Life cycle inventory analysis of PET bottles: energy and emissions across disposal pathways. IOP Conf Ser: Mater Sci Eng 81:012085. https://doi.org/10.1088/1757-899X/81/1/012085

Zaman AU (2015) A comprehensive review of the development of zero waste management: lessons learned and guidelines. J Clean Prod 91:12–25

Zero waste (2024) Available on https://www.slrecyclingltd.co.uk/zero-waste-to-landfill-what-does-it-mean/. Accessed on 27 Mar 25

Zhang Z, Chen Z, Zhang J, Liu Y, Chen L, Yang M, Osman AI, Farghali M, Liu E, Hassan D, Ihara I (2024) Municipal solid waste management challenges in developing regions: a comprehensive review and future perspectives for Asia and Africa. Sci Total Environ 172794

Chapter 2
Core Principles of Circularity and Creating a Vision of Sustainable Recovery of Resources

Harshita Jain, Lovepreet Singh, and Anamika Shrivastava

2.1 Introduction and Foundations of Circular Economy

The idea of the circular economy is a revolution in the process of perceiving, using, and managing resources by societies and industries. It goes beyond the constraints of the traditional one-dimensional linear economy model with the following phases of extraction, production, consumption and disposal on a restorative and regenerative model or way of thinking (Upadhayay et al. 2024). The understanding of value in the linear system is created when materials go through the system without much care of long-term effects on the environment and social economic degradation. This tends to lead to depletion of resources, enhancement of waste production and environmental stress. Contrarily, circular economy is concerned with optimizing the amount of value added to products, materials and resources in the economy over time, making them last longer in the economy and therefore limit their wastage and use of virgin resources (Kirchherr et al. 2023).

The fundamental tenets of the circular economy are waste reduction, product life extension, resource efficiency, and the development of systems that can regenerate natural capital. These ideas are put into practice by using tactics including prioritising renewable, recyclable, and biodegradable materials; implementing closed-loop supply chains; and designing goods for durability, reparability, and upgradeability (Agarwal and Jain 2024; Agarwal et al. 2025). The foundation of these strategies is systemic thinking, which acknowledges the connections between natural systems, economic activity, and social well-being. Therefore, the transition to circularity is a complete change that encompasses production techniques, consumer trends, business

H. Jain (✉) · A. Shrivastava
Amity Institute of Environmental Sciences, Amity University Noida, Noida, Uttar Pradesh, India
e-mail: hjain@amity.edu

L. Singh
Department of Chemical Engineering, Thapar Institute of Engineering and Technology, Patiala, India

H. Jain et al. (eds.), *The Circular Fallacy: Closing the Loop in a Consumerist World*, Circular Economy and Sustainability, https://doi.org/10.1007/978-3-032-22687-7_2

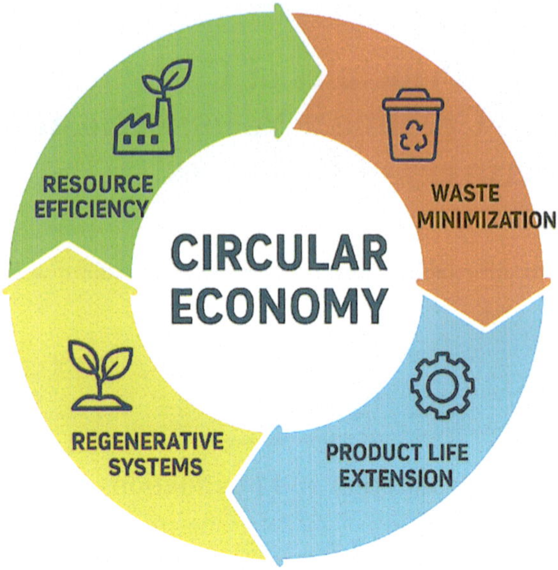

Fig. 2.1 Circular economy: a regenerative approach to resource management

models, and governance frameworks rather than a single management or technical intervention. The fundamental ideas of the circular economy, a sustainable framework that transcends the conventional linear paradigm of "take, make, dispose," are shown in Fig. 2.1. The fundamental idea is to preserve the value of resources, goods, and materials for as long as feasible.

Circularity leads to a shift in perspective on waste as temporary resource instead of its destination as part of sustainable resource recovery. Such a view creates avenues towards incorporation of advanced technologies in recovery such as mechanical, chemical and biological recovery processes that may extract, process and recreate the valuable materials as part of the production cycle. The increasing importance of digital technologies like blockchain, the Internet of Things (IoT) and artificial intelligence increases the traceability, transparency and efficiency of these processes and supports more intelligent decision-making throughout supply chains. Moreover, the process of adopting the principles of the circular economy is formed under the impact of enabling policy frameworks and economic instruments (Jain 2025; Jain et al. 2023a, b). The role of governments and international institutions is to gradually come up with regulations, incentives, and standards favoring closed-loop systems, the advancement of innovation, and the collaboration between sectors. These steps do not provide only an enabling environment of a circular transition occurrence but also ascertain that the gains are balanced on the social, environmental and economic fronts (Velenturf and Purnell 2021).

The goal of this chapter is to provide a solid conceptual framework for comprehending the circular economy and how it promotes sustainable resource recovery. It explains the fundamental ideas behind circularity, highlights the main differences between linear and circular models, and looks at the economic, regulatory, and technical forces that support its use. In order to provide academics, professionals, and policymakers with a thorough framework for addressing material loops and promoting robust, regenerative systems, it attempts to connect theory with practical insights, such as case studies and real-world implementations.

2.2 Designing for Circularity and Sustainable Resource Recovery

2.2.1 Product Design for Durability, Reparability, and Modularity

Design is the starting point of the circular economy where the features and lifecycle of the product are heavily influenced when it is in the concept and development phase. Creating durable designs will allow the products to be able to withstand the time of excessive use without seeing major drops in performance or quality and will also reduce the cost of replacing items as well as the turnover of materials that need to be destroyed (Mesa et al. 2022). Upon adding reparability to design, the life of the product can be prolonged since the damaged parts can be discharged or fixed, not necessitating tossing the whole product. Modularity increases these goals further by enabling products to be broken-down into interchangeable parts, enabling upgrading, customization and the substitution of individual modules without making the rest of the product obsolete. Collectively, these design principles minimize the impact on the environment, enable resource productive practices and offer new market ideas like product-service systems, leasing or remanufacturing. The most fundamental factors to take into consideration when it comes to the design of a circularity and the facilitation of sustainable resource recovery are demonstrated in Table 2.1. It subdivides these factors into such elements as design strategies, recovery pathways, quality enhancement measures, and enabling tools with a defined purpose, examples, and the benefits (Amend et al. 2022).

2.2.2 Resource Recovery Technologies: Mechanical, Chemical, Biological

The ability to recover and replenish valuable resources from waste streams, byproducts, and post-use products is essential to the success of circular systems. For materials like metals, polymers, and glass, where physical separation and reformation may preserve material integrity, mechanical recovery techniques including sorting, shredding, and reprocessing are often used. Complex compounds may be broken down into monomers or other basic chemicals that can be reintegrated into production using

Table 2.1 Core elements of circular design and sustainable resource recovery (Chrispim et al. 2023)

Aspect	Objective	Examples	Key benefits
Design for durability	Extend lifespan of products	Long-life appliances, heavy-duty tools	Less frequent replacement, reduced waste
Design for reparability	Enable easy repair and part replacement	Laptops with replaceable batteries, repairable kitchen equipment	Supports maintenance culture, lowers costs
Design for modularity	Allow upgrades and customization	Modular smartphones, adaptable shelving units	Extends use, reduces obsolescence
Mechanical recovery	Separate and reprocess materials physically	Metal scrap recycling, glass cullet reuse	Maintains material integrity, low processing energy
Chemical recovery	Break down materials into base components	Depolymerization of plastics, pyrolysis of biomass	Produces high-quality secondary materials
Biological recovery	Convert organic waste into useful products	Composting, anaerobic digestion	Produces bioenergy, soil enrichment
Material quality enhancement	Improve recovered material performance	Purification of recycled metals, upgraded recycled plastics	Meets product standards, wider market acceptance
Digital traceability	Track material origin and processing	Blockchain for recycled plastics, IoT-enabled sorting	Ensures transparency, supports compliance
Reintegration into manufacturing	Incorporate recovered materials into production	Recycled aluminum in automotive parts, remanufactured machinery	Reduces virgin resource demand, supports closed loops

chemical recovery processes such solvent-based extraction, pyrolysis, and depolymerisation. Organic waste may be effectively converted into bioenergy, biochemicals, or nutrient-rich soil amendments by biological processes including composting, anaerobic digestion, and enzymatic treatments. Material properties, contamination levels, and the intended quality of the recovered outputs must all be carefully taken into account when choosing the best recovery process (Ribul et al. 2021).

2.2.3 Enhancing Quality and Reintegration of Recovered Materials

Recovered materials must fulfil performance and safety requirements on par with virgin materials in order for resource recovery to be considered genuinely sustainable. Advanced purification, contaminant elimination, and material reconditioning to restore structural and functional qualities are examples of quality improvement techniques. By offering traceability and verification of material origin, composition, and processing history, digital platforms like blockchain may aid in this process. Designing processes that can accept recycled inputs without sacrificing product performance, quality, or regulatory compliance makes reintegration into production cycles easier. In addition to conserving natural resources, closing these material loops makes manufacturing processes more robust by lowering reliance on erratic international supply chains (Arun et al. 2024).

2.3 Innovations and Digital Enablers for Circularity

2.3.1 Digital Tools: IoT, AI, Blockchain for Tracking and Transparency

Circular economy strategies are becoming more and more facilitated by digital technologies and their efficient implementation. Internet of Things (IoT) enables real-time tracking of products, components, and materials during their life cycle, and companies have the opportunity to monitor the use pattern, identify performance abnormalities and plan convenient maintenance or recovery. Artificial Intelligence (AI) streamlines the process of decision-making by maximizing the flows of resources, making right predictions on the potential of recovering materials, and automating the process of sorting and classifications within the waste processing institutions (Idrissi et al. 2024). The concept of blockchain technology offers permanent records of origin and composition and history of materials processing, which is essential to the assurance of quality, compliance with regulations, and consumer confidence in harvested materials. Using these digital tools as built into one is an interconnected system that increases visibility, traceability, and efficiency within value chains. Table 2.2 summarizes the evidence from academic articles, pilot projects, and industry deployments to demonstrate the use of IoT, AI, and blockchain in the process of making things more circular, through transparency and efficiency. There will be a tangible technological counterpart to every digital tool presented, covering Wiliot IoT-powered shelf-life monitoring, all the way to advanced AI-powered sorting systems Recycleye and Cogersa, intelligent robotics such as ZenRobotics, even brand new blockchain models such as CircleChain and OpenSC. As can be seen in the table, these solutions lead to quantifiable outcomes, enhanced waste sorting, material tracking, minimizing spoilage, and enhanced stakeholder trust (Rai et al. 2024).

Table 2.2 Digital tools enabling circularity: functions, literature examples, and realized benefits (Chauhan et al. 2022)

Digital tool	Core function	Literature-based example	Realized benefit
IoT	Enables real-time tracking of products/ materials	Wiliot's battery-free printable sensors monitor supply chain items from produce to apparel, enabling waste reduction in perishables	Enhanced inventory oversight, reduced spoilage and material loss
AI (Vision and Sorting)	Automates waste identification and classification	Recycleye's AI-vision robotics sort mixed recyclables at 60 images/s with high granularity	Improved sorting precision, higher recovery yields
AI-Vision with robotics	Recognizes complex waste types via multispectral imaging	Cogersa's plastic sorter in Spain achieves 97% accuracy and 50 picks/ min using AI + hyperspectral scanning	Reliable automated sorting in high-volume settings
Combined sensors + AI	Optimizes recycling through multimodal detection	ZenRobotics' AI (ZenBrain) directs robots to differentiate plastics/construction waste with 98% accuracy	Efficient sorting, improved purity of recycled materials
AI + Autonomous pipeline	Classifies and handles textile waste	A 2024 study presents a robotic pipeline using spectral imaging, AI, and digital twin tech for textile sorting	Scalable, precise textile recycling methods
Blockchain	Records immutable data on material provenance	CircleChain's blockchain scheme tokenizes second-life components to enable traceable reuse in electronics	Trustworthy tracking, supports reuse marketplaces
Digital traceability platforms	Confirms product claims and values	OpenSC platform tracks sustainable sourcing of food via blockchain and QR codes	Strengthens consumer trust, enhances transparency

2.3.2 *Biomimicry and Nature-Inspired Innovations*

By mimicking natural processes, where waste is almost nonexistent and resource cycles are naturally regenerative, biomimicry provides a potent design paradigm for attaining circularity (Table 2.3). Nature-inspired solutions may be used in system design, where cycles of production and consumption are modelled after closed ecological loops, or in material development, such as the creation of biodegradable polymers that replicate natural degradation processes (Jain, 2024). Examples include water-purification systems based on wetlands, packaging solutions made from plant-based structures, and self-healing polymers inspired by biological tissues. Biomimicry creates designs that are effective, robust, and naturally regenerative by bringing industrial processes into line with ecological principles. This lessens the impact on the environment while improving functioning (Lebdioui 2022).

Table 2.3 Biomimicry-driven innovations supporting circular economy principles (Hayes et al. 2020)

Biomimicry approach	Natural inspiration	Example application	Key circularity benefit
Biodegradable polymers	Natural polymer degradation (e.g., chitin, cellulose)	Polylactic acid (PLA) bioplastics from corn starch	Reduces plastic waste, compostable at end-of-life
Self-healing materials	Tissue regeneration in organisms	Concrete infused with *Bacillus* bacteria to heal cracks	Extends lifespan of infrastructure, lowers repair needs
Water purification systems	Wetland filtration and plant uptake	Constructed wetlands for wastewater treatment	Low-energy treatment, nutrient recovery
Lightweight structural design	Honeycomb and bone structures	Bio-inspired lightweight composites in transport	Reduces material use and energy consumption
Thermal regulation systems	Termite mound ventilation	Passive cooling building designs	Lowers operational energy demand
Surface coatings	Lotus leaf water repellence	Superhydrophobic, self-cleaning coatings	Minimizes cleaning chemicals, prolongs material life
Packaging design	Fruit peels and shells	Edible or compostable food packaging	Eliminates non-recyclable packaging waste
Circular nutrient cycling	Forest floor nutrient loops	Integrated agroforestry systems	Restores soil health, reduces synthetic inputs

2.3.3 Emerging Recycling and Upcycling Technologies

Technological developments in upcycling and recycling are boosting the potential for recovering high-quality materials from waste streams that are becoming more complicated (Table 2.4). Materials that were previously thought to be unrecyclable may now be recovered because to innovative techniques including solvent-based fibre separation in textiles, chemical recycling of multi-layer plastics, and enhanced bio-refining of organic waste. By turning garbage into higher-value goods, upcycling ideas go beyond restoration. For example, food waste may be turned into bioplastics and biobased chemicals, or post-consumer plastics can be turned into building composites. In addition to increasing recovery yields and material quality, these cutting-edge technologies provide new markets and commercial prospects, increasing the economic viability of circular processes (Salem et al. 2023).

Table 2.4 Emerging recycling and upcycling technologies enabling high-value material recovery (Jung et al. 2023)

Technology type	Process/Method	Example application	Key circularity benefit
Chemical recycling of plastics	Depolymerization, pyrolysis, solvolysis	Breaking down PET into monomers for new food-grade bottles	Recovers high-quality feedstock, reduces virgin polymer use
Solvent-based fiber separation	Selective dissolution and recovery	Separating polyester and cotton from blended textiles	Enables recycling of previously unrecyclable fabrics
Advanced bio-refining of organics	Enzymatic or microbial conversion	Converting food waste into bioplastics or bio-based chemicals	Adds value to organic waste, reduces landfill methane
High-yield metal recovery	Electrochemical and hydrometallurgical processes	Extracting rare earths from e-waste	Reduces mining impact, recovers critical raw materials
Upcycling plastics to composites	Blending with natural fibers or additives	Turning plastic bags into construction boards	Produces durable, higher-value materials
Glass upcycling	Thermal or mechanical transformation	Turning waste glass into decorative tiles or filtration media	Expands market value of recovered glass
Rubber devulcanization	Breaking sulfur bonds in vulcanized rubber	Regenerating rubber from end-of-life tires	Extends lifecycle of rubber, reduces raw material demand
Carbon capture and material conversion	Mineralization or polymer synthesis	Converting captured CO_2 into building materials or plastics	Creates market value from emissions, closes carbon loops

2.4 Collaborative Systems and Policy Frameworks

More than only sustainable product design and technology innovation are needed for the effective adoption of a circular economy; strong stakeholder engagement, supportive legislative frameworks, and shared value-generating economic mechanisms are also essential (Fig. 2.2). In order for circular methods to become integrated, scaled, and self-sustaining within economic, social, and environmental systems, collaborative systems and supporting frameworks are necessary (Diaz et al. 2021).

2.4.1 Multi-stakeholder Collaboration and Supply Chain Integration

Circular economy implication is systemic in its nature and would involve collaboration of various actors around manufacturers, suppliers, consumers, policymakers, researchers and waste management practitioners. The solution is co-creation through multi-stakeholder cooperation which by exchanging knowledge and sharing resources allows access to more solutions than either one of them could do alone. Supply chains where information, materials, and responsibilities are shared in an open and transparent manner as an integrated supply chain are of essence in closing material loops. Indicatively, manufactures collaborating with recyclers can make sure that the recovered materials are made to design specification and logistics providers can optimize reverse flows on product take-back schemes. The industrial

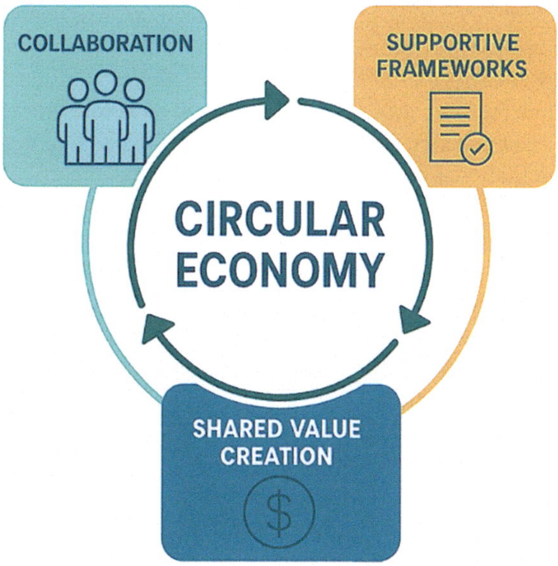

Fig. 2.2 Collaboration, supportive frameworks, and shared value creation are essential pillars for a thriving circular economy

symbiosis networks (e.g., platforms) are good examples of how collaboration can turn waste products of an industry into input and create regional resource efficiency and resilience (Inês et al. 2025).

2.4.2 Regulatory and Policy Instruments Supporting Circular Economy

By establishing goals, assigning roles, and enforcing adherence, efficient regulatory frameworks provide the legal basis for circularity. Extended producer responsibility (EPR) programs, obligatory recycling quotas, eco-design guidelines, and prohibitions on single-use items are just a few examples of the many ways that policy tools may be implemented. These actions foster accountability and encourage a shift in the market towards circular business practices. Long-term relevance is also ensured by adaptable, flexible policies that take stakeholder input and changing technology into account. For example, to speed up circular transitions across industries, the European Union's Circular Economy Action Plan combines strategic investment with legislative measures. In order to facilitate easier cross-border trading in secondary raw materials, these frameworks are also essential for standardising requirements for material quality, waste categorisation, and reporting (Dziedzic et al. 2025).

2.4.3 Economic Incentives and International Cooperation

Adopting circular solutions may be made far less expensive with the use of economic incentives including tax credits, subsidies, preferred procurement practices, and lower tariffs for circular items. These mechanisms incentivise investments in business model reform, product innovation, and recycling infrastructure. By facilitating information transfer, standardising rules, and expanding markets for recovered resources, international collaboration further increases the effect. International programs like the Ellen MacArthur Foundation's CE100 network and the Platform for Accelerating the Circular Economy (PACE) show how cross-national collaborations may exchange best practices, raise capital, and match plans with international sustainability objectives. Such collaboration guarantees that circular systems are not restricted to local or national levels but rather support robust, resource-efficient, and low-carbon global value chains in a linked economy (Van Opstal and Smeets 2023).

2.5 Case Studies and Practical Applications

The scenario where the same principles of circular economy are applied in practice and have witnessed the benefits as well as challenges is best observed through the various real-life projects aimed at altering the current regressive cycle towards regenerative systems. Through the analysis of successful experiences both on industrial and community levels, as well as the description of performed resource recovery projects, it will be possible to derive actionable insights and outline obstacles towards scaling circular solutions. The insights are also useful guides by policymakers, companies and civil society that need to incorporate circularity in both operational and strategic models (Rodríguez-Espíndola et al. 2022). Place-based projects, the circular policies in use, and environmental and economic, measurable results are described in Table 2.5. These could be a model of industrial symbiosis like in the cities of Kalundborg (Denmark) and Ulsan (South Korea), a system of community-led waste management like it is in Kamikatsu (Japan) and Surabaya (Indonesia), and novel recycling technologies such as solvent-based textile recovery in Sweden and Finland. The choice is representative of the contexts of both developed and developing countries and provides lessons on scalability, stakeholder involvement, and embedding itself within existing value chains (Jacobsen 2006). This whole-of-system review allows practitioners, policymakers and researchers to benchmark strategies, determine which best practices can be transferred and understand how to integrate circularity practices into a variety of socio-economic contexts.

2.5.1 Industrial and Community-Level Circular Initiatives

From small-scale community initiatives to massive industrial systems, circular economy ideas may be implemented at various dimensions. At the industrial level, eco-industrial parks like Denmark's Kalundborg Symbiosis demonstrate how waste may be reduced and reciprocal economic gains can be created by converting byproducts from one facility into inputs for another. Community-led projects, like Japan's Kamikatsu zero-waste villages, show how local legislation, participatory governance, and behavioural change may result in very high recycling and composting rates. Circular solutions may fulfil local requirements while lowering environmental footprints, as shown by local energy generating initiatives that use biogas from municipal trash and urban agricultural systems that use organic waste for compost (Jain and Dhupper 2025; Panda et al. 2025).

Table 2.5 Case studies and practical applications of circular economy strategies at industrial and community levels

Category	Example/ Location	Key circular practice	Outcomes and impacts	References
Industrial-scale initiatives	Kalundborg Symbiosis, Denmark	Industrial symbiosis—exchange of energy, water, and by-products between companies	Reduced CO_2 emissions by 635,000 tons/ year; cut water use by 3 million m^3/year; lowered costs	Jacobsen (2006)
	Ulsan Eco-Industrial Park, South Korea	By-product utilization in petrochemical cluster	Waste reduction by 47%; annual savings of $11 M for participating firms	Park et al. (2008)
	Rotterdam, Netherlands	Chemical recycling of mixed plastics	Recovery of high-purity polymers enabling closed-loop applications	Cafiero et al. (2025)
Community-level initiatives	Kamikatsu, Japan	Zero-waste policy with 45 waste separation categories	80% recycling rate; reduction in landfill use	Shenyoputro and Jones (2023)
	Surabaya, Indonesia	Waste-to-compost community program	Organic waste diverted from landfill; improved urban soil fertility	Farahdiba et al. (2023)
	Milan, Italy	Separate food waste collection for biogas and compost	100,000 tons/ year of food waste converted into renewable energy and fertilizer	Wang and Bai (2022)
Resource recovery projects	Sweden and Finland textile recycling plants	Solvent-based fiber separation for cotton/ poly blends	High-quality fiber recovery for apparel reuse	Peters et al. (2019)
	Pune, India	Waste pickers cooperative (SWaCH) integrating informal sector	22% diversion of municipal solid waste; improved livelihoods for 3,000 + workers	Kilby et al. (2013)
	Toronto, Canada	Construction & demolition waste recovery facility	90% material recovery rate; diversion of 800,000 tons/ year from landfill	Santos et al. (2024)

2.5.2 *Successful Resource Recovery Projects and Lessons Learned*

Numerous projects have demonstrated that targeted interventions can unlock significant value from waste streams. For instance, chemical recycling plants for plastics in the Netherlands have achieved high-purity polymer recovery from mixed waste, enabling reuse in high-performance applications. Similarly, textile recycling initiatives in Sweden and Finland have pioneered solvent-based fiber separation, reintroducing high-quality fibers into the apparel industry. Key lessons from such projects include the importance of early stakeholder engagement, the integration of recovery processes into existing supply chains, and the adoption of quality control systems to ensure market acceptance of recovered materials. Transparent communication with end-users about the performance and environmental benefits of recycled products has also proven crucial for market uptake (Trushna et al. 2024).

2.5.3 *Challenges and Opportunities for Scaling Circular Practices*

Although localised projects and successful pilots show viability, there are obstacles when attempting to scale circular methods at the national and international levels. Progress may be hampered by technical obstacles like waste stream pollution or the restricted scalability of certain recovery technology. Adoption is made more difficult by economic considerations such as competition from low-cost virgin resources, market instability for secondary minerals, and expensive initial capital expenditures. Inadequate international coordination and policy misalignment may also be obstacles to cross-border circular commerce. Nonetheless, these obstacles provide chances for creativity, cooperation, and funding. Favourable circumstances for expanding circular systems are being created by developments in digital technology, rising consumer demand for sustainable goods, and changing regulatory frameworks. Harmonised material standards, public–private collaborations, and focused financial incentives might help close the gap between promising pilots and broad adoption (Mayer et al. 2025).

2.6 Conclusion and Future Directions

A circular economy is a paradigm shift in approach to how global resources are managed since it will no longer be the conventional waste management schemes but rather the philosophy of managing resources that is devoted to making the best use of the resources, reconstruction and retention of value. As noted in this chapter, new storage and renewal technologies, networking platforms, and practical experience are enhancing the possibility and cost-effectiveness of circularity. The positive impact on the environment and socio-economic aspects of the implementation of circular solutions is supported by case studies of industrial symbiosis networks, community-based zero-waste programs and more-advanced resource recovery initiatives. Although it has come this far, there are still huge gaps in research. Scaling up Many potentially very promising technologies, including chemical recycling of complex plastics or bio-refining of organic waste, currently have a problem in scaling up and in operating at a reasonable cost. Strong life cycle assessments to analyze environmental trade-offs, quality benchmarks on secondary material that are standardized, and further adoption of digital technology to promote supply chain transparency with blockchain and AI, are also required. Moreover, social aspects such as behaviour transformation, consumer acceptance, and the provision of equitable access to the benefits of circularity should be studied further in interdisciplinary aspects. In the future, the challenge of integrating circularity in international management of resources would demand a multi-faceted strategic plan. Making cross-border comparisons of regulatory practice, policymakers also need to promote consistency in policies to ensure that markets in secondary materials are stable, and there is cross-sector cooperation. Companies ought to consider design-reuse and closed-loop manufacturing as well as make investments in innovation that enhances the recovery of materials and the durability of the products. Skills trainings, awareness and education shall play very crucial roles in the development of a team of people able to make this transition happen. To conclude, the future of circularity is linked with the combination of technologies and system-level policy changes and culture change. The global shift toward the circular economy has the potential to go beyond its niche use and become a mainstream force that can steer sustainable growth and planetary well-being into the future by aligning economic incentives with environmental stewardship, building international collaboration, and promoting innovations.

References

Agarwal S, Jain H (2024) Investigating the potential of floral waste as a vermicompost and dual-functional biosorbent for sustainable environmental management. Water Air Soil Pollut 235(5):322

Agarwal S, Jain H, Dhupper R, Mathur A (2025) Optimizing methyl orange dye removal from aqueous solutions using white chrysanthemum floral waste-derived bioadsorbent: a study of kinetics, thermodynamics, isotherms, and RSM optimization. Int J Environ Res 19(3):104

Amend C, Revellio F, Tenner I, Schaltegger S (2022) The potential of modular product design on repair behavior and user experience—evidence from the smartphone industry. J Clean Prod 367:132770

Arun M, Barik D, Chandran SSR (2024) Exploration of material recovery framework from waste—a revolutionary move towards clean environment. Chem Eng J Adv 18:100589

Cafiero LM, De Angelis D, Tuccinardi L, Tuffi R (2025) Current state of chemical recycling of plastic waste: a focus on the Italian experience. Sustainability [Online] 17

Chauhan C, Parida V, Dhir A (2022) Linking circular economy and digitalisation technologies: a systematic literature review of past achievements and future promises. Technol Forecast Soc Change 177:121508
Chrispim MC, Mattsson M, Ulvenblad P (2023) The underrepresented key elements of circular economy: a critical review of assessment tools and a guide for action. Sustain Prod Consumption 35:539–558
Diaz A, Schöggl J-P, Reyes T, Baumgartner RJ (2021) Sustainable product development in a circular economy: implications for products, actors, decision-making support and lifecycle information management. Sustain Prod Consumption 26:1031–1045
Dziedzic R, Pondicherry P, Dziedzic M (2025) Review of national policy instruments motivating circular construction. Resour Conserv Recycl 215:108053
Farahdiba AU, Warmadewanthi IDAA, Fransiscus Y, Rosyidah E, Hermana J, Yuniarto A (2023) The present and proposed sustainable food waste treatment technology in Indonesia: a review. Environ Technol Innov 32:103256
Hayes S, Desha C, Baumeister D (2020) Learning from nature–biomimicry innovation to support infrastructure sustainability and resilience. Technol Forecast Soc Change 161:120287
Idrissi ZK, Lachgar M, Hrimech H (2024) Blockchain, IoT and AI in logistics and transportation: a systematic review. Transp Econ Manag 2:275–285
Inês A, Diniz A, Moreira AC (2025) Driving eco-innovation in supply chains through multi-stakeholder collaboration: a review and research agenda. J Open Innov Technol Market Complex 11(1):100472
Jacobsen NB (2006) Industrial symbiosis in Kalundborg, Denmark: a quantitative assessment of economic and environmental aspects. 10(1–2):239–255
Jain H (2024) From pollution to progress: groundbreaking advances in clean technology unveiled. Innov Green Dev 3(2):100143
Jain H (2025) Synergizing carbon reduction technologies and nature-based solutions for a climate-neutral future. Mitig Adapt Strat Glob Change 30(6):42
Jain H, Dhupper R (2025) Holistic strategies for sustainable buildings and their impacts on soil and environmental health. J Build Pathol Rehabil 11(1):10
Jain H, Dhupper R, Shrivastava A, Kumar D, Kumari M (2023a) Leveraging machine learning algorithms for improved disaster preparedness and response through accurate weather pattern and natural disaster prediction. 11
Jain H, Dhupper R, Shrivastava A, Kumari M (2023b) Enhancing groundwater remediation efficiency through advanced membrane and nano-enabled processes: a comparative study. Groundwater Sustain Dev 23:100975
Jung H, Shin G, Kwak H, Hao LT, Jegal J, Kim HJ, Jeon H, Park J, Oh DX (2023) Review of polymer technologies for improving the recycling and upcycling efficiency of plastic waste. Chemosphere 320:138089
Kilby P (2013) Waste recycling and the household economy: the case of the pune waste-pickers' response to the changing 'Rules of the Game.' In: Elias J, Gunawardana SJ (eds) The global political economy of the household in Asia. Palgrave Macmillan UK, London
Kirchherr J, Yang N-HN, Schulze-Spüntrup F, Heerink MJ, Hartley K (2023) Conceptualizing the circular economy (Revisited): an analysis of 221 definitions. Resour Conserv Recycl 194:107001
Lebdioui A (2022) Nature-inspired innovation policy: Biomimicry as a pathway to leverage biodiversity for economic development. Ecol Econ 202:107585
Mayer J, Schittenhelm H, Pott PP (2025) Opportunities and challenges of implementing circular economy in the medical device sector in Germany. Sustain Prod Consumption 56:657–670
Mesa JA, Gonzalez-Quiroga A, Aguiar MF, Jugend D (2022) Linking product design and durability: a review and research agenda. Heliyon 8(9):e10734
Panda D, Haque S, Frishammar J, Parida V (2025) Modularity for circular economy: four circularity pathways for industrial firms. J Clean Prod 519:146001

Park H-S, Rene ER, Choi S-M, Chiu ASF (2008) Strategies for sustainable development of industrial park in Ulsan, South Korea—from spontaneous evolution to systematic expansion of industrial symbiosis. J Environ Manag 87(1):1–13

Peters GM, Sandin G, Spak B (2019) Environmental prospects for mixed textile recycling in Sweden. ACS Sustain Chem Eng 7(13):11682–11690

Rai HM, Shukla KK, Tightiz L, Padmanaban S (2024) Enhancing data security and privacy in energy applications: integrating IoT and blockchain technologies. Heliyon 10(19):e38917

Ribul M, Lanot A, Tommencioni Pisapia C, Purnell P, McQueen-Mason SJ, Baurley S (2021) Mechanical, chemical, biological: moving towards closed-loop bio-based recycling in a circular economy of sustainable textiles. J Clean Prod 326:129325

Rodríguez-Espíndola O, Cuevas-Romo A, Chowdhury S, Díaz-Acevedo N, Albores P, Despoudi S, Malesios C, Dey P (2022) The role of circular economy principles and sustainable-oriented innovation to enhance social, economic and environmental performance: evidence from Mexican SMEs. Int J Prod Econ 248:108495

Salem KS, Clayson K, Salas M, Haque N, Rao R, Agate S, Singh A, Levis JW, Mittal A, Yarbrough JM, Venditti R, Jameel H, Lucia L, Pal L (2023) A critical review of existing and emerging technologies and systems to optimize solid waste management for feedstocks and energy conversion. Matter 6(10):3348–3377

Santos G, Esmizadeh E, Riahinezhad M (2024) Recycling construction, renovation, and demolition plastic waste: review of the status quo, challenges and opportunities. J Polym Environ 32(2):479–509

Shenyoputro K, Jones TE (2023) Reflections on a two-decade journey toward zero waste: a case study of Kamikatsu town, Japan. 11

Trushna T, Krishnan K, Soni R, Singh S, Kalyanasundaram M, Sidney Annerstedt K, Pathak A, Purohit M, Stålsby Lundbog C, Sabde Y, Atkins S, Sahoo KC, Rousta K, Diwan V (2024) Interventions to promote household waste segregation: A systematic review. Heliyon 10(2):e24332

Upadhayay S, Alqassimi O, Khashadourian E, Sherm A, Prajapati D (2024) Development in the circular economy concept: systematic review in context of an umbrella framework. Sustainability [Online] 16

Van Opstal W, Smeets A (2023) Circular economy strategies as enablers for solar PV adoption in organizational market segments. Sustain Prod Consumption 35:40–54

Velenturf APM, Purnell P (2021) Principles for a sustainable circular economy. Sustain Prod Consumption 27:1437–1457

Wang S, Bai X (2022) Compatibility in cross-city innovation transfer: Importance of existing local experiments. Environ Innov Societal Trans 45:52–71

Chapter 3
Breaking the Linear Trap: Advancing Circular Economy Strategies for Sustainable Resource Management

Surbhi Srivastava, Shanu Singh, J. Kishore Kumar, and Dominic R. Karuthedath

3.1 Introduction

We have mass consumed our resources to such an extent that the linear economic model implies we are degrading environmental benefits and that means depletion of resources. The linear paradigm of "take, make, dispose" works under the premise of infinite natural resources, and the Earth has unlimited capacity to absorb waste (Beccarello and di Foggia 2018). On the other hand, the circular economy (CE) provides an alternative approach, focusing on the efficiency of resources use minimizing waste with a regenerative approach, The principles of CE, including recycling, reusing, and reducing, drive to deliver an economy in which materials are kept in closed-loop systems to cycle continuously within the economy (Belmonte-Ureña et al. 2021). This approach addresses pressing challenges, including climate change, resource scarcity, and waste generation, by transforming waste into valuable resources (Belmonte-Ureña et al. 2021). CE has become the backbone of sustainability strategies around the world. For example, the Circular Economy Action Plan is one of the EU leading policies on materials efficiency, better recycling infrastructures, and waste reduction to enhance resilience both economically and environmentally (Berg et al. 2018). At the same time, CE principles are being incorporated into sectors such as urban planning, industrial development and waste management in countries of the Global South, driven by the need to solve emerging environmental problems while pursuing economic development (Chigbu 2024).

S. Srivastava · S. Singh (✉) · J. Kishore Kumar · D. R. Karuthedath
Christ University, Bengaluru, Karnataka, India
e-mail: shanu.singh3110@gmail.com

S. Srivastava
e-mail: Surbhi233@gmail.com

H. Jain et al. (eds.), *The Circular Fallacy: Closing the Loop in a Consumerist World*, Circular Economy and Sustainability, https://doi.org/10.1007/978-3-032-22687-7_3

3.1.1 SDG 12 (Responsible Consumption and Production)

Sustainable development goal 12 (SDG 12) seeks to ensure sustainable consumption and production patterns and is adopted as part of the 2030 Agenda for Sustainable Development.

Sustainable development goal 12 SDG 12 entails ensuring sustainable consumption and production patterns meaning the sustainable management and efficient use of natural resources, the reduction of waste generation through prevention, reduction, recycling and reuse and encourages companies, especially large and transnational companies, to adopt sustainable practices and to integrate sustainability information into their reporting cycle (Dantas et al. 2021). The guiding principles of CE resonate closely with SDG 12 especially in terms of sustainable production and waste disposal practices. This means that for instance reduction in mineral waste due to recycling programs contributes to the achievement of SDG 12.5 which aims to substantially reduce waste generation through prevention, reduction, recycling and reuse by 2030 (Khairul Akter et al. 2022). Likewise, circular economy (CE) practices, such as eco-design and extended product life cycles, respond to the sustainable consumption challenge addressed in SDG 12 (Kravchenk et al. 2019).

Furthermore, CE practices generate pathways for more efficient economic production with a lesser ecological footprint. As an example, integrating CE in the textile sector significantly decreases the waste and resource consumption associated with fabric, which promotes achievements of SDG 12 by both mitigating leather waste and enabling innovation (Marrucci et al. 2019). CE offers a path to global sustainability ambitions, particularly SDG 12 (Moraga et al. 2019) by moving away from resources intensive production and consumption models.

3.1.2 Circular Economy and Its Core Principles

The circular economy (CE) is a modern economic model that overcomes the limits of the linear economy and focuses on sustainability, efficiency, and waste elimination. Linear traditional models, on the other hand, involve a take, make and dispose process that promotes high levels of raw material extraction, large amounts of waste, and negative environmental effects. CE changes this paradigm by developing a closed-loop system that continuously recycles, reuses and repurpose within the economy (Belmonte-Ureña et al. 2021). This model is attracting worldwide attention as of striking a balance between economic growth and sustainable environmental development. The micro-level businesses drive CE through specific sub-practices like redesigning their product and being efficient with materials to minimize waste and also improve their sustainability. Industrial symbiosis, on the meso-level, enables industries to partner and exchange material, resource, and energy to hance efficiency. Governments and international organizations at the macro-level adopt policies and frameworks that accelerate systemic uptake of CE principles throughout economies

(Belmonte-Ureña et al. 2021). The EUs Circular Economy Action Plan has shown this is how comprehensive policies can be a driver for change. The approach focuses on waste minimization, material recovery, and eco-design, evidencing the opportunities of CE to align with environmental objectives with economic resilience (Berg et al. 2018). Countries such as China have also integrated the CE concept into the content of national strategies to combat resource scarcity and environmental degradation and to promote sustainable development cycle (Dantas et al. 2021).

The basics of the circular economy are based on the 3R principles (Reduce, Reuse, Recycle). These principles will help to redirect business practices, policy and consumer direction towards sustainability:

Reduce: The first step to sustainability is to reduce resource use and waste generation. These principal stresses more on increasing efficiency and integrating sustainability across all production and consumption. Companies are utilizing light-weight materials, adopting energy-efficient technologies, and manufacturing durability-oriented design products to lessen environmental impact cycle (Dantas et al. 2021).

Reuse: The reuse principle aims at extending the life cycle of products and materials through their reuse in their original form. Such a strategy has already sprung up strongly in the electronics and textiles sectors, where refurbished devices and the second-hand market play a significant role in curbing waste (Shrivastava et al. 2024). Focusing on reuse can allow businesses and consumers to reduce resource extraction and prevent waste from going to landfills (Kravchenko et al. 2019).

Recycling: Recycling refers to taking old items, and using a process to make them new items or raw materials. This principle is key towards closing the loop of the production cycle as it keeps the materials in the economy with their value. Wastes in these broader categories can contain metals, plastics, and paper, and advanced recycling technologies are being used to recover these valuable resources and reduce dependence on virgin materials (Moraga et al. 2019).

Together, the principles of 3R contribute to an overarching systemic transformation in the way goods and services are produced and consumed in the economy, promoting sustainability of the economy in a way that is economically viable (innovation (Marrucci et al. 2019).

3.1.3 Linear Versus Circular Economic Models Comparison

For decades, the linear economic model has been the go-to for industries that use the excuse of economic development to bolster environmental destruction and irreversible damage to our planet. The ecological impacts of this model, based on extraction, disposable culture and disposal, are detrimental, leading to pollution, deforestation, depletion of resources (Belmonte-Ureña et al. 2021). CE offers an eco-friendly solution via the repeated utilization of resources and materials. Waste is considered as a resource, end product is not consumed in a circular model (Agarwal and Jain

2024; Agarwal et al. 2025). For example, the raw material created from converting by-products from industry can be used in other industries thereby creating a symbiotic relationship among industries to limit environmental impacts. In addition, CE is based on eco-design, which promotes the design and manufacture of durable, reparable and recyclable products (Ranta and Saari 2019). Such a move will not only reduce environmental burdens, but also strengthens economic sustainability as it reduced reliance on non-renewable resources (Berg et al. 2018).

3.1.4 Resource Efficiency and Waste Management Contribution

Across sectors, CE has shown potential for resource efficiency and waste management improvements. A set of principles from CE that empowers industries to study the materials they are using and the way it is being used, and minimize wastage (Kravchenko et al. 2019) conserving resources in the process. For instance, recycled concrete and modular building designs have reduced resources in the construction sector while cutting down costs (Dantas et al. 2021). CE minimizes energy reduce plus develops greenhouse gas discharges through the use of the many power effective technologies and processes. Energy conservation is mainly done through the incorporation of waste heat recovery systems in industrial operations, which is a perfect example of the contribution of CE (Kravchenko et al. 2019). CE focuses on converting waste into resources. One of these, for example, is management of food waste, where organic waste may be reused as bioenergy, compost or as feed for livestock, reducing the amount of organic waste delivered to a landfill site and sustaining environmentally efficient practices (Khairul Akter et al. 2022). CE reduces reliance on raw materials and consequently reduces waste management expenses. It also fosters innovation as firms invent new technologies and new business models to be consistent with CE principles (Moraga et al. 2019). CE helps with not only resource limitation issues but also with the climate change challenge. CE minimizes environmental degradation and contributes to global sustainability goals by means of reduced resource extraction while simultaneously promoting recycling (Bhardwaj et al. 2025; Jain et al. 2023; Jain 2025).

3.1.5 Connection Between CE and SDG 12

Sustainable Development Goal 12, (SDG 12) is the goal of the 2030 Agenda for Sustainable Development of the United Nations, which seeks to ensure sustainable consumption and production patterns. It acknowledges the need to combat the environmental effects of current production and consumption practices while promoting economic growth, and enhancing quality of life principles (Moraga et al. 2019).

3.1.5.1 Synergies Between Circular Economy Practices and SDG 12 Goals

Efficient Resource Utilization (SDG 12.2: CE focuses on utilizing all resources effectively by embedding a scientific approach (in terms of sustainable production). For instance, companies adopting eco-design practices produce goods with longer lifetimes, which leads to reduced extraction of raw material for short-lived products, aiding sustainable resource use principles (Moraga et al. 2019). Waste is considered a resource in CE and recycling; reusing and repurposing are cornerstones. There are several businesses that incorporate circular economy principles into the operation of their firms to promote more closed-loop production systems, utilize secondary raw materials, and support renewable energy usage. Such measures are consistent with SDG 12.6 because they incorporate sustainability into corporate strategies and reporting frameworks (Valverde and Avilés-Palacios 2021). CE tackles food waste from various angles, including smart solutions that can transform food surplus into animal feed or bio energies. Such activities not only prevent waste but also enhance the resource efficiency of agricultural and food systems (Moraga et al. 2019).

3.2 Research Objective

The following are the research objectives of the current chapter:

- Examine CE (Circular Economy) approaches in achieving SDG 12 (Sustainable Development Goal 12) relating to waste management, resources, and production in different sectors.
- Analyse the major factors that limit the adoption of Circular Economy and policy intervention, technological, and collaboration-based solutions to increase its adoption.

3.3 Research Problem

This chapter strives to solve the problem surrounding the ineffectiveness of traditional linear economic models (take, make, dispose). Such models tend to deplete resources exceedingly, encourage environmentally damaging practices, and promote consumption beyond sustainable levels. Traditional Circular Economies (CE) are beneficial because they seek to eliminate waste by rethinking production and consumption systems. However, they too have problems when it comes to their implementation including high initial investment costs, absence of clear policies, limited knowledge amongst consumers, and insufficient technologies. To tackle these issues, the research points out the need for technology and innovation policy reforms, and education of the consumers and the industries. Through these strategies, the Circular Economy can be made realistic and operational within SDG 12, which targets sustainable consumption and production alongside economic and environmental development.

3.4 Research Methodology

This study employs a systematic review approach to examine the circular economy (CE) and its connection to Sustainable Development Goal 12 (SDG 12). It starts with an extensive literature review, utilizing academic sources, policy documents, and industry case studies to create a theoretical basis. The research explores the principles of CE—reduce, reuse, and recycle and how they contribute to encouraging responsible consumption, minimizing waste, and enhancing resource efficiency. A case study approach examines CE implementation across sectors like metallurgy, energy, textiles, and food waste management, showcasing international best practices, encompassing EU regulations, China's CE Promotion Law, and Bangladesh's textile efforts. The study additionally contrasts linear and circular economic models, highlighting the environmental and economic advantages of CE. Policy approaches like Extended Producer Responsibility (EPR), collaborations between public and private sectors, and technological advancements such as Industry 4.0 and digital platforms are examined for their contribution to enhancing CE practices. The research highlights major challenges such as budget limitations, customer habits, and policy shortcomings, while suggesting remedies like financial incentives and awareness initiatives. The paper highlights the need for systemic changes, collaboration among stakeholders, and innovation to boost CE's role in SDG 12, calling for more robust policies, industry partnerships, and community involvement to achieve a sustainable economic shift.

3.5 Research Gap

In essence, this paper highlights several knowledge gaps in researching CE and its correspondence to SDG 12. Though it outlines the CE best practices across the different industries, there is less critical analysis over the scalability issues in developing countries with scarce resources. The interface of Industry 4.0 technological advancements within CE frameworks still lacks adequate and concrete exploration; even in old-fashioned industries, such as that of textiles or garments. Moreover, while it mentions policy mechanisms like EPR, the real-world effectiveness is not assessed based on the studies, especially for a world with no standardized regulation globally. This study also did not give so much emphasis to consumer behaviour that may influence the adoption of CE, such as willingness to pay for circular products. Lastly, the study does not adequately explore CE's interlinkages with other SDGs, such as climate action (SDG 13) and economic growth (SDG 8). Filling these gaps could make CE more practical and sustainable.

3.6 Illustrative Evidence from Different Sectors and Regions

CE practices are effective in promoting SDG 12 in different sectors and regions based on empirical studies. The implementation of the Circular Economy in textile manufacturing in Bangladesh has resulted in a massive reduction of fabric waste, conservation of water, and reduction in the use of chemicals. Therefore, sustainable production practices not only fulfil the requirement of SDG 12 but also help in achieving the larger goal of environmental sustainability by the industry (Rodríguez-Antón et al. 2022) (Table 3.1). The construction sector has adopted CE practices in Europe through the use of recycled materials like concrete and steel. Such measures helped to decrease wastage in construction and also encouraged resource efficiency as stated in targets 12.2 and 12.5 of the SDGs (Moraga et al. 2019). CE initiatives in developing regions have been directed towards preventing food waste by systems that utilize biowaste to produce bioenergy and fertilizers. They serve SDG 12.3 by alleviating food security issues (Valverde and Avilés-Palacios 2021; Shrivastava 2021). CE principles have been applied to e-waste recycling programs e.g., to some degree in the European Union). Wasteless management programs minimize environmental damage and increase resource use efficiency by (Rodríguez-Antón et al. 2022) reclaiming precious materials from obsolete electronic devices.

Table 3.1 Key Targets of SDG 12 and their alignment with circular economy practices (Author's Compilation)

SDG 12 Target	Description	Circular economy practice
SDG 12.2: Sustainable resource	Achieve sustainable management and efficient use of natural resources	Resource efficiency through material recycling
SDG 12.3: Food waste reduction	Halve global per capita food waste by 2030	Organic waste to bioenergy or compost
SDG 12.5: Waste reduction	Substantially reduce waste generation through prevention, reduction, recycling, and reuse	Adoption of the 3R framework
SDG 12.6: Sustainable business practices	Encourage companies to adopt sustainable practices and integrate sustainability into reporting	Eco-design and closed-loop manufacturing
S06 12.7: Public procurement	Promote sustainable public procurement policies aligned with national priorities	Procurement of recycled materials

3.7 Implementation Strategies for CE in Achieving SDG 12

National and Regional Policies in Support of CE Worldwide, governments have embraced policies supporting the circular economy (CE) as a pathway to SDG 12. National approaches tend to centre waste minimization, resource use efficiency, and sustainable consumption. Example: The Action Plan for the Circular Economy within the European Union is a wide-scale policy framework that covers various aspects for higher material recovery, less generation, and overall sustainable consumption and production patterns (Belmonte-Ureña,). China implemented a Circular Economy Promotion Law that obliges numerous industries to constrict their environmental footprint by enforcing circular economy (CE) practices (e.g., recycling, sharing resources) synchronously with growth (Pollice and Batocchio 2018). Likewise, India's EPR policies for industry also mandate post-consumer waste by manufacturers, promoting sustainable waste disposal and recycling. Global treaties and joint actions are the next level the multilateral aspects have also been key, to promote CE. The PAGE initiative of the United Nations converges countries to shift towards sustainable economic models. The Basel Convention on the Control of Transboundary Movements of Hazardous Wastes and their Disposal is indispensable in managing waste and recycling hazardous materials internationally (Prokic et al. 2022). These accords highlight the need for global collaboration on the harmonization and dissemination of circular economy practices. Table 3.2 shows circular economy practices across sectors reduce mining impacts.

3.7.1 *Industrial Practices*

CE of the Manufacturing Sector is altering manufacturing processes by encouraging resource efficiency and a reduction of waste. In the automobile sector, manufacturers

Table 3.2 Examples of circular economy practices across sectors reduce mining impacts (Author's Compilation)

Sector	Circular practice	Impact
Manufacturing	Use of recycled materials in automotive production	Reduces dependency on virgin materials and lowers costs
Textile	Recycling fabric scraps into new textiles	Reduces textile waste and conserves resources
Construction	Use of recycled concrete in infrastructure projects	Reduces construction waste and promotes resource reuse
Food Waste	Repurposing organic waste into bioenergy or compost	Minimizes landfill waste and supports sustainable farming
Electronics	Recycling lithium-ion batteries for raw material recovery	Enhances resource efficiency and reduces mining impacts

are progressively integrating recycled materials like steel and aluminium into vehicle fabrication, as an instance. With modular designs, products are easier to repair and refurbish, prolonging their lifespan and reducing material waste (Khairul Akter et al. 2022). Also, some companies like Renault have closed-loop manufacturing systems which turn end-of-life vehicles into new production inputs, upholding the prescripts of the goal 12 of the conservation of all for humanity. The CE principles have been well-accepted by the food industry to tackle food waste issues. This has attracted several innovations in surplus food redistribution and organic waste processing. Ex: Food waste has been converted to animal fodder, compost and bioenergy since they provide more sustainable practices and reduce the amount of waste that would otherwise go to landfills (Raman et al. 2024). Policies that foster the dissemination of such practices are also key to scaling these solutions to a global level. The textile and apparel sector has incorporated CE practices to mitigate the inefficiency and waste inherent in their practices. Production techniques are increasingly more sustainable, with companies putting into practice techniques like up-cycling, as well as reduction in water and energy used during the manufacturing process. For instance, the economy of Bangladesh has seen commendable CE practices since then; in this case, a reduction of textile is achieving through CE implementation in Bangladesh while conserving water that shows the capability of aligning CE with SDG 12 for the sector (Rodriguez-Anton et al. 2024).

3.7.2 Technological Innovations

Importance of Industry 4,0 and Digital Technologies Technological advances are a key enabler of CE practices. Notably, Industry 4.0 technologies such as loT, Al and Blockchain facilitate the real-time tracking of materials, enabling optimized resource use and enhanced waste management (Dantas et al. 2021). For instance, if we talk about the manufacturing processes, loT sensors play a critical role in the monitoring of material flows in the processes so that resources can be used efficiently while generating as little waste as possible (Panchal et al. 2021). Al algorithms underpin predictive maintenance, allowing industries to lengthen the lifespan of machinery and minimize material waste (Pasca et al. 2023). The use of Blockchain technology enables a transparent mechanism for supply chain practices where accountability and sustainability can be ensured. Digital platforms, in addition to enabling consumers to participate in the CE. For example, when you reuse products that you have purchased through renting, sharing, or reselling apps, you will simply be avoiding the consumption of new resources. They showcase the ability of tech to scale CE, coupled with achieving targets of SDG 12.

3.8 Practical Case Studies

3.8.1 Recycling of EV Batteries—for Resource Efficiency

The recycling of electric vehicle (EV) batteries has become influential in the connection between circular economy (CE) and resource efficiency. The driving force behind this race is the lithium-ion battery used in EVs, which requires one or more of the critically important and finite raw materials—lithium, cobalt, and nickel—which are highly convertible to the environment too. This not only decreases the demand for virgin materials but also lessens the environmental footprint resulting from mining and disposal (Chigbu et al. 2024). These recycling processes can recover valuable materials from end-of-life batteries for reuse in new battery production or other applications. Advanced recycling technologies, such as those developed by Umicore, enable the recovery of critical metals. Battery recycling directly relates with SDG 12.2 as it not only minimizes the wastage but also helps to improve resource efficiency and promote sustainable industries.

3.8.1.1 Quantitative Insight

According to a study, up to 95% of cobalt and 80% of lithium can be recovered from recycled lithium-ion batteries, which can further help avoid the need for new mining operations and reduce the amount of waste from directly charging up (Chigbu et al. 2024).

3.8.1.2 Bangladesh Recovery: Managing Textile and Apparel Waste

Textile and apparel industry is one of the largest waste-generating sectors worldwide and there are severe environmental and social consequences involved. Bangladesh, known for being one of the top producers in textiles globally, adopted CE practices for the textile sector due to the waste and inefficiency in this area as people often demand new clothing every 3 months. These include recycling Fibers, efficient water dyeing, and waste-to-resource (Khairul Akter et al. 2022). Factories within Bangladesh, as an example, are starting to use technology to recycle materials from scrapped fabrics back into new types of cloth. It not only minimizes material waste but also saves water and energy. The collaboration between manufacturers and NGOs, for example, has also resulted in sustainable production, including zero-waste apparel lines.

3.8.1.3 Qualitative Insight

These initiatives clearly showcase how the application of CE practices can minimize the environmental impact of the textile sector while generating economic opportunities for local communities that aligns with SDG 12.5 on waste reduction and SDG 12.6 on sustainable business practices.

3.8.2 Circular Approaches in EU Member States

The European Union (EU) is at the forefront of this movement with comprehensive policies and initiatives seeking to make sustainability a reality across sectors. These include policies to enhance waste management systems, eco-design policies, and the development of industrial symbiosis initiatives, whereby one industry's waste is used as a resource for another (Rodriguez-Anton et al. 2019). As an example, initiatives like "Green Deal Circular Textiles" in the Netherlands favour the recycling of post-consumer textile waste into new products, with the aim to decrease on virgin fibers whilst simultaneously tackling a waste problem. For example, with the use of circular practices in the construction sector, Germany has dramatically raised the input of recycled metals and concrete into new infrastructure (Schroeder et al. 2019).

3.8.2.1 Quantitative Insight

Since the implementation of CE policies and practices, 55% of the Municipal Waste across the EU in 2020 was recycled. Resource efficiency through circular practices is illustrated by over 90% material recovery almost entirely realized by the construction and demolition sector (Rodríguez-Antón et al. 2022).

3.9 Challenges and Opportunities

Despite increasing recognition of the potential that the circular economy (CE) offers, there are still barriers to its widespread implementation. Others stem from financial, behavioural, and policy challenges. Adopting CE practices often involves high initial investments in new technologies, infrastructure and processes. Closed-loop manufacturing systems and advanced recycling technologies, for instance, that must pay high capital costs may put a break on industries seeking to move to more sustainable systems of production, particularly regarding small and medium enterprises (SMEs) (Marrucci et al. 2019). Further, the feared uncertainties surrounding the introduction of novel CE practices (even if provisional ones) also constrain resources available from potential investors and other accompanying stakeholders. Consumer behaviour continues to be a major constraint when it comes to implementing CE. Consumers

Table 3.3 Challenges to circular economy adoption (Author's Compilation)

Challenge	Description	Impact
Financial constraints	High initial investment costs for new technologies and infrastructure	Limits the scalability of CE practices
Consumer behaviour	Resistance to purchasing recycled or refurbished products due to misconceptions about quality	Creates demand gaps for circular products
Policy gaps	Inconsistent regulations and lack of standardized CE metrics	Hinders effective monitoring and enforcement
Technological limitations	Limited access to advanced recycling and resource recovery technologies	Restricts resource efficiency improvements
Lack of education and training	Insufficient awareness and skill development for operating in circular industries	Slows down CE implementation at operational levels

are often reluctant to purchase these products due to a lack of awareness or myths regarding their quality. As an example, this high preference for new product translates into a demand gap for circular products in the textile and electronics sectors, where recycles are seen as inferior to virgin material (Panchal et al. 2021). The ability to scale up such CE initiatives is heavily reliant on the buy-in of consumers which is sorely lacking. Policy gaps at the national and international levels are another reason for inadequate or disparate policy frameworks that interfere with CE adoption. Some areas have not established corresponding policies that promote CE, such as extended producer responsibility (EPR) policies that incorporating the stability of various sectors (Panchal et al. 2021). Difficulties such as the absence of standardised metrics and indicators for CE performance, also complicate tracking progress and compliance (Table 3.3).

3.9.1 *Opportunities for Innovation*

However, this does not mean that we should abandon the circular economy, and there are still many opportunities for innovation, be it in the field of design, supply chain management or public engagement. Eco-design CE emphasizes the production of product varieties with basic properties of durability, repairability, and recyclability. When CE principles are incorporated at the design stage by using modular parts in the production of products such as electronic gadgets, they can be more easily repaired and upgraded, possibly extending the life of these products and reducing waste (Kravchenko et al. 2019). Sustainable supply chain CE aims to establish sustainable supply chains through resource sharing, waste valorisation, and the use or re-use of secondary raw materials. An example of a CE sustainable supply chain is industrial symbiosis, where one industry's waste serves as the raw material of a neighbouring industry. This reduces both costs and wastage and is linked to the objectives of SDG 12 in the form of increased resource efficiency. CE practices are beneficial only to the extent that they do not require consumers to sacrifice convenience, raises their quality of life in present times by reducing the entry price to this market, and creates a country and improving air quality Indoor air (Valverde and Avilés-Palacios 2021). Public awareness and engagement training public awareness and educational strategies play a decisive synergy role in laying the necessary foundation for consumer acceptance of CE products and services. In both marketing materials, the design process pushes the players to think of realistic lines of development that allude to the positive environmental and economic benefits of responsible consumption. This sets off positive consumer habits that help create a market for CE services and products (Valverde and Avilés-Palacios 2021).

3.9.2 *Importance of Education and Skill Development*

Education and skill development are important for the transition to a circular economy. For example, vocational training has enabled the workforce in developing countries to participate in CE initiatives such as solid waste management. Training programs are also necessary to equip employees with the right skills, such as mastering advanced recycling technologies or operating within circular industries, such as in the field of industrial symbiosis or sustainable manufacturing. Additionally, social networks can enable exchanges of capacity building and training opportunities. Educational institutions can be used as researchers in CE modeling. Research funds encourage technological development, and continuous research by universities.

Collaboration between universities, industries, and governments enhances research and practice.

3.10 Policy Recommendations and Future Directions

One of the primary roles of the policymakers is to enhance the contribution of a circular economy (CE) to SDG No. 12. It is done by the creation of comprehensive frameworks and collaboration with other stakeholders. To achieve this goal, the following policy recommendations can be suggested. Governments should design and implement a CE-based national policy to incorporate principles into their economic, industrial, and environmental policy in compliance with SDG No. 12 (Raman et al. 2024). Fiscal measures, such as tax benefits, subsidies, and low-interest loans, are critical to enable companies to implement circular practices. Operations of businesses that use production methods that are not CE should be subject to taxes while taking punitive measures against those that manifest high levels of wasted products will contribute to a CE. Through subsidies and low-interest loans, companies that exhibit CE practices will be motivated to apply and increase their capacity (Valverde and Avilés-Palacios 2021). Implementing Extended Producer Responsibility (EPR) that emphasizes producers must be responsible for final product disposal full cost promotes the reusing and recycling these products. EPR needs to be expanded to cover more industries and should be made more efficient by setting more restrictive and orderly compliance conditions in charging and penalties in non-compliance cases (Raman et al. 2024). Governments, combined with industries and academia, need to form collaborative platforms to deliver key tools to support joint research ventures for innovations. The governments should also facilitate interactions between these agencies to work seamlessly on CEs (Valverde and Avilés-Palacios 2021). Awareness needs to be raised on CE and SDG No. 12 through consumer education forums. Public relations solutions can be implemented to persuade the public on the essentials of a CE and how it can support daily consumption trends that are environmentally friendly. This will increase the preference for circular goods by creating a customer base for CE. National governments should adopt CE metrics in measuring their contribution toward achieving SDG No. 12. Examples of different metrics to be employed include material efficiency, recycling, and resource productivity. This will drive evidence-based policy

3.10.1 Future Research Areas

One of the major hurdles in moving CE forward is the absence of common and widely adopted measures. To guide future research, we consider that comprehensive indicators should be developed to evaluate the environmental, economic, and social impacts of CE practices, Material flow analysis, carbon footprint reduction, and life-cycle assessments represent some examples of quantitative metrics that can be instrumental in assessing the effectiveness of CE, ultimately informing the implementation and adaptation of policy (Valverde and Avilés-Palacios 2021). Such research must also examine the possibilities for aligning CE metrics with SDG 12 indicators, where the same unit of measurement can be used to monitor pathways toward sustainable consumption and production goals. The applicability of these metrics across sectors can be validated through case studies and/or pilot projects (Raman et al. 2024). CE contributions to SDG 12 do not stop there, and they can impact other SDGs, through their interlinkages with SDG 12. CE supports climate mitigation by cutting emissions due to resource efficiency and waste management. CE creates sustainable business models with green jobs and innovation that support SDG 8 targets. CE promotes sustainable industrialization and strives to build resilient infrastructure. This interlinkage ought to be researched in the future to better understand the integrated influence of CE on the sustainability agenda. It especially consists of an analysis of trade-offs, synergies, and potential policy conflicts between different

SDGs and the identification of "best practices" for integrated approaches (Valverde and Avilés-Palacios 2021).

3.11 Results and Findings

The chapter explores the circular economy (CE) as a transformative solution to global sustainability challenges, particularly in relation to Sustainable Development Goal 12 (SDG 12), which emphasizes responsible consumption and production. CE is presented as an alternative to the traditional linear model of production and consumption, where resources are extracted, used, and discarded. The chapter highlights how CE principles, such as reducing, reusing, and recycling, can address the ecological degradation caused by resource depletion and waste generation. This shift is vital for tackling climate change, reducing resource scarcity, and promoting more sustainable production and consumption practices. The findings demonstrate how CE has been operationalized across various industrial sectors, with real-world examples showing its effectiveness in improving resource efficiency. For instance, the chapter discusses successful CE practices in lithium-ion battery recycling, textile manufacturing, and food waste management. These case studies illustrate how industries are moving towards closed-loop systems that minimize waste and extend the lifecycle of materials, thus reducing their environmental impact while creating economic opportunities. The textile industry in Bangladesh, for example, has significantly reduced fabric waste, conserved water, and minimized chemical use through CE practices, contributing not only to environmental sustainability but also to the achievement of SDG 12. Similarly, CE initiatives in the construction sector in Europe have led to the use of recycled materials such as concrete and steel, promoting resource efficiency and waste minimization. Despite the potential benefits, the chapter identifies several challenges that hinder widespread CE implementation. High initial investment costs, particularly for small and medium-sized enterprises (SMEs), pose financial barriers. Additionally, consumer behavior remains a critical obstacle, as many consumers are reluctant to purchase recycled or refurbished products due to misconceptions about their quality. Policy gaps at both national and international levels also complicate the adoption of CE, with inconsistent regulations and a lack of standardized metrics making it difficult to track progress and enforce CE practices. Technological innovations, including the use of IoT, AI, and blockchain, are highlighted as critical enablers for scaling up CE. These technologies facilitate real-time tracking of materials, optimize resource use, and enhance waste management. The paper underscores the importance of public-private partnerships, government policies such as extended producer responsibility (EPR), and fiscal incentives to encourage companies to adopt CE practices. Additionally, the research calls for the development of standardized metrics to evaluate the environmental, economic, and social impacts of CE, ensuring alignment with SDG 12 targets.

3.12 Conclusion

The transition to a circular economy is more than just a shift in production and consumption practices, it is a fundamental transformation necessary for achieving SDG 12 on responsible consumption and production. By replacing the traditional linear model with regenerative principles of reducing, reusing, and recycling, the circular economy enhances resource efficiency, minimizes waste, and mitigates environmental harm. Case studies across industries such as manufacturing, food waste management, and textiles demonstrate its practical benefits, proving that circular strategies are not only viable but essential for long-term sustainability. The journey towards a fully circular economy requires coordinated efforts from all stakeholders, governments must create supportive policies, businesses must integrate circular models into their operations, and consumers must adopt responsible consumption habits. Innovation in eco-design, sustainable supply chains, and standardized CE metrics will be key drivers of progress. The circular economy is not just an environmental solution but a pathway to sustainable economic growth, social well-being, and climate resilience. By fostering collaboration and commitment at all levels of society, we can accelerate the transition to a circular future, one that not only achieves SDG 12 but also contributes to a more sustainable and equitable world.

References

Agarwal S, Jain H (2024) Investigating the potential of floral waste as a vermicompost and dual-functional biosorbent for sustainable environmental management. Water Air Soil Pollut 235(5):322

Agarwal S, Jain H, Dhupper R, Mathur A (2025) Optimizing methyl orange dye removal from aqueous solutions using white chrysanthemum floral waste-derived bioadsorbent: a study of kinetics, thermodynamics, isotherms, and rsm optimization. Int J Environ Res 19(3):104

Beccarello M, Di Foggia G (2018) Moving towards a circular economy: economic impacts of higher material recycling targets. Mater Today Proc 5:53154 Elsevier Ltd

Belmonte-Ureña LJ, Plaza-Úbeda JA, Vazquez-Brust D, Yakovleva N (2021) Circular economy, degrowth and green growth as pathways for research on sustainable development goals: a global analysis and future agenda. Ecol Econ 185:107050

Belmonte-Ureña LJ, Plaza-Úbeda JA, Vazquez-Brust D, Yakovleva N (n.d.) Circular economy, degrowth and green growth as pathways for research on sustainable development goals: a global analysis and future agenda

Berg A, Wilts H, Götz T (2018) Circular economy for sustainable development

Bhardwaj LK et al (2025) Resilience in the digital age: Emerging technologies for climate adaptation. In: Choudhury M, Dixit G, Majumdar S (eds) Sustainable synergy: harnessing ecosystems for climate resilience. climate change management. Springer, Cham. https://doi.org/10.1007/978-3-031-77957-2_11

Chigbu BI (2024) Advancing sustainable development through circular economy and skill development in EV lithium-ion battery recycling: a comprehensive review. Front Sustain. Frontiers Media SA.

Dantas TET, Kirchherr J, Cavalcanti RN, Viana FL (2021) How the combination of circular economy and Industry 4.0 can contribute towards achieving the sustainable development goals. Sustain Prod Consum 27:213227. Elsevier BV

Jain H (2025) Synergizing carbon reduction technologies and nature-based solutions for a climate-neutral future. Mitig Adapt Strateg Glob Change 30(42) https://doi.org/10.1007/s11027-025-10237-6

Jain H, Dhupper R, Shrivastava A, Kumar D, Kumari M (2023) AI-enabled strategies for climate change adaptation: Protecting communities, infrastructure, and businesses from the impacts of climate change. Comput Urban Sci 3(1):25. https://doi.org/10.1007/s44212-023-00044-6

Khairul Akter MM, Sultana S, Rahman M (2022) Textile-apparel manufacturing and material waste management in the circular economy: a conceptual model to achieve sustainable development goal (SDG) 12 for Bangladesh. Clean Environ Syst 4:100064

Kravchenko M, McAloone TC, Pigosso DCA (2019) Implications of developing a tool for sustainability screening of circular economy initiatives. Procedia CIRP 80:625–630. Elsevier B.V

Marrucci L, Daddi T, Iraldo F (2019) The integration of circular economy with sustainable consumption and production tools: Systematic review and future research agenda. J Clean Prod 240:118268. Elsevier Ltd

Moraga G, Huysveld S, Mathieux F, Blengini GA, Alaerts L, van Acker K, De Meester S, Dewulf J (2019) Circular economy indicators: What do they measure? Resour, Conserv Recycl 146:456–461

Panchal R, Singh A, Diwan H (2021) Does circular economy performance lead to sustainable development?—a systematic literature review. J Environ Manag 293: 112846. Academic Press

Pasca MG, Mihai C, Dobrota G (2023) How does the circular economy contribute to the pursuit of Agenda 2030? a systematic literature review. E3S Web of Conf 365:04006. EDP Sciences

Pollice F, Batocchio A (2018) The new role of procurement in a circular economy system

Prokic D, Jovanovic M, Markovic M (2022) The role of circular economy in food waste management in fulfilling the United Nations' sustainable development goals. AlimentAriA

Raman R, Lathabai HH, Nedungadi P (2024) Sustainable development goal 12 and its synergies with other SDGs: identification of key research contributions and policy insights. Discov Sustain 5:12

Ranta V, Saari UA (2019) Circular economy: enabling the transition towards sustainable consumption and production. Sustain Prod Consum 19:1–12

Rodríguez-Antón JM, Rubio-Andrés M, Alonso-Almeida MDM (2022) From the circular economy to the sustainable development goals in the European union: an empirical comparison. Int Environ Agreements Politics Econ 22(1):67–95

Rodriguez-Anton JM, Rubio-Andrés M, Alonso-Almeida MDM, CelemínPedroche MS (2019) Analysis of the relations between circular economy and sustainable development goals. Int J Sustain Dev World Ecol 26(8):708–720

Schroeder P, Anggraeni K, Weber U (2019) The relevance of circular economy practices to the sustainable development goals. J Ind Ecol 23(1):77–95

Shrivastava A (2021) Dietary arsenic exposure: Sources and risks. In: Arsenic toxicity: challenges and solutions, Singapore, Springer Singapore, pp 95–125

Shrivastava A, Abhishek K, Gupta AK, Jain H, Kumari M, Patel M, Sharma P (2024) Removal of micro-and nano-plastics from aqueous matrices using modified biochar–a review of synthesis, applications, interaction, and regeneration. J Hazard Mater Adv 16:100518

Valverde JM, Avilés-Palacios C (2021) Circular economy as a catalyst for progress towards the sustainable development goals: A positive relationship between two self-sufficient variables. Sustainability 13(22):12580 (Switzerland)

Chapter 4
Social and Psychological Consumer Conundrum, Product Obsolescence, and Sustainable Approaches

Tajudeen Alaburo Abdulsalam

Abbreviations

SPLCs	Sustainable Product Lifecycle
SLR	Systematic Literature Review
R&D	Research and Development
GPP	Green Public Procurement
EPR	Extended Producer Responsibility
RoHS	Restriction of Hazardous Substances
TAM	Technology Acceptance Model
TOE	Technology–Organization–Environment
PLCs	Product Lifecycle
UTAUT	Unified Theory of Acceptance and Use of Technology
CDT	Cognitive Dissonance Theory
TRA	Theory of Reasoned Action
TPB	Theory of Planned Behavior
IDT	Innovation Diffusion Theory
C-UTAUT	Compatibility UTAUT

T. Alaburo Abdulsalam (✉)
Department of Management and Marketing, KFUPM Business School, Dhahran, Saudi Arabia
e-mail: alaburotajudeenabdulsalam@gmail.com

H. Jain et al. (eds.), *The Circular Fallacy: Closing the Loop in a Consumerist World*,
Circular Economy and Sustainability, https://doi.org/10.1007/978-3-032-22687-7_4

4.1 Introduction

Over the past two decades, the scholarly debate on the socio-psychological drivers of consumption behavior has gained momentum, impacting theory and policy despite the criticism of planned obsolescence practices (Bharti et al. 2022; Osei-Frimpong et al. 2022). For millennia, human societies have shown that sustainable living in harmony with the environment is attainable (Jordan and Kristjánsson 2017). Firms capitalize on consumers' status-seeking, cultural values, and self-identity through planned obsolescence to meet market demand, competitive pressure, and profit margins (Rivera and Lallmahomed 2016; Proske and Finkbeiner 2020). This not only harms well-being through overconsumption but also generates 50 million tons of electronic waste worldwide, with only 20% formally recycled annually (United Nations Institute for Training and Research 2024). The electronics sector, for example, shortens product lifecycles to boost repurchase rates, contributing to 53.6 million metric tons of e-waste in 2019, with just 17.4% recycled (Forti et al. 2020). The psychological appeal of novelty products fosters unsustainable consumption, worsening resource depletion and hindering environmental goals because consumers are driven by short-term gratification, rather than long-term utility, that may trigger product turnover, increased waste generation, and inefficient use of natural resources (Bocken et al. 2016). Consumption has become central to the decision-making process, influenced by technological advancement and changing consumer preferences since the industrial revolution.

Products once considered essential are now quickly replaced, often before the end of their functional life, arguing the necessity for sustainable production and marketing approaches. Managers view socio-psychological consumption drivers as essential for developing effective strategies, especially during recessions when competition becomes intense (Weerasinghe et al. 2024; Zhang et al. 2019). Consumer behavior now reflects how well a company's strategy meets market demand, as marketing begins and ends with the consumer since consumers play roles as users, payers, and buyers. Accordingly, firms must encourage sustainable production processes, approaches, and purchasing behaviors in their daily activities (Hinterhuber and Khan 2025; Sepe et al., 2025).

The conundrum arises when consumers, driven by socio-cultural values and psychological pressures, prioritize novelty over necessity when making purchase decisions (Mukherjee and Datta 2025). Although the art of scarcity, social proof, and emotional appeals often creates a perceived need for newer products, triggering consumers toward frequent purchases. This phenomenon affects purchase behavior and raises ethical concerns regarding deceptive practices that exploit consumers' vulnerabilities (Havitz and Dimanche 1997; Bowden et al., 2015). As they increasingly seek validation, status, and self-expression through their purchases, consumers may fall victim to cycles of consumption that lead to dissatisfaction. The argument for and against product obsolescence is not new to scholarly discourse, but its effect on sustainability has gained heightened attention from policymakers in recent years (Echegaray 2016; Roberts et al. 2023; Bisschop et al. 2022; Scott and

Weaver 2018). Accordingly, planned obsolescence, where products are designed with a limited lifecycle, motivates frequent repurchases and perpetuates a throw-away culture. Contrarily, perceived obsolescence, driven by changing trends and social pressures, compels consumers to replace products that are still functional but no longer deemed fashionable and up-to-date.

This affects established brand image and contributes to environmental degradation, financial strain, and a diminished sense of value in owned products (Kuppelwieser et al. 2019; Yamamoto and Murakami 2021). As fashion cycles become increasingly fast-paced, emerging sectors of the fashion industry have adopted unsustainable production techniques to keep up with competitive pressure amidst demand shortages (Bhandari et al. 2022). In response to global interest in sustainability and its related ethical domain, food and beverage sectors have begun to offer sustainable options in their product lines (Phelan et al. 2022). However, consumers have been reluctant to adopt sustainable changes to their consumption choices, a phenomenon common to industries offering sustainable products in a market based on rapid turnover of goods. Fashion producers across sectors attempt to change their unsustainable nature but deem it feasible in the long term if consumers support sustainable fashion by purchasing it. Consumers are said to increasingly care about ethical behavior, but this attitude does not always translate into behavior toward fashion items (Teerakapibal and Schlegelmilch 2025). This presents a conundrum for marketers in an industry defined by rapid turnover of trends and associated disposal of 'unfashionable' apparel (McNeill and Moore 2015). Accordingly, the growing strain on psychological and social well-being and the potential to alleviate this strain deem the growing call for sustainable development in consumption research necessary.

Design has gained traction in recent years due to its role in sustainability (MacDonald and She 2015; Cicconi, 2020). The growing interest in sustainability and their association with production and consumption has led companies to examine how they handle environmental challenges of planned obsolescence (Maitre-Ekern and Dalhammar 2016). Eco-design, the rising concept of circular economy, is now an essential strategy for achieving a sustainable industry (Yousif and Moalosi 2024). This offers a way to support sustainable economic growth amid resource scarcity. As a precursor to the circular economy, sustainable development is a well-accepted model that aims to shift unsustainable consumption and production to sustainable practices (Liu et al. 2020). This shift shows a connection between sustainable development and consumer protection policy. Sustainable development emphasizes the social and collective aspects of consumer laws and policies, offering consumer law a chance for positive change (Lorek and Spangenberg 2014; Mensah 2019). Sustainability is now key to promoting consumption and reducing the effect of product obsolescence by involving ethical, informed, and environmentally conscious decision-making that balances personal needs with collective responsibilities. By encouraging product durability, reparability, and reusability, sustainable approaches can break the obsolescence cycle and reduce waste. Going by the proposition of the triple bottom line model (Hammer and Pivo 2017), education and awareness of sustainable consumption help consumers make choices that support long-term well-being and ecological sustainability (Sheth et al. 2011; Lim et al. 2025). As global awareness of environmental issues grows, manufacturers are focusing on the environmental performance of their products and services (Wong et al. 2012), highlighting the importance of sustainable approaches in the product lifecycle. Essential gaps persist in understanding the antecedents (A), drivers (D), and outcomes (O) of sustainable production practices along with policies (P), processes (P) and outcomes of consumer laws towards sustainable consumption. Consequently, the study proposed the Antecedent (A), Driver (D), and Outcome (O) framework by providing a comprehensive review of literature on social and psychological consumer conundrums, product obsolescence, and sustainable approaches that minimize the impact of product obsolescence on achieving sustainable development and a green economy by 2030.

The study makes theoretical (model development) and practical (managerial insights) contributions to circular economy and sustainable product lifecycle literature. First, the chapter enriches the pre-existing knowledge of product lifecycle approaches for sustainable production and consumption by introducing the ADO and PPO model as a relevant framework for explaining the strategic factors that foster crafting sustainable production approaches against planned obsolescence practices. By doing so, the chapter contributes to the study of product lifecycle and sustainable development on associated stimulus, strategic organism, and consumer decision journey that gears sustainable approaches toward the production and consumption, especially in the manufacturing and fashion industries. This deepens the reader's understanding of the relationship between the product lifecycle and sustainability in an ever-evolving competitive environment. Second, our study elucidates the potential mechanisms by which planned obsolescence impacts firms' competitive advantage in the digital landscape. This is accomplished by recognizing how social and psychological factors cause decision dilemmas for consumers and how they affect firms' production and marketing approaches amidst their innovation and change management. Third, our chapter analyzes the boundary conditions for the influence of planned obsolescence on consumers' behavior and immersive experiences, causing a sense of dissatisfaction and disloyalty. This offers a more profound understanding of the phenomenon being researched. Overall, the study demonstrates the significance and dark side of planned obsolescence toward sustainable development in the manufacturing and fashion industries by conceptually and theoretically demonstrating their role as a roadblock in achieving sustainable development by 2030.

4.2 Theoretical Background

The evolution of marketing theory shifted scholars' focus from social to behavioral science, improving the global understanding of human behavior (Achrol and Kotler 2012). This led to exploring psychological and sociological theories on human and organizational behavior. The consumer behavior school is a key example within the rational framework for identifying marketing schools that explain consumer behavior models (Ho et al. 2006; Shaw and Jones 2005; Mishra and Mishra 2023). Their proposition addresses essential arguments on personality, perception, motivation, cognition, affective needs, information search, processing, semiotics, and reasoned action. Earlier scholarly literature focused on emotional-cognitive models, known as the affective-cognitive perspective, seen in the theory of reasoned action (TRA) and planned behavior (TPB). Individual intentions and actual behavior are influenced by subjective norms, attitudes, and perceived behavioral control. Ajzen and Fishbein (1980) describe subjective norms as an individual's view of social pressure to engage in or avoid specific behavior. This argues that family and friends' opinions shape motivation and willingness to conform even though subjective norm depends on normative belief and motivation to comply regardless of personal opinion. Prior research suggests consumer ethnocentrism acts as a proxy for subjective norms (Ortega-Egea and García-de-Frutos 2021). Amidst the cultural shift toward environmental awareness in across the sectors, dissonance skews consumer decision-making. Cognitive Dissonance Theory (CDT) (Festinger 1957) suggests dissonance is psychological discomfort from conflicting thoughts, beliefs, and attitudes and it occurs when consumers like a product but feel its consumption is not environmentally sustainable. Dissonance also arises when a consumer holds an attitude but acts against it (Marikyan et al. 2023). Festinger (1957) argued CDT focuses on easing discomfort by changing beliefs, seeking information that matches convictions while ignoring contradictions, and interpreting ambiguous details to fit personal views. By reducing conflicting beliefs' importance, consumers make purchasing decisions without further dissonance.

Behavior is seen as volitional, systematic, and rational, with three boundary conditions: intention stability, volitional control, and specificity in measurement (Chatzisarantis et al. 1997; Koole et al. 2023). Scholarly literature criticizes TRA and TPB for neglecting morality, voluntariness, cognitive deliberation, habit, and survey misunderstanding of scale items towards the cognitive factors (Sniehotta et al. 2014; Ajzen 2011; Hagger 2015). The Technology Acceptance Model (TAM), a derivative of TRA, eliminates subjective norms and explains adoption intention through perceived usefulness, ease of use, and attitude toward technology usage. Despite being widely used for human adoption intention and behavior, Davis et al. (1989) link perceived usefulness and ease of use to behavioral intention. This aligns with the Technological Determinism Model, which suggests technology drives urban societal change (Vergragt and Brown 2007). However, TAM's emphasis on technology usage overlooks social, political, and economic forces influencing the adoption intention across the management and marketing domain (Luo and Remus 2014; Taherdoost, 2018). Despite often being ignored, human agency is also essential for successful implementation and societal acceptance of human behavior depending on systemic factors empowering decision-making processes (Cvitanovic et al. 2016). According to Chiu and Cho (2022), the goal-directed behavior suggests that an individual's intention is influenced by motivation (desire for the behavior), volition (attitude and subjective norm), non-volition (perceived behavioral control), emotion (anticipated positive and negative feelings), and habit (past behavior frequency). This combines desire, anticipated emotions, and past behavior to better explain intentions and actions, which suggested three key improvements to the planned behavior theory (Perugini and Bagozzi 2004).

After a few years of TAM's introduction, two studies proposed the Extended Technology Acceptance Model (ETAM) to improve its adaptivity, explanatory power, and specificity. Maillet et al. (2015) modeled antecedents of perceived usefulness and behavior intention as TAM2, adding social influence (image, subject norms, and voluntariness) and cognitive factors (result demonstrability, job relevance, and output quality). TAM2 aimed to enhance perceived usefulness in voluntary and mandatory environments but was outperformed, except for subjective norms, which influence mandatory but not voluntary settings. Taherdoost (2018) identified antecedents of perceived ease of use, categorizing them as adjustments and anchors. General beliefs about usage, such as enjoyment and objective usability, were in the anchor group, while direct experience-based beliefs, including external control, self-efficacy, anxiety, and playfulness, were in the adjustment set.

Despite these efforts, scholars continuously criticize TAM and ETAM's applicability in various domains, including the transport service industry. Venkatesh et al. (2003) introduced the Unified Theory of Acceptance and Use of Technology (UTAUT) by comparing eight models within information systems, all rooted in sociology, psychology, and communications. The model integrates performance expectancy, effort expectancy, social influence, and facilitating conditions but neglects compatibility beliefs. Bouten et al. (2008) incorporated compatibility beliefs from Karahanna et al. (2006) into UTAUT, proposing Compatibility UTAUT (C-UTAUT) to enhance explanatory power. Although aimed at assessing the relationship between compatibility beliefs and behavioral perceptions, the theory did not prioritize measuring actual usage behavior. Since UTAUT did not test different time periods, Venkatesh et al.'s (2008) proposition on experience could not be replicated, highlighting a criticism. By incorporating planned behavior theory, which introduced perceived behavioral control alongside three cognitive factors, UTAUT remains the most comprehensive adoption intention model in technology usage (Taherdoost 2018). It mirrors TAM's focus on performance and effort expectancy while adding social influence and facilitating conditions, addressing previously overlooked social and organizational factors. This aligns with social exchange theory, social cognitive theory, and system theory, which emphasize interdependent components within a larger system, especially in organization behavior, where technology adoption is recommended (Oparaocha 2016).

The Innovation Diffusion Theory (IDT) explains how stakeholders react to technological advancements and predicts innovation adoption based on user patterns and beliefs (Al-Rahmi et al. 2021). It includes five key attributes, relative benefits, complexity, compatibility, observability, and trialability, that shape public responses to new technologies. However, daily behaviors contribute to environmental damage, and individuals can take pro-social or pro-environmental actions to reduce harm. For forty years, scholars have used norm activation theory to explain these actions (Thiermann and Sheate 2020). Also called the norm activation model, it includes knowledge of consequences, ascription of responsibility, and personal norm, leading to two interpretations (Han 2014; Confente and Scarpi 2021; Dalvi-Esfahani et al. 2017). First, it follows a sequence where personal norm, a direct driver of altruistic action, is triggered by awareness of consequences through ascription of responsibility.

Second, norm activation is a process where awareness of consequences and attribution of responsibility directly evoke personal norms, leading to pro-environmental behavior. The value-belief-norm theory expands on this model, showing that pro-environmental behavior is influenced by value orientations, ecological worldview, consequences for valued objects, ascribed responsibility, and a sense of obligation (Rezaei-Moghaddam et al. 2020). Scientific literature highlights internal, external, and interpersonal factors in consumption decisions. The Goal-Framing Theory (GFT) by Lindenberg and Steg (2007) includes gain, hedonic, and normative goals to explain sustainable behavior. Sheth et al. (2011) argue that balancing profitability and comfort with normative goals supports sustainable behavior. When normative objectives align with gain and hedonic goals, the likelihood of sustainable behavior increases (Tang et al. 2020). Thus, the theorization of consumer behavior towards the consumption conundrum demands analyzing how consumers process information, manage risk, and resolve internal conflicts due to socio-psychological factors that affect consumer self-lifestyle and concept to arrive at a decision.

4.3 Material and Method

The study synthetically reviews scholarly literature on consumer behavior, product obsolescence, and sustainable product lifecycle to analyze the socio-psychological consumption drivers and how planned obsolescence causes decision dilemmas in demand for sustainable production approaches. To achieve this objective, the study relied on secondary literature by way of systematic literature review (SLR), coupled with case study illustrations, guided by interpretivist philosophy. A systematic review is a research technique that summarizes and synthesizes existing knowledge in a specific field to facilitate academic research (Tranfield et al. 2003; Whittemore et al. 2014). Researchers identify areas for further investigation by comprehensively reviewing the prior literature (Gupta et al. 2018; Gough and Thomas 2016), facilitating the establishment of themes, patterns, and inconsistencies across multiple studies, contributing to robust theoretical frameworks. According to Linnenluecke et al. (2020), SLR value stems from its ability to integrate qualitative reasoning while highlighting the significance of grounding ideas and actions in practicality and real-world applicability as a measure of their validity and importance. A thematic analysis was conducted to systematically identify and construct themes from the literature. This analytical method is ideal for a systematic examination and documentation of prevalent patterns within the data, providing deeper insights into the core area of the study, which is the development of the ADO and PPO framework.

The study uses a SLR technique to examine articles and peer-reviewed articles published in the English language between 2010 and 2025 thematically using two databases (Web of Science (WOS) and Scopus) as the main sources due to their extensive coverage and reputation as comprehensive scientific repositories (Mongeon and Paul-Hus 2016). Authors developed an advanced search query with 16 keywords to focus on using planned obsolescence and sustainable product lifecycle in consumer behavior domain. The query is as follows: *TITLE-ABSKEY: ("socio-psychological consumption drivers" OR "consumer decision dilemmas" OR "consumption conundrum" OR "psychology of consumption*") AND ("product obsolescence" OR "product obsolescence and sustainability*" OR "obsolescence in circular economy" OR "product obsolescence and consumer reaction") AND ("eco-friendly product design" OR "consumer attitudes towards eco-friendly product design" OR "future prospect for eco-friendly product design*" OR "eco-friendly product design for sustainable consumption and production*") AND ("sustainable product lifecycle*" OR "green product lifecycle management" OR "product lifecycle and sustainable value creation" OR "lifecycle-based approaches*").*

Researchers excluded any literature unrelated to the study for the purpose of fulfilling the inclusion criteria such as articles must be published in English language, related to the theme of the discussions, and undergone double-blinded peer reviews. However, the author discovered more pertinent resources by examining the reference lists of selected articles, thereby reducing the risk of overlooking significant research. Moreover, the author examined the titles and abstracts of the discovered papers to ensure they met the study's goals and satisfied predetermined inclusion and exclusion criteria. The inclusion criteria placed an emphasis on currency, authority, and relevance in accordance with the guidelines set forth by Atkinson et al. (2015). Currency examined continued relevance frequently indicated by citations; authority addressed the source's reputation, peer review, and professional editing; and relevance evaluated the item's contribution to the study context and domain. As a result, every article's content was carefully analyzed to extract relevant information. A mix of qualitative content analysis and recursive abstraction was used to examine the data (Akinyode and Khan 2018). Although notes were collected, they were not coded and were arranged according to themes. Keywords and phrases were frequently used to summarize the relevant information. This human summary aimed to remove unnecessary details and discrepancies, documenting justifications for omissions to ensure transparency. Summaries were combined and paraphrased for consistency and conciseness while preserving the original meaning. The themes were checked for alignment with the research purpose. The result was a clear and cohesive summary of the literature, organized chronologically to address the primary issues in the following subsections.

Guided by the intuition of Moher et al. (2009), Preferred Reporting Items for Systematic Reviews and Meta-Analyses (PRISMA) were applied as an effective method for constructing and cleaning database results to be ready for thematic analysis (Fig. 4.1). In the identification phase, authors developed the above-mentioned query and retrieved publications from databases by searching within their title, abstract, and keyword fields. This initial search yielded 80 and 143 records from WOS and Scopus, respectively. Subsequently, only English articles and reviews from 2010 to 2025 were obtained in the screening phase to ensure relevance and quality, yielding 68 and 80 records from WOS and Scopus, respectively. Afterward, records from both databases were merged during the eligibility phase, and duplicates and retracted entries were eliminated, resulting in 91 records. Finally, in the inclusion phase, we carefully double-checked all entries. Authors excluded the papers from

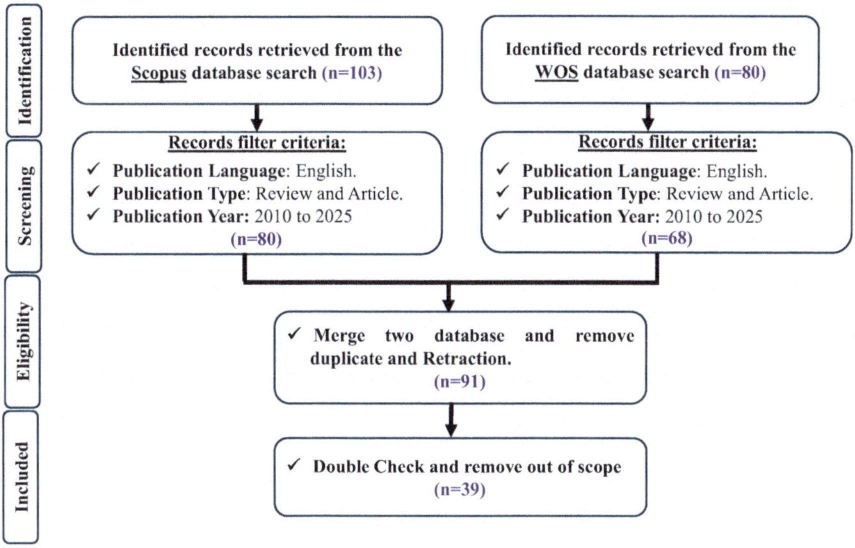

Fig. 4.1 PRISMA flowchart process (Designed by Author 2025)

the scope, resulting in 49 records used in our results analysis (see Reference). The complete selection process that consolidates the selected articles along with those excluded from screening is depicted in Fig. 4.1.

4.4 Result

4.4.1 Social and Psychological Consumer Conundrum

Fleissig (1997, p. 178) argued that the consumer consumption conundrum refers to the inconsistent results regarding non-separability and substitution between durable and nondurable goods, while the inconsistent results might arise due to the materialism, the importance a consumer attaches to worldly possessions. At the highest levels of materialism, such possessions assume a central place in a person's life and are believed to provide the greatest sources of satisfaction and dissatisfaction (Belk 1985). Consumer behavior argued how individuals make purchase decisions based on their consumption status (White et al. 2019). Their intentions to act and actual behaviors are triggered by social groups and psychological traits shaping a reference frame for future behavior (Flusberg et al. 2024). Research shows mixed findings on how socio-psychology factors drive behavior across market areas such as luxury brands, transport, food, sports marketing, and digital products (Widianti et al. 2015; Dhaliwal et al. 2025). Accordingly, little attention has been given to how specific social groups affect consumer behavior (Goldsmith et al. 2014; Goldsmith and Goldsmith 2011). Studies on sustainable products focus on the concern of broad social groups on marketing factors such as price, time, environmental concern, and green segmentation (Karman and Lipowski 2024).

Consumers buy products not only for basic needs but also to build relationships, seek status, and shape their self-image. Earlier research saw consumption as an individual activity, a view still common in industries with socially responsible qualities and high product turnover (Ruvio et al. 2020; Li et al. 2025). Studies on adoption intention recognize the role of social actors but often generalize consumer behavior instead of identifying specific groups like peers or reference groups. Research on planned behavior theory (Fishbein and Ajzen 1974) explores social norms in pro-social and pro-environmental behavior but lacks differentiation between groups. Melnyk et al. (2022) and Liu et al. (2021) show social pressure influences positive sentiments but do not distinguish between different social groups.

Rolling et al. (2021) argued conundrum as the internal and external conflicts consumers face when reconciling desires, needs, and available choices. It stems from option overload, competing priorities, and uncertainty about product performance and satisfaction. Fleissig (1997) discussed conundrum as a dissonance-laden scenario where tension between cognitive evaluations and emotional responses delays decisions, causes regret, and leads to post-purchase dissatisfaction. Sharma and Bhaduri (2013) analyzed it as a state where consumers are torn between social influences and personal preferences, leading to indecision and buyer's remorse. The discussions establish conundrum as a state of confusion or conflict during purchasing decisions, triggered by specific conditions: information overload (Lee et al. 2017), choice overload (Reutskaja et al. 2022), conflicting preferences (Ma et al. 2021), social pressure (McNeill and Moore 2015), and emotional conflict (Fischbach and Herbert 1997). When overloaded with information, consumers struggle to process it and leads to confusion and indecision because a vast array of options can paralyze consumers by increasing post-purchase regret.

When needs and wants do not align, choosing between practical and hedonic benefits creates conflict (André et al. 2018). In collectivist societies, family preferences may override personal choices and then create tension between affordability and sustainability (Lien et al. 2018). Consumers' perception of brands influences choices, and a conundrum arises when perceived quality does not match expectations or when brand image conflicts with personal identity (Thompson et al. 2006; Voorhees et al. 2015). Thus, consumer conundrum showcased the conflicts consumers face before, during, and after consumption decisions, shaped by socio-psychological factors.

The consumer conundrum complicates and extends the purchase journey when social and psychological factors intersect. Social identity theory (Tajfel et al. 1979) posits that individuals derive self-concept from group memberships, influencing brand choices as consumers select products reinforcing their identity. The theory of planned behavior (Ajzen 1991) integrates social norms and attitudes, viewing behavior as shaped by perceived expectations and personal beliefs. However, social pressure fosters impulse buying and post-purchase regret as consumers prioritize approval over rationality (Ayadi and Paraschiv 2024; Gary 2025). Psychological influences such as heuristics and biases distort perceptions making rational decision maker makes irrational choices. The scarcity heuristic causes consumers to overvalue limited-edition products relative to actual need and quality (Yi et al. 2023).

Prior studies highlight how this conundrum challenges marketers in designing strategies for effective consumption. Sharma and Bhaduri (2013) contest the assumption that bottled water consumption is purely objective, emphasizing sociocultural factors, public understanding of science, and institutional trust. Although scarcity is often linked to supply limitations, prior studies argue it as being shaped by perceived unavailability and socioeconomic background, with public risk perception of tap water driving bottled water consumption (Colosimo and Kim 2016). Media coverage reinforces this perception, making bottled water quality concerns less scrutinized. Siltaloppi and Jähi (2021) identify three conundrums hindering the shift to sustainable plastics, underscoring the challenge of aligning limited supply with consumer demand despite rising environmental awareness.

Jin and Hwang (2024) argue that authenticity remains a key driver in shaping positive likability perceptions when cultural familiarity varies. Consumers with low cultural familiarity may perceive traditional ethnic food as unfamiliar and risky, influencing their choice between traditional and localized ethnic foods. Webster et al. (2021) analyze the conundrum arising from the tension between peer group influence and the pragmatics of price, quality, and convenience in a capitalist market. Consumers are drawn to co-operatives' ethical claims and the potential benefits of dividends while weighing whether co-operatives offer competitive prices. Ethical considerations compete with practical needs, with tangible dividends and social co-operative membership significantly influencing decisions. Khurshudov (2024) highlights the smart city conundrum, emphasizing the trade-off between technological convenience and personal privacy. The "privacy paradox" reveals that individuals express concern over data privacy yet contradict themselves by accepting lengthy privacy policies without reading them or using weak passwords for convenience. The "tyranny of convenience" drives individuals to prioritize ease of use and immediate benefits over privacy concerns, often exchanging personal data without fully grasping long-term implications, shaping consumer behavior in urban environments.

The consumer conundrum is common in consumption status by evidence in serial decision making dilemma and choice conflicts that demand immediate information search and processing. This includes authenticity versus social influence (Reilly and Turcan 2023a, b), affordability versus sustainability (Aronson 2015; Giller 2020), originality vs. counterfeiting (Mukherjee and Datta 2025), brand loyalty vs. new experiences (Bowden et al. 2015), and convenience versus health (Mann 2013). Consumers aspire to make purchase choices, yet eco-friendly products often come at a higher price, creating a conundrum where they must balance sustainability with budget constraints. For example, a consumer may struggle between a cheap plastic product and a pricier biodegradable alternative (Moshood et al. 2022a, b; Brockhaus et al. 2016). The fast-food industry exemplifies a conundrum driven by psychological factors.

Despite awareness of health risks, convenience, affordability, and instant gratification often override rational decision-making, leading to regret and cognitive dissonance (Cairns et al. 2022; Lawson et al. 2021). In fashion lifestyle, social media and influencer trends pressure consumers to conform, even when these trends conflict to personal style and comfort, affecting purchasing decisions (Jacobson and Harrison 2022; Slaton and Pookulangara 2022). Consumer conundrums impact decision-making by causing impaired satisfaction, irrational choices, cognitive dissonance, and delayed decisions. To gain a competitive advantage, marketers must balance social and psychological insights to create targeted campaigns that address external influences and internal drivers (Boland et al. 2020). However, manipulating psychological triggers or exploiting social pressures risks consumer mistrust and brand harm (Fisher et al. 2008; Fleissig 1997; McNeill and Moore 2015). Transparent communication, authenticity, and value-driven marketing are essential for building long-term consumer relationships. Figure 4.2 presents the Antecedent (A), Driver (D), and Outcome (O) model, illustrating how socio-psychological factors shape self-concept and perceived marketing activities leading to varied consumer decision-making processes and subsequent decision choice.

Socio-psychological influence alters attitudes and behavior through social interactions and psychological traits. Katona et al. (2011) highlight how social networks and identity shape market dynamics. Dewan et al. (2017) examined friends' influence on online music purchases, while Gong et al. (2023) found intergenerational effects on brand usage, demonstrating behavior spreads through social networks, known as 'contagion.' Pruckner et al. (2020) studied peer influence on employees' health plan choices. Niu et al. (2022) explored peer effects on adopting production technologies, and Wu and Gau (2017) analyzed portfolio investments and social learning. Ajitha and Sivakumar (2017) showed social effects in cosmetic consumption. Sociology and psychology study concepts like 'social learning,' 'herd behavior,' and 'peer effects,' often interchangeably. Herd behavior, as Kameda and Hastie (2015) explain, occurs when people follow others' decisions. Accordingly, social learning involves observing others and cognitive learning. Kameda and Hastie (2015) argue that, unlike herd behavior, social learning requires information exchange.

Behavioral dynamics theory distinguishes biased and unbiased social learning, where biased learning involves selectively imitating certain individuals (Li et al. 2023). If imitation were random, behavior distribution would remain constant, but copying successful individuals' shifts public focus. Socio-psychological factors such as motivation, personality, beliefs, culture, social class, and reference groups shape consumer behavior. In collectivist cultures, buying decisions seek community approval, whereas individualistic cultures prioritize personal choice (Malika et al. 2022). Social class influences consumption, as people in similar economic groups share lifestyles and income levels. Consumers may struggle with authenticity, and selective perception leads them to favor familiar information while ignoring contradictions (Natarajan and Periaiya 2024), posing challenges for marketers trying to reposition brands. Motivation, explained by Maslo's (1943) hierarchy of needs, influences purchasing priorities (Rojas et al. 2023). Low-income consumers focus on basic needs, while wealthier individuals seek self-actualization. However, the

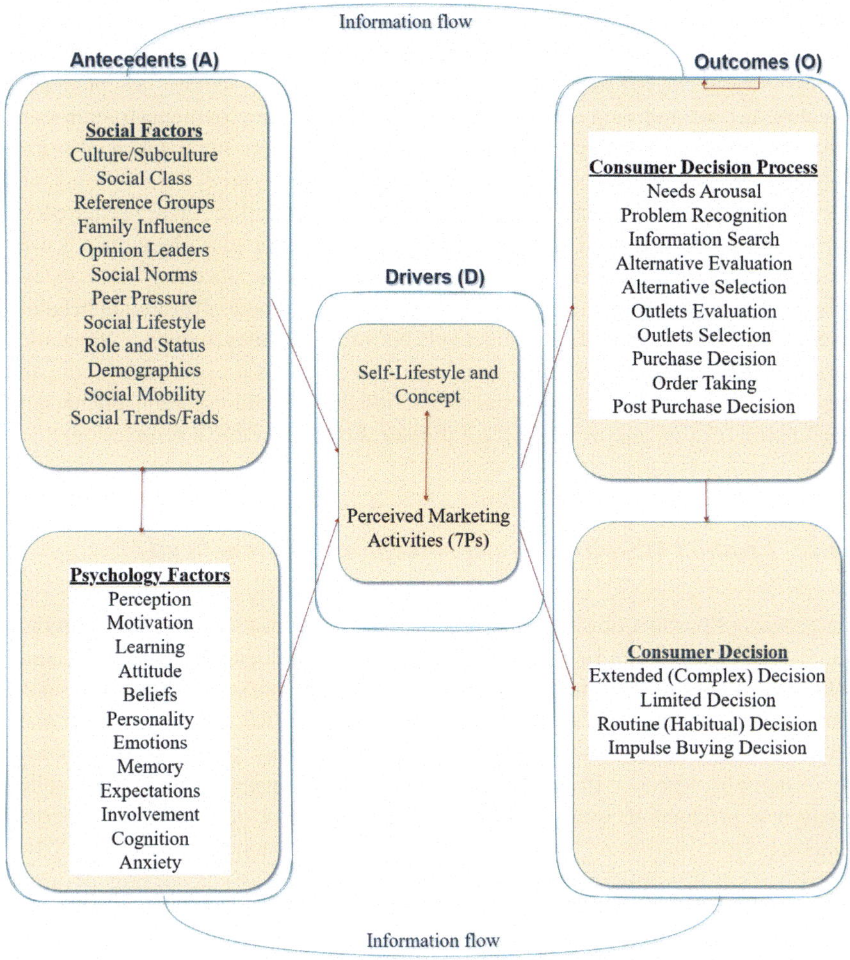

Fig. 4.2 Antecedents (A), Drivers (D), and Outcomes (O) of Consumption Decision (Adapted from Mothersbaugh et al. 2020)

theoretical proposition is criticized for assuming needs always follow a strict order (Papaleontiou-Louca et al. 2022). Overall, motivation becomes the core driver of consumer self-concept and lifestyle towards purchase decision (Fig. 4.2).

For illustration, a conundrum can be seen in the use of gig-based nursing platforms (see cases presented by Wells and Spilda 2024) by healthcare professionals that involve a registered nurse who values patient safety, continuity of care, and professional integrity. Psychologically, the nurse experiences cognitive dissonance, the discomfort that arises when their ethical beliefs (e.g., providing consistent, high-quality care) conflict with their actions (e.g., accepting short-term gig shifts in unfamiliar environments that may compromise care quality). Despite recognizing these risks, the nurse continues to participate in gig work, a behavior reinforced by normalization within the profession and peer influence. As more nurses adopt gig platforms for flexibility and higher short-term pay, the practice becomes socially accepted, even expected. Yet, rising living costs and unstable healthcare employment structures can create an environment where opting out of gig work may seem financially irresponsible. At the same time, immediate financial rewards, perceived autonomy, and loss aversion (fear of missing income opportunities) can further drive participation. The nurse rationalizes their decision by emphasizing short-term benefits ("I need this flexibility now") while downplaying long-term ethical concerns. This creates a consumption conundrum due to a discrepancy between individually rational behavior (earning income, flexibility) and collectively harmful outcomes (reduced care quality, workforce instability). Over time, this tension perpetuates a cycle where ethically questionable systems persist not because individuals fully endorse them, but because social normalization and psychological pressures override ethical reservations.

4.4.2 Product Obsolescence: Problematic or Beneficial?

The genesis of product obsolescence is rooted in limited resource usability, product failure, fashion trends, and social pressure, all of which influence consumer decision-making. Product obsolescence refers to production and marketing tactics that make a product appear one-off or becomes totally outdated (Rivera and Lallmahomed 2016). Although firms can anticipate obsolescence, many global firms intentionally design obsolescence from its conception, leading consumers to feel compelled to buy new products to replace old ones. This boosts not only firm sales' performances, but also increasing the societal waste in the environment (Echegaray, 2016). In 2019, about fifty million tons of electronic waste were produced, largely due to planned obsolescence (United Nations Institute for Training and Research 2024; Forti et al. 2020). Firms use obsolescence to control market size and drive new product adoption by making existing products obsolete after a short period of usage. According to the systematic literature review (SLR), there are five instance of product obsolescence: planned, functional, technological, postponement, and psychological obsolescence. Planned obsolescence argued making products with short lifespans by intentionally pushing consumers to repurchase. Functional obsolescence occurs when manufacturers use low-quality materials that lead to quick product deterioration and malfunction.

Unlike planned obsolescence, functional obsolescence does not intentionally shorten a product's life but still reduces its longevity through inferior material components usage (Kessler and Brendel 2016). For example, manufacturers may stop producing spare parts, halt security updates, or introduce software that reduces functionality. Both planned and functional obsolescence lead to similar outcomes but through different methods (Zallio and Berry 2017). First, planned obsolescence is marked by design intended to cause early failure, while functional obsolescence involves using subpar materials or methods that naturally reduce lifespan.

Second, functional obsolescence might also be less harmful than planned obsolescence because it arises from genuine improvements in technology driven by changing consumer needs rather than deliberate strategies to shorten a product's lifespan. In this case, consumers replace products because newer versions offer meaningful enhancements in performance, efficiency, or safety, which can increase overall value and utility. By contrast, planned obsolescence intentionally limits durability or usability to drive repeat purchases, leading to unnecessary waste, higher long-term costs for consumers, and greater environmental harm. Companies create products with varying durability, often reflected in pricing, giving consumers the choice to match products to their needs and budgets if durability is clearly communicated.

Technological obsolescence occurs when a product becomes undesirable due to technological advancements (Kuppelwieser et al. 2019). A functional product becomes outdated when new technology introduces a superior model in the eyes of consumers. Unlike other forms of obsolescence, technological obsolescence results from technology progress without malicious intent from the producer. Postponement obsolescence happens when a business applies technology or improvements only to its flagship products, focusing solely on leading products instead of all offerings (Proske and Finkbeiner 2020). All these types of obsolescence share a common trait: the product becomes outdated. However, psychological obsolescence occurs when a product loses appeal due to changing trends, even if it remains functional (Mellal 2020). This arises from consumers' perception that a product is not visually pleasing. Marketers aim to create a desire for something "newer, better, sooner" than necessary, prompting consumers to buy "in style" products to avoid being associated with outdated ones. Despite being known as "obsolescence of desirability," "subjective obsolescence," "style obsolescence," or "perceived obsolescence," psychological obsolescence often exploits consumers' tendencies to follow trends and the bandwagon effect.

This occurs where individuals mimic others' actions and preferences resulted to impulsive purchase decision that give low product satisfaction and high cognitive dissonance (Trabelsi et al. 2021). This effect reflects herd mentality, leading consumer to adopt popular tastes and opinions over their own. The belief that material possessions are essential for social acceptance, status, and identity can amplify psychological obsolescence. Research suggests young consumers are more vulnerable to social trends and the bandwagon effect than older ones (Yamamoto and Murakami 2021). Although not a new concept in theory and practice, psychological obsolescence's impact is increasingly studied. Its influence varies across societies and contexts, with group pressure affecting upper class consumers more than others (Munten et al. 2021). This variability indicates that psychological obsolescence's effects depend on specific markets and consumer groups.

In a 1932 brochure titled "Ending the Depression through Planned Obsolescence", Bernard London of the University of Wisconsin, Madison first used the term "planned obsolescence" during a U.S. recession that reduced consumer purchasing power (London 2014). Due to the recession, Americans used their items longer instead of buying new ones, leading to decreased industrial production (Guillard et al. 2023). London proposed planned obsolescence to combat the recession that encourage quick product innovation. In 1954, Brook Stevens described it as "instilling in the buyer the desire to own something a little newer, a little better, a little sooner than necessary". The goal is to encourage consumers to replace purchases before truly necessary (Guiltinan 2009). Guiltinan (2009) identified two main obsolescence mechanisms: technological (design for fashion, style, or functional enhancement) and physical (limited repair, functional life, contrived durability, and aesthetics that reduce satisfaction). Satyro et al. (2018) added design for fast consumption and limited technology updates as obsolescence mechanisms.

The key question is whether product obsolescence is beneficial or harmful. Figure 4.3 offers a theoretical overview of its good-side and dark-side in the circular economy, highlighting its impact on consumer behavior. For example, product obsolescence can cause myopia, hindering consumers' decision-making processes. Myopic consumers prioritize the present over the future and struggle to resist temptations (Acikgoz et al. 2025). Limited willpower makes it difficult to delay gratification, often leading people to favor their "present self" over their "future self" while the myopic consumers may be tempted by new products or low prices by overlooking early obsolescence risks. When a product breaks down, limited willpower can affect the decision to repair or replace it. Even well-intentioned sustainable consumption efforts may falter if repairs seem difficult, expensive, or time-consuming.

Throughout history, product obsolescence has existed for millennia (Cao and Folan 2012; Gurău 2012; Khandelwal et al. 2017). In the United States, disposable shirt collars were in use by the early 1870s. In the 1880s, this method spread to prophylactics and later to other personal hygiene goods like razors, Kleenex, tampons, and sanitary pads (Sierra-Fontalvo et al. 2023). The production of short-lived goods expanded to include automobiles and lightbulbs. Product obsolescence

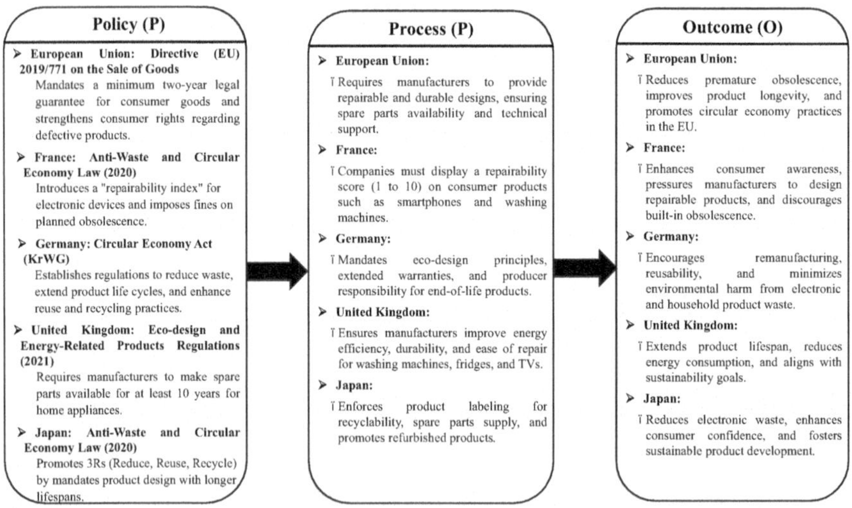

Fig. 4.3 Good and dark side of product obsolescence (Designed by Author 2025)

became a legitimate business strategy in the 1930s. Since women were believed to consume more personal goods than men, advertisements were often targeted at them. In the 1950s, Motorola produced one of the first portable radios, but it was irreparable, encouraging the purchase of new models (Brouillat 2015). According to Oraee et al. (2024), society was constantly influenced to desire things it did not have during the early stages of mass production. The light bulb is the earliest known example of product obsolescence due to its popularity in usage. In 1895, bulbs made in Ohio, USA, used Adolphe Chaillet's 1500 h-lasting filament (Alzaydi 2024). By 1924, filament technology extended the lifespan to 2500 h, but a plan was implemented to limit all bulbs to a 1000 h lifespan. Despite technology enabling more robust bulbs, manufacturers chose to shorten lifespans by altering the filament, and patents for bulbs lasting over a thousand hours were denied (Pihlajarinne 2024).

This strategy boosted light bulb sales. In the 1950 and 1960s, DuPont produced strong and durable women's nylon stockings, which were so successful that the fibers were modified to make them more brittle. The stockings began tearing more easily, requiring frequent replacement (Sierra-Fontalvo et al. 2024). Today, nylon tights rarely last more than two or three uses. Another example of product obsolescence influencing the seed market is plant breeding. Farmers used to save seeds for future seasons, but seed companies now create new, more resistant varieties to combat crop plagues (Agarwal et al. 2025). However, these seeds are often useful for shorter periods, forcing farmers to buy new seeds each season (Alzaydi 2024). The three examples highlight the social, economic, and environmental aspects of product obsolescence serving as roadblock in achieving sustainable development by 2030. According to Fig. 4.3, the dark side of product obsolescence outweighs their benefits making them problematic for the entire populace for several reasons.

Consumers generally view product obsolescence negatively, developing adverse attitudes toward its practice. Scholarly literature argued complaints that products "used to last longer," are "not made for frequent use," and that firms "cut corners to reduce costs," fueling premature obsolescence. A 2013 survey of 800 Austrian consumers found that 55% believed manufacturers intentionally shorten product lifespans (Becher and Sibony 2021). Short-life products generate negative externalities, as their disposal creates waste that harms the environment, with producers seldom internalizing these costs. Eco-design standards and extended producer responsibility address this by internalizing externalities (Lifset et al. 2023; Joltreau 2022). Planned obsolescence also relates to "time-inconsistent preferences," a market failure where consumers prioritize low prices at purchase but later wish for durability (Maitre-Ekern 2021). Consumer initially seek durable goods but shift preferences when newer models emerge. Alternatively, in technological obsolescence, consumers value durability at purchase but lose interest once functionality declines. Although scholarly literature argued consumers willingness to pay a premium for durability, their preference weakens when faced with temptations and obstacles (Goebel et al. 2018). Repair processes can further deter consumers from maintaining virtuous preferences. Market forces work well when consumers have sufficient information to compare options, but limited information leads to suboptimal choices. Firms then lack incentives to offer durable products and improve consumer access to information

(Mayol and Staropoli 2021). Overall, when traders withhold key details, information asymmetry arises, contributing to market failure, particularly concerning product lifespan.

4.4.3 Policy and Legal Standards Against Product Obsolescence: A Case of European Union

Despite the importance of laws and policies in reducing global unethical practices, specified laws and policies that treat product obsolescence and its dimensions are not specifically included in international legislation (Giolo et al. 2024). A prohibition on product obsolescence is argued to cause a major economic slowdown, explained by the circular economy principle and the capitalist model's reliance on expansion. Specifically, neither Quebec nor any other Canadian province has legislation addressing deliberate obsolescence (Lawlor 2015). Except for France, no national consumer protection law formally covers the issue of product obsolescence extensively and formally in alignment with country constitutional framework. Changes to the Environment Protection Code and the Consumer Code, introduced by the 2015 French law on energy transition for growth, include regulations on planned obsolescence (Bisschop et al. 2022). Article 70.v of the bill amends Article L.541-2 of the Environment Protection Code to prioritize consumer protection in waste avoidance and management.

The directive 1999/44/EC25 of May 25, 1999, implemented in all Member States, introduces a legal warranty of product conformity to the contract. In 2014, the European Economic and Social Committee (EESC) drafted an important opinion on consumer information and product lifespan (Maitre-Ekern and Dalhammar 2016). This advocates for a system ensuring a minimum product lifespan and promotes establishing European standards to assess product sustainability. The EESC also proposes expanding the warranty program for consumer goods but suggests a nuanced approach where product lifespan decisions are based on product use and optimization, not just extending lifespan uniformly (Brönneke 2017).

According to SLR, product obsolescence is not a recent development since campaigners were concerned about the effect of obsolescence in the early days of consumerism. The European Economic and Social Committee (EESC) was the first to raise the issue of premature obsolescence at the EU level, noting that "planned obsolescence has reached troubling levels over the past decade" (Mayol and Staropoli 2021). The Commission recently acknowledged early obsolescence as a "growing concern for EU consumers who often complain that purchased goods do not last very long" and committed to introducing legislation to address the problem. Increased policy interest has led EU authorities to commission numerous external studies and papers on planned obsolescence (De Franceschi 2018). EU policy texts often address product lifespan in relation to boosting the economy's competitiveness and sustainability. The European Green Deal exemplifies this commitment through legislative measures aimed at making Europe carbon–neutral by 2050 (Albrecht 2019). This outlines policies to "encourage businesses to offer, and to allow consumers to choose, reusable, durable, and repairable products" and to "curb the built-in obsolescence of devices," reflecting the Commission's dedication to tackling climate and environmental challenges (Saini et al. 2024). The New Consumer Agenda responded by declaring the "Green Transition" of EU consumer law, integrating environmental goals beyond merely protecting consumers' financial interests (Franzolin 2020). The Circular Economy Action Plan also highlights that addressing product obsolescence can generate new skills and create jobs while providing consumers with cost-saving opportunities (Gecit 2020). The EU has not hesitated to assert its regulatory influence globally, with the Commission's Sustainable Product Policy aiming to "set global standards in product sustainability" and influence product design and value chain management worldwide. This thus increase the likelihood of a Brussels Effect since their efforts argued that reforming policies would not happen in a vacuum. Accordingly, product durability is regulated by laws and tools from EU legislation, created with different objectives in mind. However, good policy and legal standards against product obsolesce must addressed three essential issues (Becher and Sibony 2021; Malinauskaite and Erdem 2021). This includes the prohibition of unfair trading practices, enhancement of consumer choice for long-term consumption, and producers must targets choice to promote durable production. Consequently, Fig. 4.4 provides Policy (P) Process (P) Outcome (O) model to summarize top five international acts that enact regulations against the product obsolescence practices worldwide.

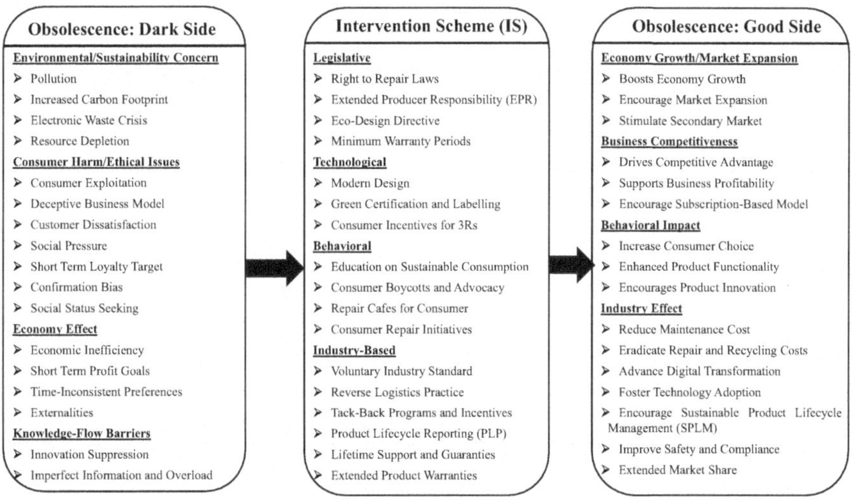

Fig. 4.4 Policy (P), Process (P) and Outcome (O) on Obsolescence. Adapted from Becher and Sibony (2021)

4.4.4 Sustainable Approaches

According to the concept of sustainable development, resources should be preserved and improved for future generations while meeting current demands (Chu and Pan 2024). Sustainable consumption poses new trajectory area of research by examines the ethical and social aspects of products in terms of their manufacturing processes and impact on consumers' well-being. Sustainable manufacturing, which promotes sustainable development, is driven by customers since they form the basis of production and marketing activities. Sustainable development, aiming for long-term economic growth that aligns with social and environmental demands, heavily relies on sustainable consumption and production (Lorek and Spangenberg, 2014). A product is considered part of a linear economy when its life cycle is complete and it is discarded as waste (Mathews and Tan 2011). Since it does not consider the pollution produced at each stage, extract, produce, consume, and dispose of, linear economy is criticized for its efficiency (Sauvea et al. 2016). According to Yang et al. (2023), the circular economy occurs when waste from an industrial process is converted into inputs for another process, water, energy, and new raw material consumption decreases, and waste disposal requirements increase. Accordingly, a circular economy must replace the linear economy to achieve sustainable consumption.

To meet the SDGs by 2030, companies must gradually alter their development strategies and employ appropriate techniques and technologies to produce more sustainable goods. For nearly 70 years, the product lifecycle (PLC) concept has been a key element in marketing theory by playing essential role across fields since its inception in 1950 and notable expansion in the 1960s (Iveson et al. 2022). Despite its benefit expands from marketing to new product creation, traditional PLCs still receive long-standing criticism from top scholars (Mitchell and Clark 2019). PLCs are usually linked to product production, starting with the extraction of raw materials and ending with recyclable, recovered, reused, or disposed of materials (Vendrell-Herrero et al. 2022). Effective PLCs management reduces waste and product obsolescence practices for materials, energy, and time.

Despite being frequently used, the PLC hypothesis has many criticism. The traditional PLC theory does not account for product reuse and repurposing, by viewing products as having only one life cycle before recycling (Bevilacqua et al. 2007). Although, iterative life cycles can be achieved through repurposing and reusing, allowing products to return to the growth phase rather than reaching end-of-life and disposal. Also, conventional PLC models are outdated and inadequate to address contemporary social, environmental, and economic challenges (Vila et al. 2015). The transition to a circular economy is hindered by the traditional PLCs, even though essential sustainability technologies and resources have advanced over the past decade. To meet the demands of today's production environment, scholarly literature advocate for sustainable product lifecycle (SPLCs) by fill the gap left by decades of criticism of traditional models. SPLCs integrate the disposal concept from early phases of design to achieve global sustainability (Zhang et al. 2020).

SPLCs, shown in Fig. 4.5, manage a product's various stages to reduce negative effects on the environment, society, and economy, the three pillars of sustainability. SPLCs promotes recycling, remanufacturing, repurposing and reuse of materials used in product development. Repurposing allows materials or products to be used for purposes other than their original design (Zhang et al. 2017). As a result, the identification of substitute applications for obsolete assets reduces material and disposal costs. Circularity can only be achieved by incorporating an SPLC approach from product design to waste repurposing and encouraging material reuse and recycling to minimize environmental impact. Arguably, the essential causes of global waste is single-use goods and materials such as the plastic packaging in the food sector which need to be phased out to increase sustainability (Badurdeen et al. 2018). Accordingly, updating traditional PLC models and incorporating sustainability into new product design is essential for firms to survive in the long-term. SPLCs considers zero-defect manufacturing, and better utilization of information to fits in the shifting customer preference. The fundamental shift in SPLC is its consideration of a product's multiple lives rather than just one (Ma et al. 2018). Historically, PLCs only included one life cycle, beginning with conception and ending with recycling or disposal. A product's recycling phase often has a considerable remaining usable life (RUL) when it reaches disposal. Consequently, products are often ended based on the user's responsibility rather than their operability (Brundage et al. 2018). Frequently, products can function normally, and even when they cannot, their components can be used in other contexts or for other products. Reusability and repurposing are fundamental concepts in the SPLC model, essential for tackling product obsolescence.

The purpose of the SPLC model is to guide sustainable design, production, lifestyle, and community-based processes (Gbededo et al. 2018). To illustrate the essential SPLCs steps, Fig. 4.5 is color-coded. The main cycle (blue) is the conventional PLC model, which has been around for a while and had a big impact on the acceptance of circularity. By eliminating the idea of disposal from the primary cycle and adding the concepts of sustainability and zero waste into the design and development stage, traditional PLCs are somewhat altered. This enhancement to the design process is required for circularity, as re-use and repurposing will not be feasible if the product is not created in a certain way. As a result of its contribution to societal trash, the disposal stage (highlighted in red) has been eliminated (Aydin and Badurdeen 2019). Since producing waste is unsustainable, discarding a product breaks the cycle and must be avoided. Second, SPLC's model is improved during the product's use period because products are often utilized, recycled, or thrown away after use. Scholars (Galati et al. 2019) introduced the repurpose and reuse cycles (shown by purple and green, respectively) to improve sustainability and utilize 100% of a product's RUL. Products are now regarded as having a single life before being recycled or thrown away, although a product or component may have several lifetimes before reaching the end of the world's supply. The SPLCs model suggests two paths when a product's first life is over (Ren et al. 2019). First, if the product is suitable, it is reused and remanufactured if needed. The product or parts of it are repurposed if they cannot function as planned or if their usage is no longer needed. Second, reusing or repurposing all of the resources used to make goods will

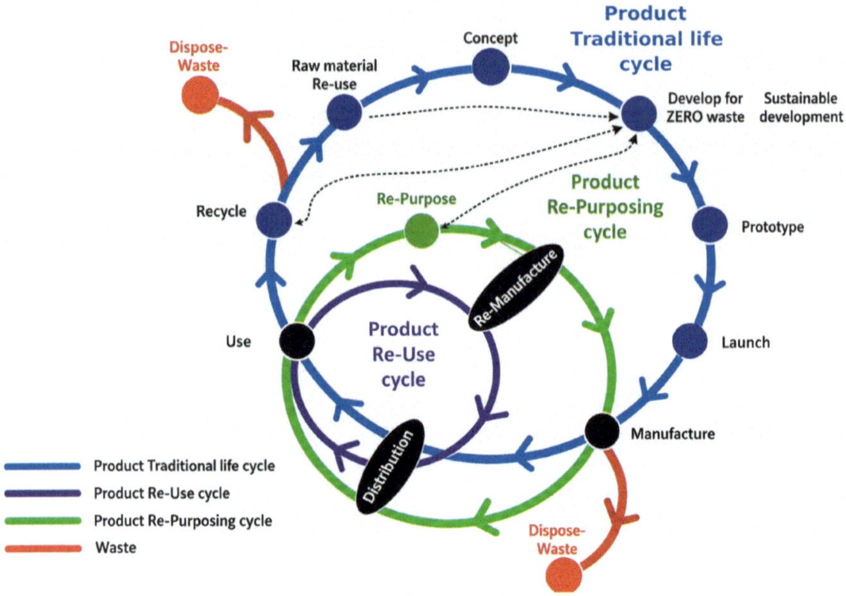

Fig. 4.5 Sustainable Product Lifecycle Management (SPLCM). Adapted from Psarommatis and May (2022)

greatly improve sustainability on a global scale. The cycle of reuse and repurposing continues until every product or component reaches its global end of life (Liu et al. 2020). Other kinds of resources, such as manufacturing time, cannot be recycled because recycling focuses on materials. Consequently, additional cycles will better utilize human labor and manufacturing time, along with natural resources (Becher and Sibony 2021). Moreover, a cycle of knowledge that emerges from the processes of use, repurpose, reuse, recycle, and product design offers valuable data for creating new sustainable goods that work with SPLCs (Rusch et al. 2023; Jensen et al. 2023). Accordingly, enhancing sustainability, the economy, the environment, and society are the three pillars that the SPLC model seeks to strengthen by making the most of a product's remaining useful life reduces material waste, optimizes resources (money and human time), and prevents waste (Shrivastava et al. 2024).

4.5 Discussion and Implication

Despite scholarly evidence linking socio-psychological factors to consumer behavior and criticizing product obsolescence, there is still limited understanding of how policy and legal standards optimize these practices. Socio-psychological drivers play a significant role in shaping modern consumption patterns, particularly within the domain of product obsolescence. As technology advances, products become obsolete at a faster pace, presenting a dilemma for consumers, businesses, and the environment. Consumers often face a conundrum between their desire for the latest products and the psychological pull of sustainability. Product obsolescence, the process where products lose relevance due to technological advancements and shifting consumer preferences, amplifies this dilemma. This strategy is prevalent in industries such as electronics, fashion, and automotive, but scholarly literature criticizes its impact on consumer decision-making by making it more complex. Psychologically, consumers experience status anxiety, where the desire to maintain social status or be seen as technologically savvy leads them to replace products prematurely. This is evident in the constant upgrading of smartphones, where people feel compelled to purchase new models despite their current devices being functional. The fear of missing out (FOMO) exacerbates the situation, as people believe not owning the latest model will lead to social exclusion or technological inadequacy. This contributes to consumer culture, where material acquisition is prioritized over longevity and sustainability. The environmental consequences of product obsolescence are dire. The constant disposal of obsolete products generates significant waste, contributing to landfill overflow and resource depletion.

In the electronics industry, for example, millions of tons of electronic waste (e-waste) are produced annually, much of which ends up in landfills or is improperly recycled, causing soil and water contamination. This unsustainable cycle leads to environmental degradation and challenges for policymakers, manufacturers, and consumers in reducing the harmful impact. SPLCs have become the center of scholarly discussions aimed at achieving sustainable production and consumption by 2030. This is due to their promotion of circular economy principles that prioritize product life extension, reparability, and recyclability. Companies can now design products with longer lifespans, offer repair services, and create recycling programs to ensure that materials are reused rather than discarded. Moreover, eco-labeling and sustainable certifications can help consumers make informed purchasing decisions, favoring environmentally responsible and ethically produced products. From a consumer perspective, educating the public about the environmental impact of their purchasing decisions is essential. Raising awareness through campaigns and social media can shift consumer values from status-driven consumption to conscious consumerism. Brands can also play a role in influencing these behaviors by marketing sustainability as a core value, offering incentives for consumers to repair rather than replace products, or encouraging product-sharing models, which reduce the need for individual ownership.

The study advances the theoretical understanding of social and psychological consumer conundrum by conducting a systematic literature review of key theories, literature, and case studies on consumption conundrum in consumer behavior domain. By focusing on the antecedent, driver and outcome, the research essentially examines the social and psychological dynamics influencing the consumer behavior, offering both a theoretical foundation for and against the consumption conundrum in general. The study's contributions are fourfold: (a) it theorize the concept of consumption conundrum by developing the ADO model to explain how social and psychological factors can drive consumer behavior towards facing consumption conundrum in their daily purchase transactions. This provides an essential understanding of socio-psychological factors influencing consumer behaviors and the challenges posed by them in the rapidly evolving sectors. (b) The study review the scholarly literature to provides valuable insights on product obsolescence, its dimensions and effect on consumer and analyze the legal policy and standard that curtail the practice of product obsolescence across the industry. This provides an essential understanding of the good and dark side of product obsolescence by answering whether their practice is problematic or beneficial for the consumer and the entire community. (c) By exploring the role of SPLCs in overcoming the problem of product obsolescence, the study offers a forward-thinking perspective that encourages further academic exploration on the concept of sustainability and product lifecycle. (d) The study synthesizes essential theories and scholarly literature on consumer behavior and circular economy to develop PPO model, benefiting both academia (for hypothesis development and testing) and practitioners (for policy formulation and implementation). This contribute to the broader discourse on the social and psychological aspects of product obsolescence and consumption conundrum. Overall, the study offers a conceptual framework for smart living research by identifying areas that require attention for achieving sustainable development by 2030 which enriches the literature and sets the stage for future research on consumer behavior, sustainability, and product obsolescence.

The reviewed articles provide valuable insights that can guide how markets, institutions, and individuals can address the tension between short-term consumption incentives and long-term sustainability goals. While cognitive dissonance is real, designing interventions should not rely solely on information provision but also on behavioral alignment mechanisms. Firms can introduce default sustainable options (e.g., repair-first recommendations, refurbished products) and use post-purchase reinforcement (e.g., dashboards that provide feedback on sustainability impact) to frame sustainable choices as identity-consistent behaviors. Doing so helps reduce the gap between consumers' stated values and actual behaviors by making sustainable choices easier, more intuitive, and psychologically rewarding. As the normalization of frequent upgrades reinforces obsolescence, there is a need to leverage social norm marketing (e.g., highlighting how many consumers choose repair or reuse). This approach can encourage status redefinition such that sustainability becomes a marker of prestige rather than novelty in a way that collective behavior can be redirected to reduce the social pressure that drives unnecessary consumption. Traditional business models often rely on planned obsolescence to sustain revenue through repeat purchases. A shift towards sustainable value creation requires the adoption of circular economy models (repair, reuse, and remanufacturing) and designing products for durability, modularity, and upgradeability. Such an approach can reduce the tension between profitability and sustainability by aligning firm incentives with stakeholder value such that consumers are better informed about product lifespan, repairability, and environmental impact. When consumers cannot fully assess a product's lifespan and repairability, providing product lifecycle labeling (e.g., durability scores, repairability indices), clear disclosure of software support timelines, and standardized sustainability certifications becomes essential. Doing that enhances consumer trust, improves decision-making quality, and empowers individuals to make choices that are consistent with both their values and long-term societal interests.

4.6 Conclusion and Future Research Call

The interplay between social and psychological consumer behaviors, product obsolescence, and sustainable approaches argued a key challenge in today's marketplace. As consumers are increasingly influenced by societal trends and psychological cues, the demand for new and upgraded products drives the cycle of obsolescence. This strains environmental resources and perpetuates unsustainable consumption patterns. However, integrating sustainable practices into product design, marketing strategies, and consumer education can help mitigate the negative effects of obsolescence. By shifting focus from planned obsolescence to long-term durability, ethical consumption, and circular economy models, businesses and consumers can foster a more responsible approach to consumption. Ultimately, the convergence of sustainability with consumer psychology offers an opportunity to reshape market dynamics, leading to a future where innovation and responsibility coexist, benefiting both society and the environment. To make significant progress toward sustainable development by 2030, industrialized nations must reduce household trash resulting from unsustainable consumption practices. Although product obsolescence is often criticized, it has still long been accepted, and prior studies on their dimensions causing the "throwaway society" are lacking. Despite growing consumer concern for sustainability, their purchasing behavior often contradicts their idea and action, influenced by marketing tactics and planned obsolescence. Researchers highlight the potential of circular economy models and policy interventions, such as Green Public Procurement (GPP), Extended Producer Responsibility (EPR) and the Restriction of Hazardous Substances (RoHS) Directive. Consumer education is also vital to shift attitudes toward sustainable consumption and emphasize the long-term environmental benefits of reducing waste and extending product life cycles. Firms must involve consumers in sustainability through self-motivation and cultivate knowledge of the justification for sustainability practices in relation to their well-being.

Despite its contribution to the scholarly literature, the study has limitations that offer opportunities for future research. The study's reliance on English-language articles from Web of Science and Scopus alone, while methodologically sound, may exclude valuable insights from non-English publications or emerging platforms, limiting contextual and cultural diversity. Excluding grey literature, such as industry reports, also reduces practical insights into real-world implementation issues related to product obsolescence and SPLCs. Although the study provides case examples for examining obsolescence and consumption conundrum, it does not use primary data or real-world scenarios to evaluate the proposed model applicability experimentally. Future research should expand the methodological approach by incorporating bibliometric reviews, surveys, ethnographic methods, and simulation models. Further exploration of smart living in terms of motives, barriers, costs, and benefits can enhance the current understanding of this field. For example, García-Machado et al. (2020) and Guo et al. (2024) employed surveys to gather primary data on sustainable consumption. Collecting primary data is essential for assessing how recent smart living developments reduce deployment barriers towards realizing sustainable development by 2030. Although the study explored various theories on intention and actual behavior, more research is needed on how theories like the diffusion of innovation theory (Rogers 1995), organizational processing of information (Galbraith 1974), the Technology Acceptance Model (TAM) and its extended version (Davis 1989), and the technology–organization–environment (TOE) framework (Tornatzky and Fleischer 1982) could examine the link between barriers, drivers, and smart living in a multimodal setting. Their application could enrich the literature by identifying barriers and motives for consumer behavior across service domain such as airline, healthcare, education and transportation service industry. Although the study calls for a governance framework, future research should assess how governance policies and industry standards affect consumption and product obsolescence using suitable methodologies accordingly. For instance, the study buttress (Siltaloppi and Jähi 2021) future research call: examination of sustainable innovation and business model development, and organizational-level challenges of conundrum that hinder sustainability transitions. This call moves beyond firm-centric incremental innovations and fosters large-scale sustainable plastic solutions by integrating stakeholders across the value chain. Moreover, investigating the root causes will offer insights into overcoming internal barriers and aligning strategic and innovation management with sustainability goals. Given the tensions inherent in integrating sustainability into business practices, understanding the organizational dynamics is essential for facilitating systemic change and fostering resilience in sustainability transitions.

Moreover, a qualitative research approach is well-suited for generating new theoretical insights into intricate phenomena, such as the social, psychological, and economic dynamics influencing the shift toward sustainable consumption. However, this method does not allow for statistical generalizations regarding the broader consumer landscape (Mukherjee and Datta 2025). Future research is essential to examine the relative significance of various barriers that hinder consumers' adoption of sustainable behaviors and the nuanced ways in which the obstacles shape decision-making across different contexts. Specifically, further studies could explore how psychological and social factors contribute to resistance against sustainable alternatives and how companies can mitigate product obsolescence through strategic innovation. Also, future studies should implements the research call of Aronson (2015) and Giller (2020) on how firms can integrate collaborative research and development (R&D), branding strategies, and infrastructural advancements to promote circular economy models. Finally, research should investigate how businesses within consumer goods industries can leverage multi-tier collaborations to scale up sustainable product solutions and address deep-rooted consumer conundrum.

References

Achrol RS, Kotler P (2012) Frontiers of the marketing paradigm in the third millennium. J Acad Mark Sci 40:35–52. https://doi.org/10.1007/s11747-011-0255-4

Acikgoz F, Borulu B, Bölen MC (2025) How does obsolescence risk influence consumer resistance to smartwatches? Inf Technol People 38(7):2684–2709. https://doi.org/10.1108/ITP-08-2023-0843

Agarwal S, Jain H (2024) Investigating the potential of floral waste as a vermicompost and dual-functional biosorbent for sustainable environmental management. Water Air Soil Pollut 235:322. https://doi.org/10.1007/s11270-024-07144-y

Ajitha S, Sivakumar VJ (2017) Understanding the effect of personal and social value on attitude and usage behavior of luxury cosmetic brands. J Retail Consum Serv 39:103–113

Ajzen I, Fishbein M (1980) Understanding attitudes and predicting social behavior. Englewood Cliffs, NJ: Prentice-Hall

Ajzen I (1991) The theory of planned behavior. Organ Behav Hum Decis Process 50(2):179–211

Ajzen I (2011) The theory of planned behaviour: reactions and reflections. In: Psychology and health. Taylor & Francis, pp 1113–1127

Akinyode BF, Khan TH (2018) Step by step approach for qualitative data analysis. Int J Built Environ Sustain 5(3)

Albrecht E (2019) Planned obsolescence: a case under torts law as intentional damage contrary to public policy (Art. 826 German Civil Code). In: Sustainable global value chains, pp 689–708

Al-Rahmi WM, Yahaya N, Alamri MM, Alyoussef IY, Al-Rahmi AM, Kamin YB (2021) Integrating innovation diffusion theory with technology acceptance model: supporting students' attitude towards using a massive open online courses (MOOCs) systems. Interact Learn Environ 29(8):1380–1392

Alzaydi A (2024) Balancing creativity and longevity: the ambiguous role of obsolescence in product design. J Clean Prod 445:141239

André Q, Carmon Z, Wertenbroch K, Crum A, Frank D, Goldstein W, Huber J, Van Boven L, Weber B, Yang H (2018) Consumer choice and autonomy in the age of artificial intelligence and big data. Cust Needs Solut 5:28–37

Aronson N (2015) Making personalized medicine more affordable. Ann N Y Acad Sci 1346(1):81–89

Atkinson KM, Koenka AC, Sanchez CE, Moshontz H, Cooper H (2015) Reporting standards for literature searches and report inclusion criteria: making research syntheses more transparent and easy to replicate. Res Synth Methods 6(1):87–95

Ayadi N, Paraschiv C (2024) Consumption-related affect and consumer well-being. Fostering consumer well-being: theory, evidence, and policy, pp 289–313

Aydin R, Badurdeen F (2019) Sustainable product line design considering a multi-lifecycle approach. Resour Conserv Recycl 149:727–737

Badurdeen F, Aydin R, Brown A (2018) A multiple lifecycle-based approach to sustainable product configuration design. J Clean Prod 200:756–769

Becher SI, Sibony A-L (2021) Confronting product obsolescence. Colum J Eur L 27:97

Belk RW (1985) Materialism: trait aspects of living in the material world. J Consum Res 12(3):265–280

Bevilacqua M, Ciarapica FE, Giacchetta G (2007) Development of a sustainable product lifecycle in manufacturing firms: a case study. Int J Prod Res 45(18–19):4073–4098

Bhandari N, Garza-Reyes JA, Rocha-Lona L, Kumar A, Naz F, Joshi R (2022) Barriers to sustainable sourcing in the apparel and fashion luxury industry. Sustain Prod Consumption 31:220–235

Bharti M, Suneja V, Chauhan AK (2022) The role of socio-psychological and personality antecedents in luxury consumption: a meta-analytic review. Int Mark Rev 39(2):269–308

Bisschop L, Hendlin Y, Jaspers J (2022) Designed to break: planned obsolescence as corporate environmental crime. Crime Law Soc Chang 78(3):271–293

Bocken NM, De Pauw I, Bakker C, Van Der Grinten B (2016) Product design and business model strategies for a circular economy. J Ind Prod Eng 33(5):308–320. https://doi.org/10.1080/21681015.2016.1172124

Boland WA, Martin IM, Mason MJ (2020) In search of well-being: factors influencing the movement toward and away from maladaptive consumption. J Consum Aff 54(4):1178–1194

Bouten M (2008) Compatibility and technology acceptance: consolidating, validating and extending concept. In: Faculty of economics and business administration [Preprint]

Bowden JLH, Gabbott M, Naumann K (2015) Service relationships and the customer disengagement–engagement conundrum. J Mark Manag 31(7–8):774–806

Brockhaus S, Petersen M, Kersten W (2016) A crossroads for bioplastics: exploring product developers' challenges to move beyond petroleum-based plastics. J Clean Prod 127:84–95

Brönneke T (2017) Premature obsolescence: suggestions for legislative counter-measures in German and European sales and consumer law. J Eur Environ Plann Law 14(3–4):361–372

Brouillat E (2015) Live fast, die young? Investigating product life spans and obsolescence in an agent-based model. J Evol Econ 25(2):447–473

Brundage MP, Bernstein WZ, Hoffenson S, Chang Q, Nishi H, Kliks T, Morris KC (2018) Analyzing environmental sustainability methods for use earlier in the product lifecycle. J Clean Prod 187:877–892

Cairns HM, Ritch EL, Bereziat C (2022) Think eco, be eco? the tension between attitudes and behaviours of millennial fashion consumers. Int J Consum Stud 46(4):1262–1277

Cao H, Folan P (2012) Product life cycle: the evolution of a paradigm and literature review from 1950–2009. Prod Plann Control 23(8):641–662

Chatzisarantis NLD, Biddle SJH, Meek GA (1997) A self-determination theory approach to the study of intentions and the intention–behaviour relationship in children's physical activity. Br J Health Psychol 2(4):343–360

Chiu W, Cho H (2022) The model of goal-directed behavior in tourism and hospitality: a meta-analytic structural equation modeling approach. J Travel Res 61(3):637–655

Chu C-H, Pan J-K (2024) A systematic review on extended reality applications for sustainable manufacturing across the product lifecycle. Int J Precision Eng Manuf-Green Technol 11(3):1017–1028

Cicconi P (2020) Eco-design and Eco-materials: an interactive and collaborative approach. Sustain Mater Technol 23:e00135

Colosimo MF, Kim H (2016) Incorporating innovative water management science and technology into water management policy. Energy Ecol Environ 1(1):45–53

Confente I, Scarpi D (2021) Achieving environmentally responsible behavior for tourists and residents: a norm activation theory perspective. J Travel Res 60(6):1196–1212

Cvitanovic C, McDonald J, Hobday AJ (2016) From science to action: principles for undertaking environmental research that enables knowledge exchange and evidence-based decision-making. J Environ Manag 183:864–874

Dalvi-Esfahani M, Ramayah T, Rahman AA (2017) Moderating role of personal values on managers' intention to adopt Green IS: examining norm activation theory. Ind Manag Data Syst 117(3):582–604

Davis FD (1989) Perceived usefulness, perceived ease of use, and user acceptance of information technology. MIS Q 319–340

De Franceschi A (2018) Planned obsolescence challenging the effectiveness of consumer law and the achievement of a sustainable economy: the apple and samsung cases. J Eur Consum Market Law 7(6)

Dewan S, Ho Y-J, Ramaprasad J (2017) Popularity or proximity: characterizing the nature of social influence in an online music community. Inf Syst Res 28(1):117–136

Dhaliwal A, Singh DP, Paul J (2025) The consumer behavior of luxury goods: a review and research agenda. J Strateg Mark 33(1):66–92

Echegaray F (2016) Consumers' reactions to product obsolescence in emerging markets: the case of Brazil. J Clean Prod 134:191–203

Festinger L (1957) A theory of cognitive dissonance. Stanford, CA: Stanford University Press
Fischbach RL, Herbert B (1997) Domestic violence and mental health: correlates and conundrums within and across cultures. Soc Sci Med 45(8):1161–1176
Fishbein M, Ajzen I (1974) Attitudes towards objects as predictors of single and multiple behavioral criteria. Psychol Rev 81(1):59
Fisher JD, Johnson DS, Marchand J, Smeeding TM, Torrey BB (2008) The retirement consumption conundrum: evidence from a consumption survey. Econ Lett 99(3):482–485
Fleissig AR (1997) The consumer consumption conundrum: an explanation. J Money Credit Bank 177–192
Flusberg SJ, Holmes KJ, Thibodeau PH, Nabi RL, Matlock T (2024) The psychology of framing: how everyday language shapes the way we think, feel, and act. Psychol Sci Public Interest 25(3):105–161
Forti V, Baldé CP, Kuehr R, Bel G (2020) The global e-waste monitor 2020: quantities, flows, and the circular economy potential. United Nations University (UNU), International Telecommunication Union (ITU), and International Solid Waste Association (ISWA). Accessed from https://ewaste monitor.info/wp-content/uploads/2020/11/GEM_2020_def_july1_low.pdf
Franzolin CJ (2020) Planned obsolescence resulting from electrical and electronic equipment: waste rights and Brazil's national solid waste policy. In: Sustainable consumption: the right to a healthy environment, pp 463–477
Galati F, Bigliardi B, Petroni A, Pinna C, Rossi M, Terzi S (2019) Sustainable product lifecycle: the role of ICT. In: Sustainability. MDPI, p 7003
Galbraith JR (1974) Organization design: an information processing view. Interfaces 4(3):28–36
García-Machado JJ, Figueroa-García EC, Jachowicz A (2020) Sustainable consumption behaviour in Poland through a PLS-SEM model. In: Perspectives on consumer behaviour: theoretical aspects and practical applications, pp 147–172.
Gary J (2025) Information push strategies in e-commerce: emotional and cognitive mechanisms shaping purchase decisions: multi-method approaches. J Market Anal 1–39
Gbededo MA, Liyanage K, Garza-Reyes JA (2018) Towards a life cycle sustainability analysis: a systematic review of approaches to sustainable manufacturing. J Clean Prod 184:1002–1015
Gecit BB (2020) Planned obsolescence: a keyword analysis. J Manag Market Logistics 7(4):227–233
Giller KE (2020) The food security conundrum of sub-Saharan Africa. Glob Food Sec 26:100431
Giolo C Jr, Saraiva JS, Aylon LL (2024) Sustainable public policy approaches to overcome planned obsolescence: models of circular economy innovations. Beijing L Rev 15:1430
Goebel P, Reuter C, Pibernik R, Sichtmann C, Bals L (2018) Purchasing managers' willingness to pay for attributes that constitute sustainability. J Oper Manag 62:44–58
Goldsmith EB, Goldsmith RE (2011) Social influence and sustainability in households. Int J Consum Stud 35(2):117–121
Goldsmith RE, Flynn LR, Clark RA (2014) The etiology of the frugal consumer. J Retail Consum Serv 21(2):175–184
Gong W, Jiang LC, Guo Q, Shen F (2023) The role of family communication patterns in intergenerational COVID-19 discussions and preventive behaviors: a social cognitive approach. BMC Psychol 11(1):290
Gough D, Thomas J (2016) Systematic reviews of research in education: aims, myths and multiple methods. Rev Educ 4(1):84–102
Guillard V, Le Nagard E, de Campos Ribeiro G (2023) A typology of consumers regarding perceived obsolescence: the paradox of eco-conscious consumers. J Clean Prod 412:137202
Guiltinan J (2009) Creative destruction and destructive creations: environmental ethics and planned obsolescence. J Bus Ethics 89:19–28
Guo X, Wang L, Guo Y, Liu Y (2024) The impact of sustainable consumption behaviour on natural resource conservation in China: a cross-sectional analysis. Resour Policy 89:104610
Gupta S, Rajiah P, Middlebrooks EH, Baruah D, Carter BW, Burton KR, Chatterjee AR, Miller MM (2018) Systematic review of the literature: best practices. Acad Radiol 25(11):1481–1490

Gurău C (2012) A life-stage analysis of consumer loyalty profile: comparing Generation X and Millennial consumers. J Consum Mark 29(2):103–113
Hagger MS (2015) Retired or not, the theory of planned behaviour will always be with us. In: Health psychology review. Taylor & Francis, pp 125–130
Hammer J, Pivo G (2017) The triple bottom line and sustainable economic development theory and practice. Econ Dev Q 31(1):25–36
Han H (2014) The norm activation model and theory-broadening: Individuals' decision-making on environmentally-responsible convention attendance. J Environ Psychol 40:462–471
Havitz ME, Dimanche F (1997) Leisure involvement revisited: conceptual conundrums and measurement advances. J Leis Res 29(3):245–278
Hinterhuber A, Khan O (2025) What drives sustainable procurement? Insights from the theory of planned behavior. Int J Oper Prod Manag 45(13):28–52
Ho TH, Lim N, Camerer CF (2006) Modeling the psychology of consumer and firm behavior with behavioral economics. J Mark Res 43(3):307–331
Iveson A, Hultman M, Davvetas V (2022) The product life cycle revisited: an integrative review and research agenda. Eur J Mark 56(2):467–499
Jacobson J, Harrison B (2022) Sustainable fashion social media influencers and content creation calibration. Int J Advert 41(1):150–177
Jensen SF, Kristensen JH, Adamsen S, Christensen A, Waehrens BV (2023) Digital product passports for a circular economy: data needs for product life cycle decision-making. Sustain Prod Consumption 37:242–255
Jin H, Hwang J (2024) The dual lens of authenticity and likability: the effect of cultural familiarity on consumer responses to ethnic foods. Int J Hosp Manag 117:103655
Joltreau E (2022) Extended producer responsibility, packaging waste reduction and eco-design. Environ Resour Econ 83(3):527–578
Jordan K, Kristjánsson K (2017) Sustainability, virtue ethics, and the virtue of harmony with nature. Environ Educ Res 23(9):1205–1229
Kameda T, Hastie R (2015) Herd behavior. In: Emerging trends in the social and behavioral sciences: an interdisciplinary, searchable, and linkable resource, pp 1–14
Karahanna E, Agarwal R, Angst CM (2006) Reconceptualizing compatibility beliefs in technology acceptance research. MIS Q 781–804
Karman A, Lipowski M (2024) Switching to sustainable products: the role of time, product, and customer characteristics. Psychol Mark 41(5):1082–1099
Katona Z, Zubcsek PP, Sarvary M (2011) Network effects and personal influences: the diffusion of an online social network. J Mark Res 48(3):425–443
Kessler T, Brendel J (2016) Planned obsolescence and product-service systems: linking two contradictory business models. J Competence-Based Strateg Manag 8:29–53
Khandelwal M, Hill ME Jr, Greenough P, Anthony J, Quill M, Linderman M, Udaykumar HS (2017) Why have improved cook-stove initiatives in India failed? World Dev 92:13–27
Khurshudov A (2024) The smart city conundrum: technology, privacy, and the quest for convenience. Smart Sustain Built Environ [Preprint]
Koole SL, Jostmann NB, Baumann N (2023) From intentions to action: an integrative review of action control theory and research. Adv Exp Soc Psychol 68:291–375
Kuppelwieser VG, Klaus P, Manthiou A, Boujena O (2019) Consumer responses to planned obsolescence. J Retail Consum Serv 47:157–165
Lawlor R (2015) Delaying obsolescence. Sci Eng Ethics 21(2):401–427
Lawson SJ, Gleim MR, Hartline MD (2021) Decisions, decisions: variations in decision-making for access-based consumption. J Market Theory Practice 29(3):358–374
Lee SK, Lindsey NJ, Kim KS (2017) The effects of news consumption via social media and news information overload on perceptions of journalistic norms and practices. Comput Hum Behav 75:254–263
Li J, Zhao J, Huang J (2025) Social avoidance needs boost AI's nonsocial attribute valuation in secret consumption. Technol Soc 81:102817

Li S, Hong Y-C, Craig SD (2023) A systematic literature review of social learning theory in online learning environments. Educ Psychol Rev 35(4):108

Lien NH, Westberg K, Stavros C, Robinson LJ (2018) Family decision-making in an emerging market: tensions with tradition. J Bus Res 86:479–489

Lifset R, Kalimo H, Jukka A, Kautto P, Miettinen M (2023) Restoring the incentives for eco-design in extended producer responsibility: the challenges for eco-modulation. Waste Manag 168:189–201

Lim WM, Das M, Sharma W, Verma A, Kumra R (2025) Gamification for sustainable consumption: a state-of-the-art overview and future agenda. Bus Strateg Environ 34(1):1510–1549

Lindenberg S, Steg L (2007) Normative, gain and hedonic goal frames guiding environmental behavior. J Soc Issues 63(1):117–137

Linnenluecke MK, Marrone M, Singh AK (2020) Conducting systematic literature reviews and bibliometric analyses. Aust J Manag 45(2):175–194

Liu H, Jayawardhena C, Osburg V-S, Yoganathan V, Cartwright S (2021) Social sharing of consumption emotion in electronic word of mouth (eWOM): a cross-media perspective. J Bus Res 132:208–220

Liu Y, Zhang Y, Ren S, Yang M, Wang Y, Huisingh D (2020) How can smart technologies contribute to sustainable product lifecycle management? J Clean Prod 249:119423

London B (2014) Ending the depression through planned obsolescence. Revue Du MAUSS 44(2):47–50

Lorek S, Spangenberg JH (2014) Sustainable consumption within a sustainable economy–beyond green growth and green economies. J Clean Prod 63:33–44

Luo MM, Remus W (2014) Uses and gratifications and acceptance of Web-based information services: an integrated model. Comput Hum Behav 38:281–295

Ma J, Harstvedt JD, Dunaway D, Bian L, Jaradat R (2018) An exploratory investigation of additively manufactured product life cycle sustainability assessment. J Clean Prod 192:55–70

Ma J, Tu H, Zhang P, Fan L, Cheng B, Ma J (2021) Can work–family conflict influence purchase preference? experiential versus material consumption. J Bus Res 135:620–632

MacDonald EF, She J (2015) Seven cognitive concepts for successful eco-design. J Clean Prod 92:23–36

Maillet É, Mathieu L, Sicotte C (2015) Modeling factors explaining the acceptance, actual use and satisfaction of nurses using an Electronic Patient Record in acute care settings: an extension of the UTAUT. Int J Med Inform 84(1):36–47

Maitre-Ekern E (2021) Re-thinking producer responsibility for a sustainable circular economy from extended producer responsibility to pre-market producer responsibility. J Clean Prod 286:125454

Maitre-Ekern E, Dalhammar C (2016) Regulating planned obsolescence: a review of legal approaches to increase product durability and reparability in Europe. Rev Eur Comp Int Environ Law 25(3):378–394

Malika M, Maheswaran D, Jain SP (2022) Perceived financial constraints and normative influence: discretionary purchase decisions across cultures. J Acad Market Sci 1–20

Malinauskaite J, Erdem FB (2021) Planned obsolescence in the context of a holistic legal sphere and the circular economy. Oxf J Leg Stud 41(3):719–749

Mann N (2013) Human evolution and diet: a modern conundrum of health versus meat consumption, or is it? Animal Prod Sci 53(11):1135–1142

Marikyan D, Papagiannidis S, Alamanos E (2023) Cognitive dissonance in technology adoption: a study of smart home users. Inf Syst Front 25(3):1101–1123

Mathews JA, Tan H (2011) Progress toward a circular economy in China: the drivers (and inhibitors) of eco-industrial initiative. J Ind Ecol 15(3):435–457

Mayol A, Staropoli C (2021) Giving consumers too many choices: a false good idea? a lab experiment on water and electricity tariffs. Eur J Law Econ 51(2):383–410

McNeill L, Moore R (2015) Sustainable fashion consumption and the fast fashion conundrum: fashionable consumers and attitudes to sustainability in clothing choice. Int J Consum Stud 39(3):212–222

Mellal MA (2020) Obsolescence–a review of the literature. Technol Soc 63:101347
Melnyk V, Carrillat FA, Melnyk V (2022) The influence of social norms on consumer behavior: a meta-analysis. J Mark 86(3):98–120
Mensah J (2019) Sustainable development: meaning, history, principles, pillars, and implications for human action: literature review. Cogent Soc Sci 5(1):1653531
Mishra AS, Mishra RR (2023) Marketing schools of thought and their present day relevance. FIIB Bus Rev 12(4):351–361
Mitchell S-L, Clark M (2019) Reconceptualising product life-cycle theory as stakeholder engagement with non-profit organisations. J Mark Manag 35(1–2):13–39
Moher D, Liberati A, Tetzlaff J, Altman DG, PRISMA Group*, t (2009) Preferred reporting items for systematic reviews and meta-analyses: the PRISMA statement. Ann Internal Med 151(4):264–269
Mongeon P, Paul-Hus A (2016) The journal coverage of Web of Science and Scopus: a comparative analysis. Scientometrics 106:213–228
Moshood TD, Nawanir G, Mahmud F, Mohamad F, Ahmad MH, AbdulGhani A (2022a) Biodegradable plastic applications towards sustainability: a recent innovations in the green product. Clean Eng Technol 6:100404
Moshood TD, Nawanir G, Mahmud F, Mohamad F, Ahmad MH, AbdulGhani A, Kumar S (2022b) Green product innovation: a means towards achieving global sustainable product within biodegradable plastic industry. J Clean Prod 363:132506
Mothersbaugh D, Hawkins D, Kleiser S (2020) Consumer behavior: building marketing strategy 14th Edition
Mukherjee S, Datta B (2025) Counterfeit conundrum: Insights into bottom of the pyramid consumer behavior in India. J Retail Consum Serv 82:104114
Munten P, Vanhamme J, Swaen V (2021) Reducing obsolescence practices from a product-oriented PSS perspective: a research agenda. Recherche Et Appl En Market (English Edition) 36(2):42–74
Natarajan M, Periaiya S (2024) Revealing consumer review attitude through online review and website cues. J Consum Mark 41(3):261–280
Niu Z, Chen C, Gao Y, Wang Y, Chen Y, Zhao K (2022) Peer effects, attention allocation and farmers' adoption of cleaner production technology: taking green control techniques as an example. J Clean Prod 339:130700
Oparaocha GO (2016) Towards building internal social network architecture that drives innovation: a social exchange theory perspective. J Knowl Manag 20(3):534–556
Oraee A, Pohl L, Geurts D, Reichel M (2024) Overcoming premature smartphone obsolescence amongst young adults. Clean Responsible Consumption 12:100174
Ortega-Egea JM, García-de-Frutos N (2021) Mapping the influence of country-of-origin knowledge, consumer ethnocentrism, and perceived risk on consumer action against foreign products. J Consum Behav 20(5):1164–1178
Osei-Frimpong K, McLean G, Islam N, Otoo BA (2022) What drives me there? the interplay of socio-psychological gratification and consumer values in social media brand engagement. J Bus Res 146:288–307
Papaleontiou-Louca E, Esmailnia S, Thoma N (2022) A critical review of Maslow's theory of spirituality. J Spirituality Mental Health 24(4):327–343
Perugini M, Bagozzi RP (2004) The distinction between desires and intentions. Eur J Soc Psychol 34(1):69–84
Phelan AA, Meissner K, Humphrey J, Ross H (2022) Plastic pollution and packaging: corporate commitments and actions from the food and beverage sector. J Clean Prod 331:129827
Pihlajarinne T (2024) Shifting normality–regulation of planned and premature obsolescence in Europe. In: Promoting sustainable innovation and the circular economy. Routledge, pp 183–200
Proske M, Finkbeiner M (2020) Obsolescence in LCA–methodological challenges and solution approaches. Int J Life Cycle Assess 25:495–507
Pruckner GJ, Schober T, Zocher K (2020) The company you keep: health behavior among work peers. Eur J Health Econ 21(2):251–259

Psarommatis F, May G (2022) Achieving global sustainability through sustainable product life cycle. In: IFIP international conference on advances in production management systems. Springer, pp 391–398

Reilly JE, Turcan RV (2023a) Leadership: facing the authenticity conundrum. In: The emerald handbook of authentic leadership. Emerald Publishing Limited, pp 559–576

Reilly JE, Turcan RV (2023b) Leadership: The conundrum of authenticity. In: The emerald handbook of authentic leadership. Emerald Publishing Limited, pp 3–37

Ren S, Zhang Y, Liu Y, Sakao T, Huisingh D, Almeida CMVB (2019) A comprehensive review of big data analytics throughout product lifecycle to support sustainable smart manufacturing: a framework, challenges and future research directions. J Clean Prod 210:1343–1365

Reutskaja E, Cheek NN, Iyengar S, Schwartz B (2022) Choice deprivation, choice overload, and satisfaction with choices across six nations. J Int Mark 30(3):18–34

Rezaei-Moghaddam K, Vatankhah N, Ajili A (2020) Adoption of pro-environmental behaviors among farmers: application of Value–Belief–Norm theory. Chem Biol Technol Agric 7:1–15

Rivera JL, Lallmahomed A (2016) Environmental implications of planned obsolescence and product lifetime: a literature review. Int J Sustain Eng 9(2):119–129. https://doi.org/10.1080/19397038.2015.1099757

Roberts H, Milios L, Mont O, Dalhammar C (2023) Product destruction: exploring unsustainable production-consumption systems and appropriate policy responses. Sustain Prod Consumption 35:300–312

Rogers EM (1995) Diffusion of innovations, 4th The Free Press. New York

Rojas M, Méndez A, Watkins-Fassler K (2023) The hierarchy of needs empirical examination of Maslow's theory and lessons for development. World Dev 165:106185

Rolling V, Seifert C, Chattaraman V, Sadachar A (2021) Pro-environmental millennial consumers' responses to the fur conundrum of luxury brands. Int J Consum Stud 45(3):350–363

Rusch M, Schöggl J, Baumgartner RJ (2023) Application of digital technologies for sustainable product management in a circular economy: a review. Bus Strateg Environ 32(3):1159–1174

Ruvio A, Bagozzi RP, Hult GTM, Spreng R (2020) Consumer arrogance and word-of-mouth. J Acad Mark Sci 48:1116–1137

Saini J, Gupta AK, Shrivastava A, Dhupper R (2024) Study of urban warming in Chandigarh union territory, India: geospatial approach: J. Saini et al. Theor Appl Climatol 155(8):8245–8261

Satyro WC, Sacomano JB, Contador JC, Telles R (2018) Planned obsolescence or planned resource depletion? a sustainable approach. J Clean Prod 195:744–752

Sauvé S, Bernard S, Sloan P (2016) Environmental sciences, sustainable development and circular economy: alternative concepts for trans-disciplinary research. Environ Dev 17:48–56

Scott KA, Weaver ST (2018) The intersection of sustainable consumption and anticonsumption: repurposing to extend product life spans. J Public Policy Mark 37(2):291–305

Sepe F, Valerio M, Anna P, Mario T (2025) Fashion and sustainability: evidence from the consumption of second-hand clothes. Corp Soc Responsib Environ Manag 32(1):947–962

Sharma A, Bhaduri S (2013) Consumption conundrum of bottled water in India: an STS perspective. Bull Sci Technol Soc 33(5–6):172–181

Shaw EH, Jones DGB (2005) A history of schools of marketing thought. Mark Theory 5(3):239–281

Sheth JN, Sethia NK, Srinivas S (2011) Mindful consumption: a customer-centric approach to sustainability. J Acad Mark Sci 39:21–39

Shrivastava A, Abhishek K, Gupta AK, Jain H, Kumari M, Patel M, Sharma P (2024) Removal of micro-and nano-plastics from aqueous matrices using modified biochar–a review of synthesis, applications, interaction, and regeneration. J Hazard Mater Adv 16:100518

Sierra-Fontalvo L, Gonzalez-Quiroga A, Mesa JA (2023) A deep dive into addressing obsolescence in product design: a review. Heliyon 9(11)

Sierra-Fontalvo L, Ruiz-Pastor L, Gonzalez-Quiroga A, Mesa JA (2024) Decoding product obsolescence: a taxonomic approach from product design attributes. J Clean Prod 475:143635

Siltaloppi J, Jähi M (2021) Toward a sustainable plastics value chain: core conundrums and emerging solution mechanisms for a systemic transition. J Clean Prod 315:128113

Slaton K, Pookulangara S (2022) Collaborative consumption: an investigation into the secondary sneaker market. Int J Consum Stud 46(3):763–780

Sniehotta FF, Presseau J, Araújo-Soares V (2014) Time to retire the theory of planned behavior. In: Health psychology review. Taylor & Francis, pp 1–7

Taherdoost H (2018) A review of technology acceptance and adoption models and theories. Procedia Manuf 22:960–967

Tajfel H, Turner JC, Austin WG, Worchel S (1979) An integrative theory of intergroup conflict. Organ Identity Reader 56(65):9780203505984–16

Tang Y, Chen S, Yuan Z (2020) The effects of hedonic, gain, and normative motives on sustainable consumption: multiple mediating evidence from China. Sustain Dev 28(4):741–750

Teerakapibal S, Schlegelmilch BB (2025) Mind the gap: identifying the chasm between environmental awareness, attitudes and action in fast fashion consumption. Int Mark Rev 42(1):103–127

Thiermann UB, Sheate WR (2020) Motivating individuals for social transition: the 2-pathway model and experiential strategies for pro-environmental behaviour. Ecol Econ 174:106668

Thompson CJ, Rindfleisch A, Arsel Z (2006) Emotional branding and the strategic value of the doppelgänger brand image. J Mark 70(1):50–64

Tornatzky LG, Klein KJ (1982) Innovation characteristics and innovation adoption-implementation: a meta-analysis of findings. IEEE Trans Eng Manag (1):28–45

Trabelsi I, Zolghadri M, Zeddini B, Barkallah M, Haddar M (2021) Prediction of obsolescence degree as a function of time: a mathematical formulation. Comput Ind 129:103470

Tranfield D, Denyer D, Smart P (2003) Towards a methodology for developing evidence-informed management knowledge by means of systematic review. Br J Manag 14(3):207–222

United Nations Institute for Training and Research (UNITAR) (2024) The global e-waste monitor 2017: quantities, flows, and resources. E-Waste Monitor; United Nations Institute for Training and Research (UNITAR). https://ewastemonitor.info/

Vendrell-Herrero F, Gomes E, Opazo-Basaez M, Bustinza OF (2022) Knowledge acquisition throughout the lifecycle: product and industry learning frameworks. J Knowl Manag 26(6):1633–1647

Venkatesh V, Morris MG, Davis GB, Davis FD (2003) User acceptance of information technology: toward a unified view. MIS Q 425–478

Venkatesh V, Brown SA, Maruping LM, Bala H (2008) Predicting different conceptualizations of system use: the competing roles of behavioral intention, facilitating conditions, and behavioral expectation1. MIS Q 32(3):483–502

Vergragt PJ, Brown HS (2007) Sustainable mobility: from technological innovation to societal learning. J Clean Prod 15(11–12):1104–1115

Vila C, Abellán-Nebot JV, Albiñana JC, Hernández G (2015) An approach to sustainable product lifecycle management (Green PLM). Procedia Eng 132:585–592

Voorhees CM, White RC, McCall M, Randhawa P (2015) Fool's gold? Assessing the impact of the value of airline loyalty programs on brand equity perceptions and share of wallet. Cornell Hospitality Q 56(2):202–212

Webster A, Wilson JF, Wong ND (2021) Commerce with a bit of ethics or ethics with a bit of commerce? the conundrum of British consumer co-operation 1863–1990. J Manag Hist 27(3):318–338

Weerasinghe AS, Rasheed EO, Rotimi JOB (2024) Environmental and socio-psychological drivers of building users' behaviours: a case study of tertiary institutional offices in Auckland. J Facil Manag 22(4):564–587

Wells KJ, Spilda FU (2024) Uber for nursing: how an AI-powered gig model is threatening health care

White K, Habib R, Hardisty DJ (2019) How to SHIFT consumer behaviors to be more sustainable: a literature review and guiding framework. J Mark 83(3):22–49

Whittemore R, Chao A, Jang M, Minges KE, Park C (2014) Methods for knowledge synthesis: an overview. Heart Lung 43(5):453–461

Widianti T, Sumaedi S, Bakti IGMY, Rakhmawati T, Astrini NJ, Yarmen M (2015) Factors influencing the behavioral intention of public transport passengers. Int J Qual Reliabil Manag 32(7):666–692

Wong CWY, Lai K, Shang K-C, Lu C-S, Leung TKP (2012) Green operations and the moderating role of environmental management capability of suppliers on manufacturing firm performance. Int J Prod Econ 140(1):283–294

Wu W-L, Gau Y-F (2017) Home bias in portfolio choices: social learning among partially informed agents. Rev Quant Financ Acc 48:527–556

Yamamoto H, Murakami S (2021) Product obsolescence and its relationship with product lifetime: an empirical case study of consumer appliances in Japan. Resour Conserv Recycl 174:105798

Yang M, Chen L, Wang J, Msigwa G, Osman AI, Fawzy S, Rooney DW, Yap P-S (2023) Circular economy strategies for combating climate change and other environmental issues. Environ Chem Lett 21(1):55–80

Yi C, Jiang Z, Zhou M (2023) Investigating the effects of product popularity and time restriction: the moderating role of consumers' goal specificity. Prod Oper Manag 32(9):2723–2739

Yousif T, Moalosi R (2024) The role of industrial designers in achieving the green economy through recycling. J Eng 2024(1):7291504

Zallio M, Berry D (2017) Design and planned obsolescence. theories and approaches for designing enabling technologies. Des J 20(sup1):S3749–S3761

Zhang A, Hochman Z, Horan H, Navarro JG, Das BT, Waldner F (2019) Socio-psychological and management drivers explain farm level wheat yield gaps in Australia. Agron Sustain Dev 39(1):10

Zhang X, Zhang L, Fung KY, Bakshi BR, Ng KM (2020) Sustainable product design: a life-cycle approach. Chem Eng Sci 217:115508

Zhang Y, Ren S, Liu Y, Sakao T, Huisingh D (2017) A framework for big data driven product lifecycle management. J Clean Prod 159:229–240

Chapter 5
Ethical Marketing and Consumer Buying Behaviour: Do Young Consumers in the GCC Truly Care About Marketing Ethics?

Ghaith Al-Abdallah

5.1 Introduction

In today's globalised culture, consumers pay significant attention to sustainability claims made by products and brands, as companies strive to meet customer expectations (Eid and Al-Abdallah 2024). This growing awareness is gradually shifting market demand towards goods and services that prioritise environmental responsibility and public health (Ciasullo et al. 2017). However, a growing body of scholarly research highlights a clear gap between consumers' positive attitudes towards sustainability and ethical behaviour and their actual purchasing decisions, underscoring the need to understand the mechanisms driving this discrepancy. Ethics, being a fundamental component of sustainability (de Paula and Cavalcanti 2000) contribute to the overall success of the organisation and foster a positive reputation for both the compnay and its employees in the eyes of the customers, which increased the interest of both customers and organisation managers in the ethical dimension of work (Gillett and Tennent 2020). Petersen (2013) defines business ethics as "the principles, norms, and standards that guide an organisation's activities, internal relations, and interactions with external stakeholders." Marketing ethics, a subset of business ethics, can be defined generally as standards of behaviour and moral decisions that are applied to marketing practices (Gaski 1999). According to Koozechian et al. (2021), multinational corporations are becoming increasingly aware of the significance of business and marketing ethics and their contribution to the company's overall success and achievement of financial goals. Most modern enterprises recognise the need to implement tools for social and ethical marketing, which enhance awareness and

G. Al-Abdallah (✉)
University of Kurdistan Hewlêr, Erbil, Iraq, and Catholic University in Erbil, Erbil, Iraq
e-mail: ghaith.abdallah@outlook.com

H. Jain et al. (eds.), *The Circular Fallacy: Closing the Loop in a Consumerist World*, Circular Economy and Sustainability, https://doi.org/10.1007/978-3-032-22687-7_5

consumer protection. At its core, social and ethical marketing implements the principles of consumer orientation, innovation, value creation, awareness of the social mission, and corporate social responsibility. Increasingly, companies are willing to be guided by marketing ethics principles, such as consumer and producer freedom, minimising potential damage, meeting basic needs, economic efficiency, innovation, consumer education and awareness, and consumer protection (Voronov et al. 2021).

Every business requires an ethical climate that sharpens its behaviour and guides its operations. The American Marketing Association (AMA) (2023) stipulates that organisations must interact with customers in a transparent, trustworthy, and courteous manner. In addition, it is necessary to establish organisational policies to protect customers privacy during business transactions and information exchange. Iglesias et al. (2020) state that ethics are essential for the development of business exchange, as ethical interaction increases the likelihood of developing relationships and the durability and sustainability of business relationships. Amine et al. (2012) acknowledge that unethical behaviour has extremely negative effects on the productivity, image, and reputation of an organisation. Ellemers and Gilder (2022) conclude that it would be too expensive to compensate the aggrieved party for the negative consequences of unethical behaviour. According to Ndubisi et al. (2014), any ethical issues that arise between an organisation and its customers will be widely publicised through the media, which will have a negative impact on the organisation. According to Guerola-Navarro et al. (2022), business objectives have shifted from focusing on acquiring new customers to focusing on customer retention practices and extending the duration of the organisation-customer relationship, which cannot be achieved without ethical business and marketing. Amine et al. (2012) conclude that while ethics play a significant role in the management of an organisation and have a positive impact on its performance, they are also the foundation of maintaining customer relationships, as the ethical climate of the organisation will influence the behaviour of employees and decision-makers, which will be reflected in the organisation's overall performance (Jain et al. 2023). Ethical behaviour of the organisation is the basis for a successful relationship between the company and the customer through the maintenance of mutual trust, which has a significant impact on the enhancement of the organisation's performance (Bag et al. 2024). The effective implementation of ethical business models relies on the individual ethical marketer who acts as an interface between the market, the consumer, and the practices of organisations. Sustainable marketing has to be implemented through the actions of individual marketers, customers, and the institutional system to be successful (Ardley and May 2020). From the consumer side, consumer ethics refers to the moral principles and standards that guide consumers in determining whether certain consumption behaviours are ethically right or wrong, in which cultural and personal dimensions are crucial factors influencing individual ethical attitudes and behaviours (Lu et al. 2014). Minton et al. (2018) reach the conclusion that all individuals from diverse religious backgrounds concur that individual actions must be guided by ethical principles. The definition of the word "right" varies from source to source; right can be defined according to one's interests, whether rewards or punishments, or in accordance with society's opinions and expectations. In general, an individual's behaviour is governed by his or her moral philosophy, which is shaped by their culture, education, society, and environment (Lokhvytska et al. 2022). Accordingly different markets are expected to behave differently in this regard, reflecting the importance of this study in the selected markets.

While issues of perceived ethical importance were noted by some scholars as highlighted above, the past two decades have seen this subject gain importance among both consumers and companies, particularly in the context of the hyper-competitive business environment (Lee 2020). Schlegelmilch and Öberseder (2010) note that the degree of business commitment to ethical conduct is multifaceted, varying according to factors such as culture, education, experience, and normative ethical standards. Moreover, several researchers have examined the role of perceived ethical importance in steering individual behaviour. For example, Al-Rashid et al. (2023) demonstrate that a greater emphasis on ethics is associated with higher individual involvement, while Barnett and Valentine (2004) describe perceived ethical importance as a state-like construct, closely linked to behavioural intention and predictive of individual decision-making. This heightened focus has contributed to the development and acceptance of the "ethical marketing" concept, a subset of business ethics, which seeks to motivate both consumers and organisations to adopt practices that enhance collective well-being and quality of life through a balanced integration of moral and economic principles (Laczniak and Murphy 2019; Saini et al. 2024). Kotler et al. (2022) contend that an ethical approach to marketing can achieve a tripartite objective: improving customer satisfaction, fostering more defensible business success, and promoting social welfare. This rationale helps explain why perceived ethical importance has increasingly become central to marketing strategies, with a stronger focus on consumer needs and preferences. Ethics in marketing activity is the application of fairness standards or moral rights and wrongs to marketing decision-making, behaviour, and practice in the organisation (Al-Hakimi et al. 2022). Disputes about marketing ethics and socially responsible behaviour by firms are fundamentally debatable, and over three decades of research continue to provide academics and practitioners with opposing and contradictory perspectives on the value of ethical marketing in various regions of the world. Ethical marketing practices have a significant impact on product value-adding and customer-value brand relationships (Tanveer et al. 2021; Tan and Salo 2023). This underscores the critical role of ethical perceptions in shaping behavioural choices. A corporate marketing mix strategy with respect to ethical issues is crucial for generating consumer–brand relationships and enhancing perceived product quality, which are influenced by corporate brand loyalty (Lee and Jin 2019). Additionally, Kadic-Maglajlic et al. (2019) argue that a stronger perception of ethical importance increases the likelihood of adopting morally desirable marketing practices. Collectively, these findings suggest that ethics is a key variable in directing individual behavioural choices. However, despite the growing academic and professional focus on this concept, as reflected in the widespread use of terms such as "ethical consumer" and "ethical buying behaviour", it remains uncommon for either consumers or companies to exhibit completely ethical behaviour, as other factors are also taken into account such as cultural factors, macro and micro environments and markets, and geographical area

(Sciarelli 2002; Newholm and Shaw 2007; Bairrada et al. 2024; Kadic-Maglajlic et al. 2019; Bucic et al., 2012). As a matter of fact, some scholars believe that consumers are generally not behaving ethically (Eckhardt et al. 2010; Carrigan and Attalla 2001), while others believe that even consumers with ethical values are not behaving ethically when it comes to their purchasing habits. Others believe that despite their ethical intentions, ethically minded consumers rarely purchase ethical products or pay attention to business and marketing ethics (Carrington et al. 2010, 2014, 2016; Hassan et al. 2016). In the same vine, Johnstone and Tan (2015) identify several barriers to ethical consumption behaviour within the context of green and sustainable consumption. Besharat et al. (2024) state that efforts to promote ethical consumer behaviour in emerging markets often encounter significant consumer resistance. One reason for this is that consumers in these markets may struggle to prioritise ethical, other-focused attributes, such as Fair Trade or eco-friendly options, particularly when these attributes do not offer direct personal benefits. The influence of marketing ethics on consumer purchasing behaviour is a contentious topic in this field. Intuitively, embracing marketing ethics should attract consumers to the firm's offering, whereas failing to do so should repel them; however, how true is this assertion? How much influence do such ethical practices have on consumer decisions, particularly purchasing decisions? According to López-Fernández (2020), models and theories of consumer behaviour have, with a few notable exceptions, largely ignored the role of ethics. These intention–behaviour gaps are important to researchers and industry, yet they are poorly understood in different setting and require further investigation, which is a gap this study attempts to bridge.

According to Aguirre et al. (2020), culture plays a crucial role in defining the ethical standards that any company operating within its country's borders must adhere to. Moreover, as organisations expand their market operations beyond their domestic market, they will face an increasing number of ethical issues due to cultural diversity, making it essential to evaluate the impact of such ethical practices on consumers in different cultures and markets. Oe and Yamaoka (2022) report that ethical marketing by producers had a non-significant impact, whereas communicating sustainability had a significant impact in Thailand. In general, rapid economic growth and the rise of consumerism in emerging markets have exerted substantial pressure on both the environment and consumers (Al-Abdallah and Al-Salim 2021; Anas et al. 2024). While well-researched ethical consumption strategies may prove effective in Western contexts, they often cannot be directly applied to emerging markets due to institutional and socio-cultural differences (Li et al. 2024). Despite the economic, social, and environmental significance of emerging markets, most research on ethical consumerism has predominantly focused on developed market contexts, limiting the understanding of the concept in emerging markets. This highlights the need to expand empirical knowledge to better understand the peculiarities, dynamics, social and cultural variations associated with ethical consumerism in emerging markets (Kutaula et al. 2024; Prikshat et al. 2024). While some research has been conducted in certain emerging markets, the results can vary significantly across such markets based on social and cultural capital (Prikshat et al. 2024). Building on Ali and Wisniesk (2010), who indicate that culture plays a role in shaping consumers' orientations and attitudes, it has been observed that, unlike in other contexts, the consumerism scale is positively and significantly correlated with the consumer ethics scale in Islamic markets, where reference groups have a significant effect on young consumers (Al-Rawabdeh et al. 2021). Nayak et al. (2024) explore how ethical consumers' involvement with social networking sites affects their intentions to purchase ethical products, thereby elucidating the impact on the purchase intentions of ethical consumers in India. Huang et al. (2012) examined Chinese consumers' ethical beliefs in relation to five selected demographic characteristics (gender, age, religion, family income and education) and found that more ethical Chinese consumers tend to be younger, more religious, and to have a lower family income. Similarly, in the Gulf Cooperation Council (GCC) countries, the needs and desires of the young consumer market are largely shaped by their firm faith in Islam and its values (Saeed et al. 2023). The GCC region, for instance, has a unique socio-cultural dimension; GCC countries score highly on the cohesive measure of Ronen and Shenkar's (2013) clustering index, thanks to their religious, linguistic, and cultural similarity. The primary reason for such an occurrence is that the GCC market has been identified as one of the most promising and relatively untapped emerging markets (Shaikh et al. 2018; Halari et al. 2019). The countries in this region are Bahrain, KSA, Kuwait, Oman, Qatar, and the UAE. Efforts to conceptualise religion's influence on marketing and consumer behaviour have been scarce, which is the main gap that the current study is addressing in one of the most important emerging markets in the world (Ustaahmetoğlu 2020). Therefore, the purpose of this study is to assess the impact of marketing ethics specifically on the purchasing behaviour, of young consumer, who are supposed to be well informed about the importance of sustainability in the relatively rich GCC countries (see Alsuwaidan 2023; Pradeep and Pradeep 2023; Al-Nuaimi 2024; Alotaibi and Aljaafari 2024; Jayabalan et al. 2024).

5.2 Consumer Purchasing Behaviour

Consumer behaviour is a form of human behaviour which, in the literature, is defined in a variety of ways. Consumer behaviour can be defined as all the actions involved in obtaining, consuming and disposing of products and services, along with the decisions that precede and determine these actions (Blackwell et al. 2007). It is the study of the processes involved when individuals or groups select, purchase, use and dispose of products, services, ideas or experiences to satisfy their needs and desires (Solomon 2019). Consumer behaviour includes mental, physical and emotional activities undertaken by people when they wish to select, purchase, use or discard the product or service that fulfils their needs and demands (Eshra and Beshir 2017). According to the AMA, consumer behaviour is the dynamic interaction of affect and cognition, behaviour and the environment by which human beings conduct the exchange aspects of their lives (AMA—American Marketing Association Dictionary 2016). Consumers will go through several steps whereby they search for information about a product, make a decision as to whether to buy or not, use the product or service, and then, in the post-purchase phase, evaluate and dispose of the product or service (Al-Abdallah and Bataineh 2018). The individual consumer purchase behaviour affects societies at large because it determines the demand for raw materials, logistical inputs, the workforce, and all other stakeholders needed to produce a product or service that a consumer wishes to purchase (Al-Abdallah et al. 2018).

There are many ways to model consumer behaviour, depending on the goal of the research, as there are several models that have explored and discussed consumer purchasing behaviour. Among the most significant models is the black box model, which shows how stimuli, consumer characteristics, decision processes and consumer responses interact (Paz and Vargas 2023). This model offers valuable insights into how consumers make purchasing decisions while taking into consideration the role of marketing and other external factors. While it may be challenging to pinpoint exactly why someone makes a specific purchase, the model helps to identify relevant patterns in consumer thinking. Businesses can utilise this knowledge to design products and services that meet customer needs and encourage repeat purchases. It blends elements from psychology, sociology and economics, attempting to understand the buyer decision-making process both individually and in groups. It studies characteristics of individual consumers, such as demographics and behavioural variables, in an attempt to understand people's wants, while also assessing influences on the consumer from groups such as family, friends, reference groups and society in general (Liu et al. 2022; Abbott et al. 2023). The black box model is related to the black box theory of behaviourism, where the focus is not only on the internal processes within a consumer but on the relationship between the stimuli and the consumer responses (Singh 2021). The essence of the model is that it suggests consumers will respond in particular ways to different stimuli after they have 'processed' those stimuli in their minds. In more detail, the model proposes that external factors act as stimuli for the consumer behaviour, but that the consumer's personal characteristics and decision-making processes interact with the stimulus before a particular behavioural response is generated (Liu et al. 2022). The stimuli can be distinguished between interpersonal stimuli (between people) and intrapersonal stimuli (within people) and links the model to the black box theory of consumer behaviour, with a focus on the relationship between stimuli and consumer responses (Łatuszyńska et al. 2012; Chauhan 2013). Chauhan (2013) describes how the black box model of consumers helps marketers understand consumer interaction with stimuli, characteristics, decision processes and responses towards advertisements and marketing communication. The mechanism of marketing stimuli is determined by businesses, whereas the environmental stimuli are due to social factors such as the economic, political and cultural situation of a locality. The model presumes that the consumers' black box is a composite of buyer characteristics and their decision processes, which together define their response, as illustrated in Fig. 5.1.

The main reason why Kotler et al. (2010) consider the black box model to be a well-known framework for explaining consumer behaviour is that it elucidates the consumers' perceived exposure to stimuli and their actual purchase decisions. It assumes that consumers will respond in a specific way to various stimuli at different

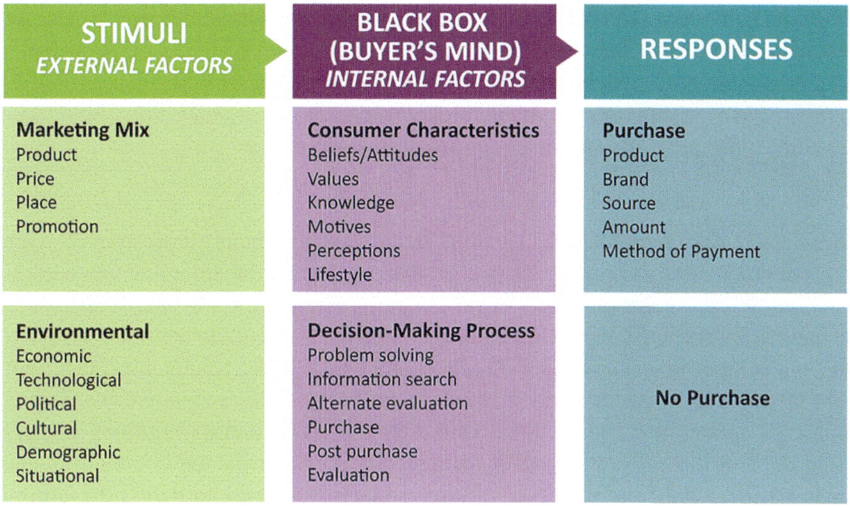

Fig. 5.1 The black box model

stages of their thought processes; however, both internal and external factors can influence consumer behaviour (Esti Susanti 2014). In the model, marketing and all other stimuli enter the customer's "black box" and produce certain responses. For instance, marketing stimuli are carefully planned and produced by firms, whereas environmental stimuli are generated by social factors based on the economic, political and cultural circumstances of a society. The inputs or buying influences in this model include communications about available brands, their prices, qualities, availability, service, styles, options and images. The channels through which this information reaches the buyer include advertising media, sales personnel, acquaintances, the buyer's family and personal observations. However, the buyer's black box also contains the buyer's inherent characteristics and decision-making processes, which will ultimately determine the buyer's response (Chauhan 2013). Therefore, in order to understand buyer behaviour, one must focus on the factors that determine the buyer's characteristics within the black box model, as these characteristics influence how stimuli are perceived, and the decision-making process determines the subsequent purchasing behaviour undertaken (Madhavan and Kaliyaperumal 2015). After processing the inputs in the model, the buyer's responses are manifested in decisions regarding product choice, dealer selection, purchase quantities, and frequency of purchase (Chauhan 2013). While this model simplifies the understanding of the complex mental processes occurring within consumers, it nevertheless highlights the importance of external factors in influencing purchasing decisions (Paz and Vargas 2023). Therefore, marketers utilise this model to tailor stimuli, recognising that although consumers' internal processes are not directly observable, they can be inferred by studying responses to external influences. Despite some of its limitations, the black box model remains a foundational framework for analysing and influencing consumer behaviour (Keegan and Rowley 2017) and was found to be the best fit for measuring purchasing behaviour in this study.

5.3 Young Consumers in the GCC

According to Kuźniar et al. (2021), young consumers are typically those aged 28 years or younger. Gentina and Tang (2024) state that youth materialism influences adolescents' unethical consumer beliefs by investigating the relationships between materialism and Generation Z teenagers' consumer ethics, with culture acting as a moderator in a comparative study between China and France. The results indicate significant differences in young consumer behaviour between the two countries. Huang et al. (2012) state that more ethical Chinese consumers tend to be younger, more religious, and have a lower family income. Similarly, in the GCC countries, the needs and desires of the young consumer are largely shaped by their firm faith in Islam and its values (Saeed et al. 2023). The young consumer in the GCC market is regarded as a segment with considerable potential for further engagement by marketers (Al-Rawabdeh et al. 2021). For example, the financial services decisions of the young generation in the UAE are influenced by consumer expectations regarding the conformity of banks' operations with Islamic principles (Shome et al. 2018).

The GCC region is comprised of the Gulf Cooperation Council members—namely, the UAE, Bahrain, Saudi Arabia, Oman, Qatar and Kuwait—where the vast majority of the population are Muslims, primarily Sunni, sharing similar cultural principles and foundations (Pollock 2019). Over the last decade, high prices for oil and natural gas have contributed to the rapid economic growth of these nations. This economic growth has altered both societies and economies, and concurrently, the population has undergone substantial change. Over the past two decades, the total population of the GCC region has nearly doubled, rising from 26.2 million in 1995 to 56.4 million in 2021. Despite the fact that the majority of this population growth is attributable to large-scale external migration to support economic and social development, there have also been substantial changes in the local populations. In this context, the term 'local population' refers to citizens of the GCC nations. Between 2010 and 2021, the population of several GCC states grew by over 19%. All GCC nations have young native populations; in 2021, more than fifty per cent of the local population was under the age of 25, while the proportion of senior citizens remained low, with approximately 5% of the population being over 65 years old in some countries (Al Flaiti 2023). The young population is beginning to affect the labour market, as an increasing number of young locals enter the labour force, thereby enhancing their purchasing power (Al-Abdallah et al. 2021).

According to a recent report by Deeb (2022), the average consumer in the GCC is well informed about the ongoing changes affecting their purchasing patterns including sustainability and ethics. Additionally, the average young consumer is observant and eager to enjoy a satisfying customer experience. While many argue that young consumers in the Gulf region are well informed about the importance of sustainability and ethics (Alsuwaidan 2023; Pradeep and Pradeep 2023; Al-Nuaimi 2024; Alotaibi and Aljaafari 2024; Jayabalan et al. 2024), others are reporting that a new type of young consumer is emerging; one who is more price-sensitive, digitally savvy but socially responsible as well (Iftahy et al. 2022), showing significant changes in the GCC countries. Such developments raise the question of the role of ethical marketing practices in the purchasing behaviour of young consumers in the GCC.

5.4 Ethical Marketing

Murphy et al. (2005) define ethical marketing, or marketing ethics, as "the systematic study of how moral standards are applied to marketing decisions, behaviours and institutions." Laczniak and Murphy (2019) consider marketing ethics as a subfield of applied ethics that focuses on the moral principles underlying the operation and regulation of marketing. This approach provides a framework for applying fairness in decision-making and guiding organisational behaviour. Al-Hakimi et al. (2022) conclude that sound marketing practices frequently result in the development of long-term customer relationships. Furthermore, the application of ethics in marketing pertains to each component of the marketing mix, including product, price, distribution and promotion (Ndubisi et al. 2014). According to Aguirre et al. (2020), marketing ethics is an organisation's responsibility and a prerequisite for sustaining long-term, mutually beneficial relationships and establishing trust. Ndubisi et al. (2014) emphasise the significance of constructing and reorganising relationships between organisations and clients on the basis of ethical communication. AMA promotes excellence in marketing by endorsing a collective conception of what is considered desirable, important and morally proper. In this framework, values serve as the criteria for evaluating the actions of others. The AMA has adopted a code of ethics that includes the principles of honesty, responsibility, equity, transparency and citizenship, which can be used as an ethical framework for analysing particular marketing issues (Ferrell et al. 2021; American Marketing Association 2023).

The AMA ethical values represent the collective understanding of what the community finds appropriate, desirable, and morally acceptable (American Marketing Association 2023). These ethical values should govern and direct the behaviour of marketers and companies, serving as the criteria for evaluating both their own actions and those of others. One of the AMA's three guiding principles is to 'do no harm', which entails not only consciously avoiding harmful actions or omissions but also striving to benefit all stakeholders and society at large. Consequently, marketers must exhibit high ethical standards and, at a minimum, comply with all applicable laws and regulations when making marketing decisions or implementing marketing strategies. In addition, they are expected to promote and maintain integrity by pursuing openness and equity in every aspect of the marketing ecosystem, and to embrace ethical values by building strong relationships and enhancing stakeholder confidence through the affirmation of core values. These core values of the AMA code of ethics are detailed as follows (American Marketing Association 2023):

- **Honesty**: Uphold truthfulness consistently in all situations and with all stakeholders, rejecting any manipulation or coercion that might undermine trust; provide valuable solutions that align with intentions expressed in communications; and honour both explicit and implicit commitments and promises.
- **Responsibility**: Recognise the social responsibilities associated with heightened marketing and economic influence, integrate environmental and societal stewardship into decision-making processes, aim to exceed industry or legal standards in safeguarding private or sensitive information, and acknowledge the consequences of marketing decisions and strategies.
- **Equity**: Encourage diversity within the marketing workforce and promote inclusive practices by appreciating and embracing individual, cultural and ethnic differences; recognise the needs and commitments of vulnerable market segments and those historically disadvantaged; and avoid stereotyping or negatively portraying any group, thereby ensuring respectful treatment of buyers, suppliers, intermediaries and distributors from diverse backgrounds.
- **Transparency**: Foster openness and clear communication, remain responsive to feedback from customers and stakeholders, address potential risks proactively with full disclosure of information that may influence marketing processes, avoid conflicts of interest, and duly recognise the contributions of consultants, employees and colleagues in maintaining professional integrity.
- **Citizenship**: Fulfil economic, legal, philanthropic and societal responsibilities to stakeholders, recognise the value of marketing in both business and society, prioritise ecological protection by communicating sustainability efforts transparently, contribute to the community through volunteerism and charitable donations, and advocate for fair trade and production practices across the marketing ecosystem.

In this study, ethical marketing is measured through the five core ethical values outlined by the AMA: namely honesty, responsibility, equity, transparency and citizenship. Based on the preceding sections, the research framework is presented in Fig. 5.2.

Based on the above the following research hypotheses can be formulated:

Ha. Ethical Marketing has a statistically significant positive effect on Young Consumer Purchasing Behaviour.

Fig. 5.2 Research framework

This hypothesis is breaking down and measured through the following sub hypotheses:

Ha1.1 Honesty has a statistically significant positive effect on Young Consumer Purchasing Behaviour.

Ha1.2 Responsibility has a statistically significant positive effect on Young Consumer Purchasing Behaviour.

Ha1.3 Equity has a statistically significant positive effect on Young Consumer Purchasing Behaviour.

Ha1.4 Transparency has a statistically significant positive effect on Young Consumer Purchasing Behaviour.

Ha1.5 Citizenship has a statistically significant positive effect on Young Consumer Purchasing Behaviour.

5.5 Research Methodology

The current study adopts a descriptive methodology, intending to provide a comprehensive account of the examined phenomena without interference or manipulation. For the purposes of this study, a quantitative approach was deemed most suitable, employing a survey strategy utilising a seven-point Likert scale questionnaire. The questionnaire was developed using insights gleaned from previous studies and was presented in both English and Arabic. It comprises three main sections: a cover letter and consent form, demographic information of the respondents, and a series of statements designed to measure the research variables. In order to reach a wider audience, the questionnaire was distributed digitally by sharing the survey link.

5.6 Research Population and Sample

As highlighted before, the GCC countries: namely the United Arab Emirates, Bahrain, Saudi Arabia, Oman, Qatar and Kuwait are known for having predominantly young local populations, with more than 50% of local citizens being under 25 years of age (Al Flaiti 2023). Consequently, the target population for this study comprises young local citizens within the GCC, estimated at approximately 28 million individuals. The study utilised a non-probability, quota snowball sampling technique, in which initially, the researcher circulated the questionnaire link on various social media platforms, specifically targeting young consumers in the GCC region, requesting that they complete the survey and share it with other young locals within their networks. Once each country reached its predetermined quota determined based on its share of the total GCC population, data collection for that country was concluded (as illustrated at the end of Table 5.1). The data collection period spanned from January to May 2024. In total, 1283 responses were received during the data collection phase, and following screening, 1266 valid and filtered questionnaires were deemed suitable for statistical analysis. The demographic details of the research sample are presented in Table 5.1.

The data presented in Table 5.1 indicate the characteristics of the study sample. The majority of respondents were male, held a bachelor's degree or higher (including higher diplomas and master's degrees), had an income level of $1,500 to less than 3,000, and predominantly represented Saudi Arabia, the largest country in the GCC. The research sample adds value to the study, as studies on the middle-income customer segments, where most young consumers belong, have been relatively rare in GCC markets (Rahman et al. 2021).

Table 5.1 Sample profile

Variable	Category	Counts	%
Gender	Male	723	57.1
	Female	543	42.9
	Total	1266	100%
Age	Less than 18	389	30.7
	18-less than 28	877	69.3
	Total	1266	100%
Educational level (currently enrolled)	High school or less	315	24.9
	Community college diploma	98	7.7
	Bachelor and above	853	67.4
	Total	1266	100%
Personal income level per month	$0–499	311	24.6
	$50–999	148	11.7
	$1,000–1,499	205	16.2
	$1,500–2,999	320	22.2
	$3,000+	282	25.3
	Total	1266	100%
Country	Saudi Arabia	783	61.9
	UAE	203	16
	Oman	98	7.7
	Kuwait	92	7.3
	Qatar	58	4.6
	Bahrain	32	2.5
	Total	1266	100%

Table 5.2 Reliability results

Variable	Cronbach's alpha value	Conclusion
Honesty	0.885	Reliable
Responsibility	0.859	Reliable
Equity	0912	Reliable
Transparency	0.919	Reliable
Citizenship	0.829	Reliable
Purchasing behaviour	0.931	Reliable

5.7 Reliability

The reliability of the data was assessed using Cronbach's alpha. The results are presented in Table 5.2.

Table 5.2 presents the Cronbach's alpha values for the research variables, ranging from 0.829 to 0.931. These values indicate that the constructs possess adequate reliability, as suggested by Nunnally (1978).

5.8 Construct Validity

The construct validity was assessed through correlation coefficients. The results are presented in Table 5.3.

Table 5.3 displays the correlation values between the questionnaire items. The highest possible correlation value is 1, while the lowest accepted value is set at 0.40 (Laher 2010). According to the findings presented above, all correlation values fall

Table 5.3 Correlation coefficients test

	Independent xariables				
	Honesty	Responsibility	Equity	Transparency	Citizenship
1	0.795	0.861	0.735	0.745	0.775
2	0.827	0.819	0.724	0.714	0717
3	0.718	0.829	0.716	0.731	0.699
	Dependent variable				
	Purchasing behaviour				
1	0.862				
2	0.899				
3	0.798				
4	0.780				
5	0.811				

Table 5.4 The mean, standard deviation, MI, and the level of importance of study variables

Variables	Mean	SD	MI	Level
Independent variables				
Honesty	4.927	1.449	70.4%	High
Responsibility	4.693	1.658	67%	High
Equity	4.916	1.445	70.2%	High
Transparency	5.506	1.063	78.7%	V. High
Citizenship	4.431	1.462	63.3%	Moderate
Dependent variable				
Purchasing Behaviour	5.12	1.316	73.1%	High

Means description (1–1.86 not important, 1.87–2.73 v. low, 2.74–3.6 low, 3.61–4.47 moderate, 4.48–5.34 high, 5.35–6.21 v. high, 6.22–7 extremely high)

within the range of greater than 0.40 and less than 1, indicating acceptable construct validity.

5.9 Descriptive Analysis of the Study Variables

A descriptive analysis was made to the study variables to determine the means, standard Deviation and the degree of importance degree. Table 5.4 demonstrates the result.

Table 5.4 indicates that participants hold positive views across all items. Transparency is rated as the most important variable, followed by Honesty and Equity, both of which are considered highly important. Responsibility also holds a high level of importance, while Citizenship is regarded as moderately important. Additionally, Purchasing Behaviour is perceived as highly important.

5.10 Normality and Multicollinearity

Before testing the hypotheses, two fundamental assumptions of linear regression must be examined: the normality of data distribution and the degree of multicollinearity among the independent variables (dimensions) (Cohen et al. 2003). Normality is assessed using skewness and kurtosis, while multicollinearity is evaluated using the variance inflation factor (VIF). The results of these tests are presented in Table 5.5.

Table 5.5 presents the VIF, skewness, and kurtosis values. The data is considered normally distributed if the skewness values and the kurtosis values lie between (–1) and (+1), all the obtained values fall between –1 and + 1, indicating excellent symmetry of a data distribution and reflect a normal distribution (Hair et al. 2022, p. 66). According to Maddala (1992) a VIF value that is more than 10 indicates a

Table 5.5 Normality indictor and the VIF test for multicollinearity

	Skewness	Kurtosis	VIF
Independent variables			
Honesty	0.402	0.194	1.780
Responsibility	–0.524	–0.543	2.723
Equity	–0.616	–0.180	2.645
Transparency	0.697	0.337	2.290
Citizenship	–0.347	–0.263	2.634
Dependent variable			
Purchasing behaviour	0.886	0.716	–

serious multicollinearity problem that prevent the application of regression analyses, a value between 5 and 10 reflects a moderate problem, while a value less than 5 reflects low multicollinearity problem. All obtained values were less than 3, which means that there is no significant multicollinearity between the independent variable dimensions i.e. very low correlation among the independent variable dimensions.

Based on the results above, regression analyses can be safely applied.

5.11 Testing the Hypothesis

This research is based on one main hypothesis: Ha. Ethical Marketing has a statistically significant positive effect on Young Consumer Purchasing Behaviour. This hypotheses is tested through the its five dimension in the following five sub hypotheses:

Ha1.1 Honesty has a statistically significant positive effect on Young Consumer Purchasing Behaviour.

Ha1.2 Responsibility has a statistically significant positive effect on Young Consumer Purchasing Behaviour.

Ha1.3 Equity has a statistically significant positive effect on Young Consumer Purchasing Behaviour.

Ha1.4 Transparency has a statistically significant positive effect on Young Consumer Purchasing Behaviour.

Ha1.5 Citizenship has a statistically significant positive effect on Young Consumer Purchasing Behaviour.

To examine the research hypothesis, multiple regression analysis was employed, and the outcomes are detailed in Table 5.6.

Table 5.6 reveals that the regression model is significant, as evidenced by the significant f-value (68.241), where the significance level (Sig. f) is less than 0.05. This indicates that ethical marketing has a significant positive impact on the purchasing behaviour of young consumers. The results further indicate that ethical marketing

Table 5.6 Multiple regression results

Variable	β	t value	Sig. (t)	f value	p-value (Sig. f)
Honesty	0.349	6.669	0.000	68.241	0.000
Responsibility	0.317	4.072	0.000		
Equity	0.294	1.527	0.002		
Transparency	0.399	6.817	0.000		
Citizenship	0.217	6.524	0.008		
Adjusted $R^2 = 0.355$					

Table 5.7 Summary of hypotheses testing

Hypothesis	Statement	Sig. (t)	Result
Ha1.1	Honesty has a statistically significant positive effect on Young Consumer Purchasing Behaviour	0.000	Supported
Ha1.2	Responsibility has a statistically significant positive effect on Young Consumer Purchasing Behaviour	0.000	Supported
Ha1.3	Equity has a statistically significant positive effect on Young Consumer Purchasing Behaviour	0.002	Supported
Ha1.4	Transparency has a statistically significant positive effect on Young Consumer Purchasing Behaviour	0.000	Supported
Ha1.5	Citizenship has a statistically significant positive effect on Young Consumer Purchasing Behaviour	0.008	Supported
Main Hypothesis	*Statement*	*Sig. (f)*	*Result*
Ha	Ethical Marketing has a statistically significant positive effect on Young Consumer Purchasing Behaviour	0.00	Supported

accounts for 35.5% of the variance in young consumers' purchasing behaviour ($R^2 = 0.355$) which is a moderate level of interpretation.

Moreover, all ethical marketing dimensions exhibit significance, as evidenced by Sig. (t) values being less than 0.05 for each dimension. Among these dimensions, transparency emerges with the highest impact ($\beta = 0.399$), followed by honesty ($\beta = 0.349$), responsibility ($\beta = 0.317$), equity ($\beta = 0.294$), and finally, citizenship ($\beta = 0.217$). In light of these findings, the research main and sub hypotheses results are presented in Table 5.7.

5.12 Discussion and Conclusion

This study aimed to investigate the impact of ethical marketing on the purchasing behaviour of young consumers in the GCC markets. The ethical dimensions examined were derived from the new code of conduct of the AMA, encompassing honesty, responsibility, equity, transparency, and citizenship. The empirical results indicate that ethical marketing influences the purchasing behaviour of young consumers in the GCC; however, the effect size is relatively modest, with ethical marketing accounting for only 35.5% of the variance in young consumers' purchasing behaviour. This suggests that 64.5% of the variance in behaviour is attributable to other factors. It is plausible that, while young consumers in the GCC appreciate ethical marketing practices that align with their religious beliefs, they also place substantial value on other tangible factors such as price, brand, promotion, and personal perception.

The findings underscore transparency as the most influential dimension of ethical marketing for young consumers in the GCC. These consumers value companies that engage in open and clear communication, respond effectively to customer feedback, and address potential risks through the full disclosure of information affecting marketing processes. This is in line with research by Cambier and Poncin (2020), Yang and Battocchio (2021) and Chopdar and Paul (2023), which emphasised the significant impact of marketing and communication transparency on consumer empowerment, responsiveness, and branding.

Honesty emerges as the second influential dimension of ethical marketing, indicating that young consumers in the GCC prioritise truthfulness in their interactions with companies. They reject manipulation and coercion, valuing organisations that honour both explicit and implicit commitments and promises. This emphasis on honesty is consistent with Islamic ethical standards, making it highly esteemed by young consumers in the GCC, a finding that aligns with Abbas et al. (2020), who highlighted the importance of honesty in fostering customer satisfaction.

Responsibility is the third influential dimension of ethical marketing, reflecting the consideration of social responsibilities by young consumers in the GCC. They appreciate companies' environmental and societal stewardship, responsible behaviour in protecting private information, and the acknowledgement of the consequences stemming from marketing decisions and strategies. These observations are in agreement with the bibliometric analysis conducted by Kamila and Jasrotia (2023) and the findings of Al-Abdallah and Ahmed (2018) on the impact of responsibility in generating customer loyalty.

Equity is the fourth influential dimension of ethical marketing, indicating that young consumers in the GCC value diversity and inclusive practices in marketing. They appreciate companies that embrace individual, cultural and ethnic differences, recognise commitments to vulnerable market segments, avoid negative stereotyping and exhibit respectful treatment for all consumers from diverse backgrounds. This is particularly relevant in the GCC, which attracts a significant foreign workforce, and reflects the view of young consumers towards companies' ethical marketing practices. This observation is consistent with the perspectives of Abbas et al. (2020) and Kamila and Jasrotia (2023) regarding the importance of equity in shaping consumer behaviour.

Citizenship is the fifth influential dimension of ethical marketing impacting young consumers in the GCC, suggesting that fulfilling economic, legal, philanthropic and societal responsibilities is important to them which is consistent with the findings of Andersen and Johansen (2021). They value companies that prioritise ecological protection, communicate sustainability efforts transparently, contribute to the community through volunteerism and charitable donations, and adhere to fair trade and ethical production practices. However, the impact of this dimension appears relatively low among young consumers in the GCC, which may indicate that they place greater emphasis on other ethical marketing dimensions. This could be attributed either to a lack of awareness of these activities or an insufficient appreciation of their importance.

In summary, while ethical marketing does exert an influence on the purchasing behaviour of young consumers in the GCC, its overall impact is moderate, with transparency, honesty, responsibility, equity and citizenship playing varying roles. Although young consumers appear to value ethical marketing practices, other factors also significantly contribute to their purchasing decisions.

5.13 Recommendations

Drawing from the study's outcomes, the researcher recommends that companies targeting young consumers in the GCC integrate a substantial emphasis on ethical marketing practices when developing their marketing offers and value propositions. It is crucial for these companies to articulate these values prominently in their communications with both consumers and the wider public. Specifically, there should be a strong focus on transparency, given its identified as having the highest impact on the purchasing behaviours of young consumers in the GCC, accompanied by an equally robust emphasis on honesty. Moreover, companies should actively recognise and fulfil their responsibilities towards society, environmental conservation, and individual well-being. Incorporating responsible business practices into day-to-day operations and ensuring transparent communication about these efforts is paramount.

Additionally, there is a need to educate consumers on the significance of marketing ethics for the overall well-being of society and the future welfare of consumers. Joint marketing campaigns between companies that embrace ethical marketing and official recognition by relevant authorities can serve as effective means to achieve this objective. In the same vein, the study suggests that authorities and policymakers should underscore the importance of ethical marketing practices for young consumers. Such an emphasis could contribute to raising general awareness about the implications of ethical practices on society at large, thereby potentially amplifying their impact on consumer purchasing intentions in the future. Furthermore, companies that adopt ethical marketing should be formally acknowledged by authorities and consumer protection groups. This formal recognition is essential, given the pivotal role these practices play in safeguarding consumers and contributing to societal well-being. Such a comprehensive communication strategy could further enhance the perceived value of ethical marketing among both consumers and the broader community.

5.14 Limitations and Future Research Orientation

The study has certain limitations that warrant consideration for future research endeavours. Firstly, the research exclusively focuses on young consumers; it is recommended that similar studies be conducted across different age groups to enable a comprehensive comparison of results. Additionally, the investigation confines its exploration of ethical marketing practices to those outlined by the AMA. Future research could extend this analysis by examining alternative ethical marketing frameworks to provide a more thorough understanding of prevailing ethical practices.

Furthermore, while the study identifies ethical marketing as a factor influencing young consumers' purchasing behaviour, it acknowledges the potential impact of other variables such as price, branding, and social as well as personal preferences. Future research should undertake a comparative analysis of these factors to ascertain their relative significance in shaping the perceptions and behaviours of young consumers in the GCC. Moreover, since the study's scope is not sector-specific, it is recommended that future research explore the nuances of ethical marketing practices within specific industry sectors. This sector-focused investigation could offer a more in-depth understanding of how ethical marketing principles are applied and perceived across diverse business contexts.

Lastly, the research is confined to the GCC markets; expanding the study to include other emerging markets would provide an opportunity to compare results and enhance the generalisability of the findings. This comparative approach across different regions could reveal variations in the impact of ethical marketing on consumer behaviour, taking into account cultural, economic, and social differences.

References

Abbott R, Sin R, Pedersen C, Harris T, Beck T, Nilsson S, Dong T, Wang Y, Li Y (2023) The role of dark pattern stimuli and personality in online impulse shopping: an application of S-O-R theory. J Consum Behav 22(6):1311–1329

Abbas A, Nisar QA, Mahmood MAH, Chenini A, Zubair A (2020) The role of Islamic marketing ethics towards customer satisfaction. J Islam Mark 11(4) 1001–1018. https://doi.org/10.1108/JIMA-11-2017-0123

Aguirre GC, Hyman MR, Jones JL (2020) Dedicated marketing ethics course: design and test. Mark Educ Rev 30(3):177–194

Al Flaiti A (2023) Demographics of the GCC—understanding the changes in the local population' [poster session]. In: 64th ISI world statistics congress. Ottawa, Canada

Al Hakimi MA, Saleh MH, Borade DB, Hasan MB, Sharma D (2022) Competitor orientation and SME performance in competitive environments: the moderating effect of marketing ethics. J Entrepreneurship Emerg Econ [ahead of print]

Al Rashid MA, Harumain YAS, Goh HC, Ali Z, Nadeem M, Campisi T (2023) Perceived norms of public transport use as the determinants of older adults' social exclusion: evidence from Pakistan. Cities 137:104264

Al Rawabdeh H, Ghadir H, Al Abdallah G (2021) The effects of user generated content and traditional reference groups on purchase intentions of young consumers: a comparative study on electronic products. Int J Data Netw Sci 5(4):691–702

Al-Abdallah G, Khair N, Elmarakby R (2021) The impact of social networking sites on luxury vehicles purchase decision process in Gulf Cooperation Council countries. J Int Consum Mark 33(5):559–577

Al-Abdallah GM, Ahmed RS (2018) The impact of corporate social responsibility on customer loyalty in the Qatari telecommunication sector. J Bus Retail Manag Res 13(01)

Al-Abdallah GM, Al-Salim MI (2021) Green product innovation and competitive advantage: an empirical study of chemical industrial plants in Jordanian qualified industrial zones. Benchmarking: An Int J 28(8):2542–2560

Al-Abdallah GM, Bataineh AQ (2018) Social networking sites and fashion e purchasing process. J Bus Retail Manag Res 13(02)

Al-Abdallah GM, Abou Moghli AA, Al Thani AH (2018) An examination of the e-commerce technology drivers in the real estate industry

Ali AJ, Wisniesk JM (2010) Consumerism and ethical attitudes: an empirical study. Int J Islam Middle East Financ Manag 3(1):36–46

Al-Nuaimi S (2024) Sustainable consumption through education for sustainability in higher education institutes in Qatar (Doctoral dissertation, Hamad Bin Khalifa University (Qatar))

Alotaibi S, Aljaafari MAH (2024) Social commerce in Saudi Arabia: opportunities and challenges in a digital society. Sustainability 16(24):10951

Alsuwaidan MSM (2023) Green marketing and customer behaviour in Saudi Arabia: an examination of eco-friendly electric cars purchase probability (Doctoral dissertation, Victoria University)

AMA–American Marketing Association Dictionary (2016). https://www.ama.org/resources/Pages/Dictionary.aspx?dLetter=C. Accessed 11 May 2016

American Marketing Association (2023) AMA Statement of Ethics. https://www.ama.org/ama-statement-of-ethics. Accessed 22 Feb 2023

Amine MEA, Chakor A, Alaoui AM (2012) Ethics, relationship marketing and corporate performance: theoretical analysis through the mediating variables. Int Bus Res 5. https://doi.org/10.5539/ibr.v5n8p68

Anas M, Zhang W, Bakhsh S, Ali L, Işık C, Han J, Liu X, Rehman HU, Ali A, Huang M (2024) Moving towards sustainable environment development in emerging economies: the role of green finance, green tech-innovation, natural resource depletion, and forested area in assessing the load capacity factor. Sustain Dev 32(4):3004–3020

Andersen SE, Johansen TS (2021) Corporate citizenship: challenging the corporate centricity in corporate marketing. J Bus Res 131:686–699

Ardley B, May C (2020) Ethical marketer and sustainability: facing the challenges of overconsumption and the market. Strateg Chang 29(6):617–624

Bag S, Srivastava G, Gupta S, Sivarajah U, Wilmot NV (2024) The effect of corporate ethical responsibility on social and environmental performance: an empirical study. Ind Mark Manag 117:356–370

Bairrada CM, Coelho AFDM, Moreira JRM (2024) Attitudes toward ethical consumption in clothing: comparing Peruvian and Portuguese consumers. J Int Consum Mark 36(2):180–196

Barnett T, Valentine S (2004) Issue contingencies and marketers' recognition of ethical issues, ethical judgments and behavioral intentions. J Bus Res 57(4):338–346

Besharat A, Nardini G, Mesler RM (2024) Bringing ethical consumption to the forefront in emerging markets: the role of product categorization. J Bus Ethics 1–16

Blackwell R, Miniard P, Engel D (2007) The consumers' behaviour. St. Petersburg, Piter

Bucic T, Harris J, Arli D (2012) Ethical consumers among the millennials: a cross national study. J Bus Ethics 110(1):113–131

Cambier F, Poncin I (2020) Inferring brand integrity from marketing communications: the effects of brand transparency signals in a consumer empowerment context. J Bus Res 109:260–270

Carrigan M, Attalla A (2001) The myth of the ethical consumer–do ethics matter in purchase behaviour? J Consum Mark 18(7):560–578

Carrington MJ, Neville BA, Whitwell GJ (2010) Why ethical consumers don't walk their talk: towards a framework for understanding the gap between the ethical purchase intentions and actual buying behaviour of ethically minded consumers. J Bus Ethics 97:139–158

Carrington MJ, Neville BA, Whitwell GJ (2014) Lost in translation: exploring the ethical consumer intention–behaviour gap. J Bus Res 67(1):2759–2767

Carrington MJ, Zwick D, Neville B (2016) The ideology of the ethical consumption gap. Mark Theory 16(1):21–38

Chauhan NM (2013) Consumer behaviour and his decision of purchase. Int J Res Manag Pharm 2(5):1–4

Chopdar PK, Paul J (2023) The impact of brand transparency of food delivery apps in interactive brand communication. J Res Interact Market

Ciasullo MV, Maione G, Torre C, Troisi O (2017) What about sustainability? an empirical analysis of consumers' purchasing behavior in fashion context. Sustainability 9(9):1617

Cohen J, Cohen P, West SG, Aiken LS (2003) Applied multiple regression/correlation analysis for the behavioral sciences, 3rd edn. Lawrence Erlbaum Associates, Mahwah, NJ

de Paula GO, Cavalcanti RN (2000) Ethics: essence for sustainability. J Clean Prod 8(2):109–117

Deeb R (2022) Understanding the GCC customers, Interview by S. Serrano, Campaign Middle East

Eckhardt GM, Belk R, Devinney TM (2010) Why don't consumers consume ethically? J Consum Behav 9(6):426–436

Eid MAH, Al-Abdallah G (2024) Sustainable development through biomimicry: enhancing circular economy practices for environmental sustainability. Sustain Dev 32(6):6045–6056

Ellemers N, de Gilder D (2022) The moral organization: key issues, analyses, and solutions. Springer Nature

Eshra N, Beshir N (2017) Impact of corporate social responsibility on consumer buying behavior in Egypt. World Rev Bus Res 7(1):32–44

Esti Susanti C (2014) The antecedence of customer loyalty in traditional restaurants in East Java, Indonesia. Int J Process Manag Benchmarking 4(1):22–35

Ferrell OC, Hartline M, Hochstein BW (2021) Marketing strategy. Cengage Learning

Gaski JF (1999) Does marketing ethics really have anything to say?–a critical inventory of the literature. J Bus Ethics 18(3):315–334

Gentina E, Tang TLP (2024) Youth materialism and consumer ethics: do Gen Z adolescents' self concepts (power and self esteem) vary across cultures (China vs. France)? Ethics Behav 34(2):120–150

Gillett AG, Tennent KD (2020) The rise of marketing. In: The Palgrave handbook of management history, pp 841–858

Guerola Navarro V, Gil Gomez H, Oltra Badenes R, Soto Acosta P (2022) Customer relationship management and its impact on entrepreneurial marketing: a literature review. Int Entrepreneurship Manag J 1–41

Hair JF, Hult GTM, Ringle CM, Sarstedt M (2022) A primer on partial least squares structural equation modeling (PLS-SEM) (3rd ed). Sage

Halari A, Helliar C, Power DM, Tantisantiwong N (2019) Taking advantage of Ramadan and January in Muslim countries. Q Rev Econ Finance 74:85–96

Hassan LM, Shiu E, Shaw D (2016) Who says there is an intention–behaviour gap? assessing the empirical evidence of an intention–behaviour gap in ethical consumption. J Bus Ethics 136:219–236

Huang CC, Lu LC, You CS, Yen SW (2012) The impacts of ethical ideology, materialism, and selected demographics on consumer ethics: an empirical study in China. Ethics Behav 22(4):315–331

Iftahy A, Jha J, Khairallah S, Lambert M (2022) How to win in the Gulf's dynamic consumer and retail sectors. McKinsey & Company. https://www.mckinsey.com/industries/consumer-packaged-goods/our-insights/how-to-win-in-the-gulfs-dynamic-consumer-and-retail-sectors. Accessed: [insert date]

Iglesias O, Markovic S, Bagherzadeh M, Singh JJ (2020) Co creation: a key link between corporate social responsibility, customer trust, and customer loyalty. J Bus Ethics 163:151–166

Jain H, Dhupper R, Shrivastava A, Kumar D, Kumari M (2023) AI-enabled strategies for climate change adaptation: protecting communities, infrastructure, and businesses from the impacts of climate change. Comput Urban Sci 3(1):25. https://doi.org/10.1007/s44212-023-00044-6

Jayabalan SK, Albusaidi ASO, Negi GS, Iqbal MI, Abdulqader HA (2024) Consumer acceptance, social behavior, driving, and safety issues regarding electric vehicles in Oman. World Electric Veh J 15(12):549

Johnstone ML, Tan LP (2015) Exploring the gap between consumers' green rhetoric and purchasing behaviour. J Bus Ethics 132:311–328

Kadic Maglajlic S, Micevski M, Lee N, Boso N, Vida I (2019) Three levels of ethical influences on selling behavior and performance: synergies and tensions. J Bus Ethics 156:377–397

Kamila MK, Jasrotia SS (2023) Ethics and marketing responsibility: a bibliometric analysis and literature review. Asia Pac Manag Rev

Keegan BJ, Rowley J (2017) Evaluation and decision making in social media marketing. Manag Decis 55(1):15–31

Koozechian H, Sardari M, Ehsani M, Amiry M (2021) Designing and testing ethical marketing model in sports manufacturing industry. Sport Manag Stud 13(65):179–202

Kotler P, Armstrong G, Armstrong GM (2010) Principles of marketing. Pearson Education India

Kotler P, Keller KL, Chernev A (2022) Marketing management (16th ed). Pearson Education

Kutaula S, Gillani A, Gregory Smith D, Bartikowski B (2024) Ethical consumerism in emerging markets: opportunities and challenges. J Bus Ethics 1–23

Kuźniar W, Surmacz T, Wierzbiński B (2021) The impact of ecological knowledge on young consumers' attitudes and behaviours towards the food market. Sustainability 13(4):1984

Laczniak GR, Murphy PE (2019) The role of normative marketing ethics. J Bus Res 95:401–407

Laher S (2010) Using exploratory factor analysis in personality. SA J Ind Psychol 36(1):1–7

Łatuszyńska M, Furaiji F, Wawrzyniak A (2012) An empirical study of the factors influencing consumer behaviour in the electric appliances market. Contemp Econ 6(3):76–86

Lee JY, Jin CH (2019) The role of ethical marketing issues in consumer brand relationship. Sustainability 11(23):6536

Lee SH (2020) Achieving corporate sustainability performance: the influence of corporate ethical value and leader member exchange on employee behaviors and organizational performance. Fashion Textiles 7(1):25

Li CX, Liu XX, Ye J, Zheng S, Cai S (2024) Ethical pursuit or personal nirvana? unpacking the practice of Danshari in China. J Bus Ethics 191(4):675–695

Liu F, Liu S, Jiang G (2022) Consumers' decision making process in redeeming and sharing behaviors toward app based mobile coupons in social commerce. Int J Inf Manag 67:102550

Lokhvytska L, Rozsokha A, Azman C (2022) Can moral values be formed without influencing the development of an individual's worldview? Philos Cosmol 28:73–92

López Fernández AM (2020) Price sensitivity versus ethical consumption: a study of millennial utilitarian consumer behavior. J Market Anal 8(2):57–68

Lu LC, Huang YW, Chang HH (2014) Confucian dynamism, the role of money and consumer ethical beliefs: an exploratory study in Taiwan. Ethics Behav 24(1):34–52

Maddala GS (1992) Introduction to econometrics, 2nd edn. Macmillan Publishing Company, New York, p 631

Madhavan M, Kaliyaperumal C (2015) Consumer buying behavior: an overview of theory and models. St. Theresa J Humanit Soc Sci 1(1):74–112
Minton EA, Leary RB, Upadhyaya S (2018) Religion's influence on consumer response to moral versus justice message appeals. Int J Consum Stud 42(6):768–778
Murphy PE, Laczniak GR, Bowie NE, Klein TA (2005) Ethical marketing
Nayak S, Pereira V, Kazmi BA, Budhwar P (2024) To buy or not to buy? exploring ethical consumerism in an emerging market—India. J Bus Ethics 191(4):811–835
Ndubisi NO, Nataraajan R, Lai R (2014) Customer perception and response to ethical norms in legal services marketing. J Bus Res 67(3):369–377
Newholm T, Shaw D (2007) Studying the ethical consumer: a review of research
Nunnally JC (1978) An overview of psychological measurement. In: Clinical diagnosis of mental disorders: a handbook, pp 97–146
Oe H, Yamaoka Y (2022) The impact of communicating sustainability and ethical behaviour of the cosmetic producers: evidence from Thailand. Sustainability 14(2):882
Paz MDR, Vargas JCR (2023) Main theoretical consumer behavioural models: a review from 1935 to 2021. Heliyon 9(3)
Petersen A (2013) The importance of governance and business ethics. CIPS
Pollock D (2019) GCC security issues, as their publics see them: a rare look at real data. In: The dilemma of security and defense in the gulf region
Pradeep S, Pradeep M (2023) Awareness of sustainability, climate emergency, and generation Z's consumer behaviour in UAE. Clean Responsible Consumption 11:100137
Prikshat V, Patel P, Kumar S, Gupta S, Malik A (2024) Role of socio cultural capital and country level affluence in ethical consumerism. J Bus Ethics 1–15
Rahman MM, Chowdhury MAF, Haque MM, Rashid M (2021) Middle income customers and their perception of Islamic banking in Sylhet: one of Bangladesh's most pious cities. Int J Islam Middle East Financ Manag 14(2):230–246
Ronen S, Shenkar O (2013) Mapping world cultures: cluster formation, sources and implications. J Int Bus Stud 44:867–897
Saeed M, Waheed Z, Baig AK, Azmi IAG (2023) Muslim consumers' brand switching behavior in Lahore and Kuala Lumpur: a qualitative comparative case study. J Islamic Market 14(6):1577–1602
Saini J, Gupta AK, Shrivastava A, Dhupper R (2024) Study of urban warming in Chandigarh union territory, India: geospatial approach: J. Saini et al. Theor Appl Climatol 155(8):8245–8261
Schlegelmilch BB, Öberseder M (2010) Half a century of marketing ethics: shifting perspectives and emerging trends. J Bus Ethics 93(1):1–19
Sciarelli S (2002) Business quality and business ethics. Total Qual Manag 13(8):1141–1149. https://doi.org/10.1080/0954412020000009
Shaikh SA, Ismail MA, Ismail AG, Shahimi S, Mohd Shafiai MH (2018) Intertemporal consumption behaviour in OIC countries. Int J Islam Middle East Financ Manag 11(4):529–552
Shome A, Jabeen F, Rajaguru R (2018) What drives consumer choice of Islamic banking services in the United Arab Emirates? Int J Islam Middle East Financ Manag 11(1):79–95
Singh UG (2021) Investigating mobile payment adoption in Mauritius. African J Bus Econ Res 16(3):257
Solomon MR (2019) Consumer behaviour: a european perspective, 7th edn. Pearson Education
Tan TM, Salo J (2023) Ethical marketing in the blockchain based sharing economy: theoretical integration and guiding insights. J Bus Ethics 183(4):1113–1140
Tanveer M, Ahmad AR, Mahmood H, Haq IU (2021) Role of ethical marketing in driving consumer brand relationships and brand loyalty: a sustainable marketing approach. Sustainability 13(12):6839
Ustaahmetoğlu E (2020) The influence of different advertisement messages and levels of religiosity on attitude and purchase intention. Int J Islam Middle East Financ Manag 13(2):339–356

Voronov AA, Gorlachev PV, Mirzoeva EV, Rudenko VA, Popova TS (2021) Role of social and ethical marketing in improving sustainable business development. In: Cooperation and sustainable development. Springer International Publishing, Cham, pp 1249–1255

Yang J, Battocchio AF (2021) Effects of transparent brand communication on perceived brand authenticity and consumer responses. J Product Brand Manag 30(8):1176–1193

Chapter 6
Nature's Battery: How Wastewater and Plants Could Power Tomorrow?

Navya Vikraman Nair and Sanju Sreedharan

6.1 Introduction

Across the globe, escalating challenges related to wastewater management and energy security are posing serious threats to environmental sustainability and public health. Rapid urbanization, industrial expansion, and population growth are placing unprecedented pressure on water resources and energy infrastructure. According to the United Nations World Water Development Report (United Nations World Water Assessment Programme (UN WWAP) 2023), more than 80% of global wastewater is discharged into the environment untreated, leading to the degradation of freshwater ecosystems and the proliferation of waterborne diseases. Simultaneously, global energy demand is projected to increase by 25% by 2040, further straining fossil fuel-dependent systems and exacerbating their environmental and economic impacts (International Energy Agency (IEA) 2021). Traditional wastewater treatment plants (WWTPs), while effective in pollutant removal, are highly energy-intensive and often fail to recover valuable resources from waste streams. The United States Environmental Protection Agency (US EPA) (2006) estimates that WWTPs contribute nearly 3% to global electricity consumption. Furthermore, centralized wastewater treatment systems frequently struggle to reach remote or underserved communities, particularly in the Global South, where infrastructural and economic constraints persist. These limitations not only contribute to environmental pollution but also represent significant missed opportunities for resource recovery and sustainable development.

N. V. Nair (✉)
School of Environment, Enterprise, and Development (SEED), University of Waterloo, Waterloo, ON, Canada
e-mail: n9nair@uwaterloo.ca

N. V. Nair · S. Sreedharan
Department of Civil Engineering, SCMS School of Engineering and Technology, Karukutty, Ernakulam, Kerala, India

H. Jain et al. (eds.), *The Circular Fallacy: Closing the Loop in a Consumerist World*, Circular Economy and Sustainability, https://doi.org/10.1007/978-3-032-22687-7_6

The circular economy paradigm presents a transformative framework to address these intertwined challenges of waste and energy. Moving beyond the traditional linear model of "take-make-dispose," the circular economy promotes the regenerative use of materials and energy within closed-loop systems (MacArthur 2013). In the context of wastewater, this entails reimagining effluents not as liabilities but as potential sources of clean water, nutrients, and renewable energy. Circular wastewater management thus calls for the adoption of technologies that are energy-efficient, decentralized, and capable of transforming waste into value. However, many existing wastewater technologies fall short of these ideals. Conventional WWTPs primarily target pollutant removal and sludge disposal, with limited integration of energy recovery mechanisms. Although anaerobic digesters and biogas systems offer partial solutions, they are often capital-intensive and operationally complex. As the global movement toward sustainable infrastructure gains momentum, there is a pressing need for decentralized, nature-inspired technologies that embody the principles of circularity while addressing the energy-water nexus.

One such emerging technology is the Plant Microbial Fuel Cell (PMFC), a bioelectrochemical system that leverages the symbiotic interactions between plant roots and soil microorganisms to generate electricity while simultaneously treating wastewater. PMFCs function by utilizing root exudates (organic compounds released by plant roots) as substrates for electroactive bacteria (EAB) in the rhizosphere. These microorganisms oxidize the organics and transfer electrons to an anode, generating an electric current through an external circuit to a cathode, where a reduction reaction occurs (Strik et al. 2008; Helder et al. 2010). This dual-function system exemplifies circular economy principles by purifying wastewater, producing renewable energy, and supporting plant growth within a single biological framework. Unlike conventional WWTPs, PMFCs are inherently decentralized, cost-effective, and compatible with existing ecosystems (Kumar et al. 2025). They offer promising solutions for rural sanitation, agricultural runoff management, and urban green infrastructure. As such, PMFCs have the potential to become integral to regenerative design strategies within the water-energy-food nexus.

This study investigates the application of PMFCs using two plant species: Red Spinach (*Amaranthus dubius*) and Rice Paddy (*Oryza sativa*). These species were selected based on their high biomass yield, rapid growth rates, and established roles in tropical and subtropical agriculture. Red Spinach is a nutritious leafy vegetable widely cultivated for food, while Rice Paddy is a staple crop with deep-rooted cultural and economic importance in Asia. Both species exhibit robust root systems conducive to microbial colonization and biofilm development, key factors in enhancing PMFC performance. Red Spinach was chosen for its potential to support edible crop production in PMFC systems without competing with staple food crops for land and inputs. Its short lifecycle allows for rapid assessment of plant growth, nutrient uptake, and system compatibility. Rice, conversely, offers insights into PMFC operation under flooded, anaerobic conditions typical of paddy fields. Its longer cultivation period and dense rhizosphere interactions make it a suitable candidate for studying sustained microbial activity and voltage generation.

The overarching objective of this study is to assess the feasibility of PMFCs as nature-based solutions for integrated WWT and energy generation. Specifically, the study aims to: (i) compare the performance of Red Spinach and Rice Paddy in PMFC systems using synthetic dairy and domestic wastewater; (ii) evaluate bioelectricity generation through voltage trend analysis; (iii) assess WWT efficiency via parameters such as BOD, COD, turbidity, nutrient removal, and effluent quality; (iv) examine plant growth, biomass yield, and compatibility with PMFC operations; (v) analyze microbial colonization and biofilm development using Scanning Electron Microscopy (SEM); and (vi) identify practical limitations and opportunities for scaling PMFCs in real-world contexts. This research is significant in that it situates PMFCs within broader discourses on circular economy, climate resilience, and sustainable development. It contributes empirical data on plant–microbe-electrode interactions and supports the growing body of literature advocating biomimetic infrastructure as part of the global transition to low-carbon, resource-efficient systems (Saini et al. 2024).

The following sections elaborate on the scientific underpinnings of PMFCs, detailing the electrochemical mechanisms and biofilm dynamics essential to their operation. This is followed by a comprehensive description of the experimental methodology, encompassing wastewater composition, reactor configuration, plant cultivation protocols, and analytical techniques. Subsequent chapters present and analyze findings related to voltage generation, pollutant removal, plant growth, and microbial colonization. The discussion synthesizes these outcomes in the context of circular economy principles and proposes pathways for practical implementation. Finally, the chapter concludes by summarizing key insights and recommending future directions for research, policy, and application. By reframing wastewater as a valuable input and aligning engineering practice with ecological design, this study envisions PMFCs as foundational to a regenerative future where water, energy, and food systems are co-managed for resilience and sustainability.

6.1.1 Circular Economy and Nature-Inspired Waste Solutions

The circular economy is predicated on the idea of designing out waste and pollution, keeping materials in use, and regenerating natural systems (MacArthur 2013). This framework fundamentally challenges the traditional linear approach to WWT, which typically involves extracting clean water through energy-intensive processes that produce substantial sludge and rely heavily on chemical inputs. As urban populations increase and climate change continues to stress water systems, the demand for circular, decentralized, and low-impact wastewater solutions has grown increasingly urgent. Wastewater, once regarded merely as a byproduct of domestic, industrial, and agricultural activity, is now increasingly recognized as a resource. It contains organic matter, nutrients, and thermal energy that, if effectively harnessed, can contribute to energy generation, soil enrichment, and water reuse. Circular wastewater management aims to tap into these embedded values by utilizing technologies that recover resources while minimizing environmental degradation (Bhardwaj et al. 2025; Jain et al. 2023; Jain 2025). Within this context, PMFCs present a nature-based, regenerative alternative to conventional WWT systems.

Traditional WWT technologies (particularly activated sludge systems) are energy-intensive and costly to operate. The activated sludge process necessitates continuous aeration to support microbial degradation of organic matter, accounting for up to 60% of a treatment plant's energy consumption (Tchobanoglus et al. 2003). Additionally, these systems generate large volumes of excess sludge requiring further treatment, transportation, and disposal, often via environmentally detrimental methods such as incineration or landfilling. In contrast, PMFCs function passively, relying on natural interactions between plant roots and electroactive microorganisms in the rhizosphere. These microbes oxidize organic matter and release electrons, which are captured by anode electrodes and transferred to a cathode, thereby generating electricity. PMFCs accomplish both organic matter degradation and bioelectricity generation without requiring external aeration or chemical additives (Agarwal and Jain 2024; Agarwal et al. 2025). This self-sustaining functionality significantly reduces both operational costs and environmental impacts. Furthermore, PMFCs generate considerably less sludge compared to activated sludge systems. The microbial communities in PMFCs form stable, dense biofilms on electrode surfaces, which limits the proliferation of suspended biomass that typically contributes to sludge accumulation. This reduction not only simplifies sludge management but also aligns with circular economy objectives of waste minimization.

PMFCs exemplify circular economy principles by converting wastewater into multiple valuable outputs: purified water, electricity, and biomass. Root exudates from plants such as red spinach and rice paddy serve as natural substrates for microbial metabolism, while the treated effluent achieves quality levels suitable for reuse in irrigation or safe environmental discharge. Electricity generated by PMFCs, though modest in power density (~2.11 mW/cm^2), can support low-energy applications such as biosensors, lighting in sanitation units, or environmental monitoring (Kaku et al. 2008; Strik et al. 2008; Helder et al. 2010). Critically, this energy is derived from the oxidation of organic waste within the wastewater, effectively transforming pollution into a renewable energy source. In addition to clean water and energy, PMFCs enable food production by supporting plant cultivation. Leafy vegetables like red spinach can be grown directly within PMFC setups, utilizing the nutrient content of the wastewater to promote growth without jeopardizing food safety—provided treatment efficacy is maintained. The schematic diagram (Fig. 6.1) shows the plant roots in the rhizosphere, microbial activity, anode and cathode placement, electron flow, and electricity generation.

PMFCs represent a compelling example of biomimicry in environmental engineering. Biomimicry involves emulating natural ecosystems to develop sustainable technologies (Benyus 1997). In natural wetlands, interactions among plant roots, microbes, and water facilitate pollutant removal, nutrient cycling, and biodiversity support. PMFCs replicate these interactions within engineered systems, offering a

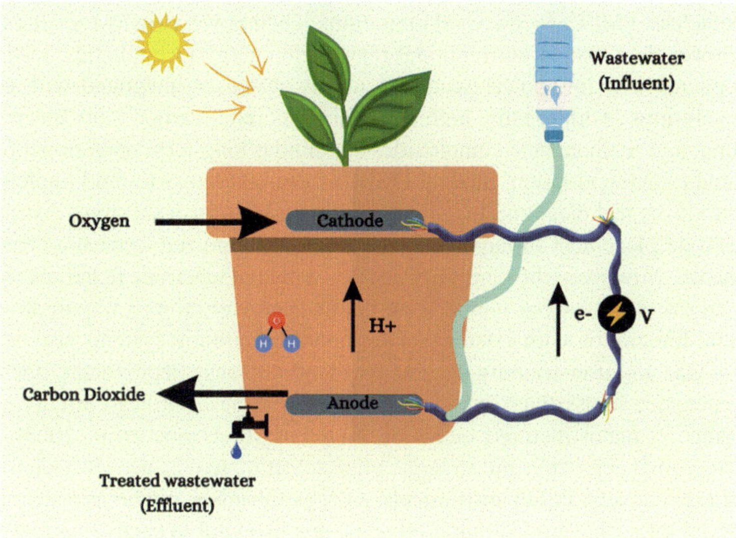

Fig. 6.1 Schematic diagram illustrating the working principle of a PMFC (Nair and Sreedharan, 2023)

model for circular water-energy solutions. The rhizosphere, or root zone, serves as a dynamic interface where plant-released organic compounds stimulate microbial activity. PMFCs exploit this interaction to simultaneously degrade pollutants and harvest energy. The use of carbon cloth as an anode mimics the conductivity of natural soil particles and roots, enhancing microbial adhesion and facilitating electron transfer. By positioning nature as both model and mentor, PMFCs challenge mechanistic, resource-intensive models of infrastructure. Instead, they advance self-regulating, self-sustaining systems that resonate with the ecological cycles in which they are embedded. A comparative analysis between PMFCs and conventional wastewater systems underscores their potential as disruptive innovations within the sector (Table 6.1).

Table 6.1 Comparison of PMFCs with conventional WWT technologies

Parameter	Activated sludge systems	Chemical treatment	PMFCs
Energy demand	High (aeration required)	Moderate to high	Low (passive operation)
Sludge generation	High	Variable	Low
Chemical use	Low to moderate	High	None
Resource recovery	Limited	Rare	High (energy, biomass, water)
Scalability	Centralized	Centralized	Decentralized
Cost	High CAPEX and OPEX	High chemical cost	Moderate, low OPEX

Nonetheless, PMFCs face several limitations. Their power density remains significantly lower than that of conventional renewable energy technologies, restricting their application to low-power scenarios unless scaled or integrated with energy storage solutions. Additionally, high material costs (e.g., carbon cloth electrodes), biofouling, and maintenance complexities can hinder long-term operational performance and broader adoption. Scaling PMFCs from laboratory to field implementations introduces technical and logistical challenges. Uniform substrate distribution and electrode placement in larger systems must be optimized to ensure consistent performance. Moreover, the durability and resistance of electrode materials to environmental stressors, such as temperature fluctuations and salinity, require thorough validation. Integration with existing sanitation or agricultural systems also necessitates modular and user-friendly designs that facilitate easy deployment, operation, and maintenance across diverse contexts. To overcome these hurdles, current research is focusing on nanostructured electrode materials (ul Haque et al. 2023), engineered microbial consortia, and strategic plant selection to enhance electron transfer and nutrient recovery. Policy instruments, such as subsidies, public–private partnerships, and pilot-scale demonstration projects, are also vital to catalyzing real-world deployment.

PMFCs hold a strategic position within the closed-loop energy-water nexus, offering a triple dividend by addressing water pollution, energy insecurity, and food production concurrently. They fulfill core circular economy goals by minimizing waste through effective pollutant removal and low sludge generation, maximizing reuse via the recycling of wastewater into irrigation-quality effluent, and recovering resources through electricity production and biomass cultivation. Their modular and decentralized architecture renders them especially suitable for peri-urban farms, household-level treatment systems, eco-industrial parks, and smart urban infrastructures. When integrated with Internet of Things (IoT) technologies or renewable energy grids, PMFCs can function simultaneously as sanitation units and environmental data nodes, contributing to adaptive resource management and real-time monitoring. Thus, PMFCs provide a promising, biomimetic solution for circular wastewater management. By transforming organic waste into usable outputs (such as clean water, energy, and plant biomass), they embody the principles of waste valorization, system integration, and ecological design. While scalability and operational longevity remain critical challenges, interdisciplinary advances in material science, microbial ecology, and policy frameworks are key to unlocking their potential. As urbanization accelerates and climate stressors intensify, PMFCs represent an innovative class of environmental infrastructure (for e.g., low-energy, low-cost, and high-impact). Situated at the confluence of biology and engineering, they exemplify how nature-inspired technologies can redefine our relationship with waste and catalyze the transition to a regenerative, sustainable future.

6.1.2 Science Behind Plant Microbial Fuel Cells (PMFCs)

Microbial Fuel Cells (MFCs) are bioelectrochemical systems that utilize the metabolic processes of microorganisms to convert organic matter into electrical energy. In a typical MFC, EAB oxidize organic substrates and transfer the resulting electrons to an anode. These electrons then flow through an external circuit to a cathode, where a reduction reaction, often involving oxygen, completes the circuit, thereby generating an electric current (Logan et al. 2006). MFCs have garnered significant interest as a sustainable technology for recovering energy from waste due to their dual functionality in WWT and electricity production (Kaku et al. 2008). However, conventional MFCs typically require external sources of organic matter and tightly controlled environmental conditions, which limit their scalability and field adaptability.

To overcome these limitations, researchers have developed PMFCs, which integrate living plants into MFCs to serve as continuous sources of organic substrates. In PMFCs, plants release organic compounds, primarily carbohydrates and amino acids, into the soil through rhizodeposition. These root exudates act as electron donors for EAB residing in the rhizosphere, the narrow zone of soil influenced by root activity (Strik et al. 2008). The bacteria metabolize these compounds, producing electrons and protons. The electrons are captured by a buried anode in the anaerobic soil zone, while the protons travel through the soil matrix or a proton exchange membrane (PEM) to the cathode. Electrons flow through an external circuit to the cathode, often exposed to air, where oxygen is reduced to water. This sequence of redox reactions enables electricity generation in PMFCs:

$$\text{Anode Reaction : Organic matter} + H_2O \rightarrow CO_2 + H^+ + e^- \tag{6.1}$$

$$\text{Cathode Reaction : } O_2 + H^+ + e^- \rightarrow H_2O$$

The voltage generated by PMFCs generally ranges from 100 to 500 mV per cell. While modest, this continuous energy output, coupled with plant growth and WWT, makes PMFCs a multifunctional and sustainable solution for environmental applications. Plants play a crucial role in PMFCs, not only as providers of carbon substrates but also as ecosystem engineers shaping rhizosphere conditions. Rhizodeposition varies by plant species, growth stage, and environmental conditions such as light, temperature, and nutrient availability (Dakora and Phillips 2002). Additionally, many plants release oxygen into the rhizosphere via radial oxygen loss (ROL), creating aerobic zones in the upper soil layers while maintaining anaerobic conditions in deeper strata. This redox stratification supports a diverse microbial community, facilitating efficient pollutant degradation and energy production. Plants with extensive root systems, high exudation rates, and tolerance to fluctuating redox conditions are especially suited for PMFC applications. This study utilizes Red Spinach and Rice Paddy for their adaptability, high biomass output, and proven support of microbial activity in saturated soils.

The rhizosphere serves as a microbial hotspot where EAB, such as Geobacter sulfurreducens, Shewanella oneidensis, and species from the Desulfuromonas and Pseudomonas genera, mediate electron transfer. These bacteria use external electron acceptors like anodes instead of traditional ones such as nitrate or sulfate. Electron transfer occurs via two main mechanisms: direct electron transfer (DET), facilitated by membrane-bound cytochromes and conductive pili (nanowires), and mediated electron transfer (MET), which relies on redox shuttles (Lovley 2008). Stable biofilm formation on the anode surface is essential for sustained electricity generation. SEM analyses from the present study confirm that densely packed biofilms with high extracellular polymeric substance (EPS) content are positively correlated with higher voltage outputs, especially in systems treating nutrient-rich synthetic dairy wastewater.

A fully functional PMFC comprises five key components: (i) Anode: Located in the anaerobic soil zone, the anode is typically made of conductive materials like carbon cloth, graphite felt or activated carbon. It serves as the terminal electron acceptor for EAB. (ii) Cathode: Positioned at or above the soil surface and exposed to atmospheric oxygen, the cathode completes the electrical circuit by facilitating the reduction reaction. (iii) Substrate/Medium: This includes the soil or growing medium that supports plant roots, microbial communities, and electron flow. It must also allow for hydraulic conductivity if used for WWT. (iv) External Circuit: Connects the anode and cathode through a resistor or electrical load, allowing for voltage and power measurements. (iv) Plant: The central biological engine of the PMFC, providing organic substrates, oxygen, and structural complexity in the rhizosphere. System configurations of PMFCs vary, including sediment-based systems, constructed wetlands, tubular setups, floating treatment wetlands, flat-plate designs, and even artificial trees designed for urban applications (Prasad and Kalla 2021). These are visually represented in Fig. 6.2, which outlines several PMFC models, each with unique structural and functional attributes:

Recent advancements in PMFC research focus on improving electrode conductivity, enhancing microbial compatibility, and increasing electrode surface area. Materials such as carbon nanotubes, graphene, and metal–organic frameworks (MOFs) have shown promise in boosting electron transfer rates and supporting biofilm formation (ul Haque et al. 2023; Feng et al. 2019). In addition, sustainable alternatives like biochar, produced via pyrolysis of organic biomass, are being explored as cost-effective and eco-friendly electrode materials. Surface modifications through heat treatment and chemical functionalization further enhance microbial adhesion and reduce activation losses. Beyond materials, plant science innovations also contribute to PMFC optimization. Researchers are exploring the use of transgenic or selectively bred plants with traits such as enhanced root exudation, increased oxygen transport, and improved root conductivity. These traits can improve substrate availability for EAB, thus elevating PMFC performance. PMFCs embody the convergence of biological and electrochemical systems, transforming waste into usable energy through the synergistic interaction of plant roots, microbial communities, and engineered electrodes. By deepening our understanding of rhizosphere dynamics, redox processes, and material interfaces, PMFCs can transition

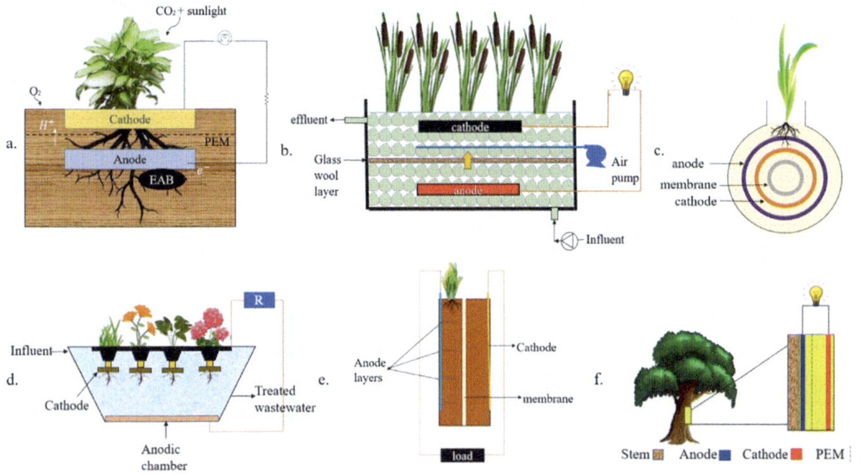

Fig. 6.2 Schematic of various PMFC models: **a** Sediment PMFC, **b** Constructed wetland-MFC, **c** Tubular PMFC, **d** Floating treatment wetland-MFC, **e** Flat plate PMFC, and **f** Power-generating trees (R = Resistance, PEM = Proton exchange membrane, EAB) (Prasad and Kalla 2021)

from experimental setups to scalable, real-world technologies. Their multifunctionality, enabling simultaneous electricity generation, WWT, and plant biomass production, positions them as key innovations within the broader framework of circular economy and sustainable development.

6.2 Research Methodology

This study investigates the feasibility of PMFCs as a dual-purpose technology for decentralized WWT and bioelectricity generation. The experimental design was structured to assess PMFC performance under different substrate conditions using two plant species: Red Spinach and Rice Paddy. Four distinct PMFC systems were established by pairing each plant type with two types of synthetic wastewater, formulated to simulate dairy and domestic effluents. The study focused on key performance metrics such as voltage output, pollutant removal efficiency, microbial colonization, and plant growth.

6.2.1 Wastewater Source and Composition

The dairy wastewater used in this study was formulated based on samples collected from the MILMA Dairy Plant at Edappally, Ernakulam, while the domestic wastewater was modeled after samples obtained from the Sewage Treatment Plant (STP) at

Table 6.2 Composition of synthetic wastewater used in PMFC reactors

Parameter/component	Dairy wastewater (mg/L)	Domestic wastewater (mg/L)
Glucose	180	–
Milk powder	1500	–
Starch	30	–
Ammonium chloride	150	–
Magnesium sulphate heptahydrate	100	100
Calcium carbonate	150	–
Potassium dihydrogen phosphate	100	100
Peptone	–	160
Meat extract	–	1100
Urea	–	30
Sodium chloride	–	150
Calcium chloride	–	150

Note Both formulations were prepared using distilled water

the SSET Campus, which treats effluents from hostels and canteens. To ensure consistency and control across experimental conditions, given the variability in raw wastewater composition, synthetic formulations were prepared to replicate the average chemical characteristics of the collected samples. These formulations captured the essential organic and nutrient loads typical of dairy and domestic wastewater streams (Tchobanoglus et al. 2003). The formulations in Table 6.2 ensured a balanced representation of organic and mineral loads, particularly carbohydrates, proteins, nitrogenous compounds, and essential minerals.

6.2.2 Experimental Setup and Reactor Design

The PMFC reactors were constructed using airtight, food-grade plastic containers (1 L vol) for initial trials and larger tapered cylindrical buckets (top diameter: 32 cm; bottom diameter: 17 cm; height: 38 cm for soil-based PMFCs. Each bucket was filled with 10 kg of oven-dried, sieved loamy soil, with no added compost or fertilizer, to minimize interference from extraneous organic sources. The uniform soil matrix facilitated consistent root development and microbial colonization across experimental units. Table 6.3 presents the experimental configuration of the four PMFC systems, combining two plant types, Red Spinach and Rice Paddy, with two wastewater types, dairy and domestic, to evaluate performance across varied biological and chemical conditions.

Figure 6.3 illustrates the functioning of PMFC systems integrating red spinach and rice paddy plants irrigated with domestic and dairy wastewater, respectively.

Table 6.3 Experimental configuration of PMFC systems

Set	Plant type	Wastewater type	Code
1	Red spinach	Dairy wastewater	RS-D
	Red spinach	Domestic wastewater	RS-H
2	Rice paddy	Dairy wastewater	RP-D
	Rice paddy	Domestic wastewater	RP-H

Each setup features a dual-chamber configuration with an anode buried in the rhizosphere and a cathode placed near the soil surface. Organic matter in the wastewater is oxidized by EAB at the anode, releasing electrons and protons. Electrons flow through an external circuit to the cathode, where they reduce oxygen, completing the bioelectrochemical circuit. Root exudates and photosynthetic activity enhance microbial metabolism and electron transfer. Simultaneously, nutrient uptake by plant roots and microbial degradation processes purify the wastewater, resulting in treated effluent. This closed-loop system exemplifies the circular economy by integrating WWT, energy recovery, and crop production in a single, sustainable framework.

Red Spinach and Paddy seeds were evenly sown in a bucket. Four germinated seedlings were then transplanted equidistantly in each reactor to ensure uniform spacing and minimize intra-species competition. A thin water layer (2–3 cm) was maintained above the soil surface in all reactors to simulate conditions typical of paddy fields and enhance root-microbe interactions (Yoshida 1981). Carbon cloth was used for both anode and cathode due to its high conductivity, large surface area, and microbial compatibility (Logan et al. 2006). The electrodes were vertically embedded in the soil with a spacing of 16 cm. The anode was buried in the anaerobic zone, while the cathode was placed near the soil surface and exposed to atmospheric

Fig. 6.3 Schematic representation of PMFC systems using domestic and dairy wastewater

oxygen. Copper wires connected the electrodes to a 100 Ω external resistor, forming a closed circuit for voltage generation.

6.2.3 Operational Methodology and Performance Monitoring

The reactors were operated in batch mode over 109 days, from January 15 to May 4, 2018, under ambient laboratory conditions with access to natural daylight. Environmental variables such as temperature and humidity were monitored using the Samsung Weather Application to maintain optimal conditions for plant growth, particularly for rice. Synthetic wastewater was introduced every 48 h, with influent volumes determined based on the soil's field capacity to prevent waterlogging and maintain aerobic and anaerobic zones conducive to microbial activity. Influent was supplied via 1-L plastic bottles connected to the reactor bases, promoting upward flow for uniform nutrient distribution. Effluent was collected in separate 1-L bottles for subsequent chemical and biological analysis. Voltage output (V) was recorded daily using a digital multimeter, with readings taken every hour from 10 a.m. to 4 p.m. This schedule captured diurnal variations in electricity generation linked to plant photosynthetic activity (Strik et al. 2008).

Effluent samples were collected on January 23 and February 1, 2018, representing the system's early operational phase. Samples were immediately stored at 4 °C and brought to room temperature prior to analysis. All effluent analyses were conducted following American Public Health Association (APHA) Standard Methods (1995). Monitored parameters included pH, turbidity, conductivity, total suspended solids (TSS), biological oxygen demand (BOD), chemical oxygen demand (COD), nitrate, phosphate, calcium, magnesium, and total hardness, providing a comprehensive assessment of treatment performance (Metcalf 2003). To evaluate microbial colonization, post-experiment SEM was conducted on electrode surfaces. SEM imaging provided visual evidence of biofilm development, microbial attachment, and EPS structures relevant to electron transfer efficiency and system stability (Schröder 2007).

Plant growth was monitored weekly by measuring the height from the soil surface to the tip of the tallest leaf. At the end of the 109-day period, plants were harvested and fresh biomass was recorded. Due to the relatively short experimental duration, rice plants did not reach grain-bearing maturity, which typically requires 120–150 days (Yoshida 1981). Therefore, plant height and biomass were used as proxies for productivity. Collected data were analyzed to compare the performance of the four PMFC configurations, highlighting differences in voltage output, pollutant removal, and plant growth. Descriptive statistics and comparative performance metrics were employed to evaluate the effects of wastewater type and plant species on PMFC functionality. This comprehensive methodology offers an integrated evaluation of PMFCs as multifunctional units within the circular economy. Through careful experimental design, consistent monitoring, and multidisciplinary analysis, the study provides empirical support for the adoption of PMFCs in sustainable wastewater management and renewable energy production.

6.3 Results and Discussion

6.3.1 Voltage Generation in PMFC Systems

Voltage generation is a primary performance indicator in PMFC systems, reflecting the bioelectrochemical activity of electroactive microbial communities in the rhizosphere and their ability to convert chemical energy into electrical energy. This study systematically monitored voltage output across four distinct PMFC configurations over a 109-day period, from January 15 to May 4, 2018. The objective was to evaluate the influence of plant species and wastewater substrate type on voltage output as a means of determining system efficiency and bioelectrochemical sustainability. Voltage trends in each configuration were influenced by a combination of plant growth stages, substrate complexity, microbial community development, and environmental factors such as diurnal light cycles. These factors collectively determined microbial respiration rates and the efficiency of electron transfer at the root–anode interface.

6.3.1.1 Voltage Trends in Red Spinach PMFCs

Red Spinach-based systems exhibited the most pronounced and consistent voltage increases over the experimental period. Figure 6.4 illustrates the daily voltage generation for RS-D and RS-H systems. Starting from an initial voltage of 3.43 mV for RS-D and 8.61 mV for RS-H on January 15, both systems showed a gradual and steady rise in voltage as microbial colonization progressed and the plants entered active growth phases. By mid-February, RS-H had already reached a voltage of approximately 97 mV, while RS-D lagged behind at 21.50 mV. The gap between the two systems widened during March and April. RS-H peaked at a maximum voltage of 258.31 mV on April 13, while RS-D reached a lower maximum of 181.94 mV on the same day. This significant difference in peak performance can be attributed primarily to substrate composition. Domestic wastewater, used in RS-H, contains simpler, more readily biodegradable organic compounds like carbohydrates, peptones, and amino acids. These compounds facilitate rapid microbial metabolism, resulting in higher electron production and efficient transfer to the anode. Dairy wastewater, on the other hand, contains complex organic matter such as milk fats, proteins, and long-chain carbohydrates that require longer microbial degradation times. This complexity likely slowed microbial adaptation and electron transfer efficiency in RS-D.

A secondary factor contributing to RS-H's performance was the sustained microbial diversity in the rhizosphere. Root exudates from Red Spinach, particularly during

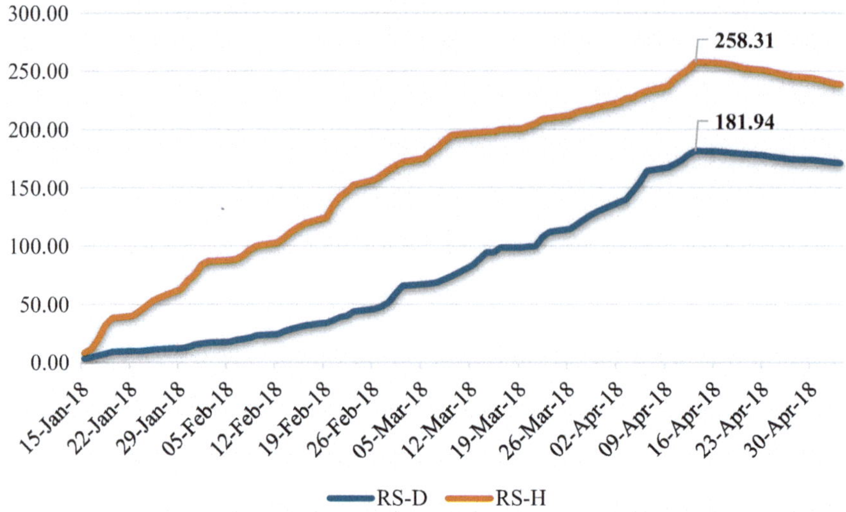

Fig. 6.4 Average daily voltage output for spinach PMFC systems (RS-D and RS-H) from January 15 to May 4, 2018. RS-H (with domestic wastewater) achieved a peak of 258.31 mV, outperforming RS-D (with dairy wastewater), which peaked at 181.94 mV

its mid-growth phase in late February and March, likely enhanced microbial density and diversity at the anode, further accelerating voltage generation. The influence of photosynthesis was also observed through diurnal voltage fluctuations, with voltage values peaking during the day and dipping at night. This trend is consistent with previous studies (e.g., Strik et al. 2008) that attribute increased microbial activity to light-induced root exudation. Interestingly, although RS-H produced a higher peak voltage, the system also exhibited a slow decline after mid-April, ending at 239.30 mV on May 4. This drop may be explained by the depletion of labile organic matter in the domestic wastewater and the onset of microbial senescence. In contrast, RS-D displayed a more gradual increase and a plateaued decline, ending at 171.60 mV. This more stable behavior may be due to the prolonged availability of complex organics in dairy wastewater, supporting sustained, albeit lower, microbial activity.

6.3.1.2 Voltage Trends in Rice Paddy PMFCs

Rice Paddy-based PMFC systems showed moderate voltage generation relative to their Red Spinach counterparts. The RP-H (Rice Paddy with domestic wastewater) and RP-D (Rice Paddy with dairy wastewater) configurations began with voltages of 2.50 and 0.61 mV, respectively, on January 15. By January 23, RP-H had reached 17.84 mV, while RP-D reached 14.49 mV, establishing the initial trend where domestic wastewater again supported more rapid microbial activation. Voltage output trends for RP-D and RP-H systems are shown in Fig. 6.5. This trend continued through February and March, with RP-H surpassing 100 mV by March 16, and RP-D crossing that threshold nearly two weeks later. RP-H peaked at 171.01 mV on April 19, whereas RP-D achieved a maximum of 125.80 mV on April 17. The temporal differences in peak voltage output support the hypothesis that simpler organics in domestic wastewater are metabolized quickly by EAB, leading to earlier voltage peaks. However, this rapid metabolism can lead to early substrate exhaustion, as evidenced by RP-H's decline to 150.41 mV by May 4.

RP-D, in contrast, exhibited a slower voltage build-up and a gentler decline. It ended the experiment with a voltage of 80.77 mV, suggesting that dairy wastewater, although initially less conducive to rapid microbial activation, provided a more enduring source of nutrients. This trend reflects the interplay between substrate complexity and microbial ecology, where slow-degrading substrates sustain microbial consortia for longer durations. Notably, RP systems recorded lower peak voltages than RS systems overall. This difference can be partially explained by the differences in root structure and rhizodeposition patterns between the two plants. Red Spinach produces fine, fibrous roots with high surface area and abundant exudates, which enhance microbe-root-electrode interactions. Rice Paddy roots, adapted for flooded conditions, are thicker and may produce fewer exudates in soil-based PMFCs, reducing the electron flux toward the anode. Moreover, paddy rice requires more water and prefers submerged conditions, which might have disrupted optimal aerobic-anaerobic gradients essential for electrogenic microbial activity in this experimental setup.

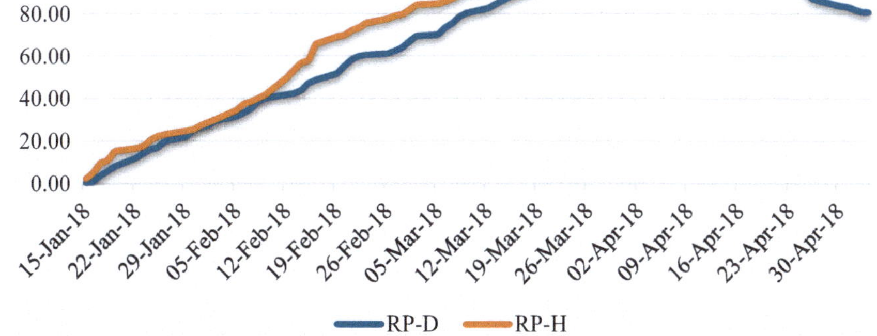

Fig. 6.5 Average daily voltage output for paddy PMFC systems (RP-D and RP-H). RP-H reached a peak voltage of 171.01 mV, while RP-D peaked at 125.80 mV, with gradual decline indicating substrate exhaustion and microbial dynamics

Voltage generation in PMFCs extends beyond its role as a simple operational metric. It encapsulates the transformative potential of converting waste into a renewable energy stream within a circular economy framework. Among the tested configurations, the RS-H system emerged as the most promising, consistently delivering the highest voltage output while simultaneously contributing to wastewater purification and nutrient recovery. This system exemplifies a closed-loop solution by integrating three vital resource cycles. First, the energy loop is closed through microbial oxidation of organic compounds in wastewater, which liberates electrons that are transferred to the anode and used to generate bioelectricity. Second, the nutrient loop is addressed as plants uptake residual nutrients, particularly nitrogen and phosphorus, thereby reducing the potential for downstream eutrophication. Third, the water loop is achieved through effective pollutant removal, resulting in treated effluent that can be reused for irrigation or other non-potable purposes. Together, these loops embody the core principles of regenerative design, where waste streams are repurposed as inputs for energy, food, and water systems.

The strong performance of RS-H underscores the viability of PMFCs in achieving multiple sustainability goals, particularly those outlined by the United Nations Sustainable Development Goals (SDGs). Specifically, PMFCs contribute to SDG 6 (Clean Water and Sanitation) by enhancing decentralized WWT capabilities; SDG 7 (Affordable and Clean Energy) through renewable electricity generation; and SDG 12 (Responsible Consumption and Production) by converting organic waste into valuable outputs. The implications are especially significant for the Global South, where access to centralized sanitation and energy infrastructure is limited. PMFCs, when optimized for material cost, microbial efficiency, and system scalability, offer a low-cost, low-tech, and environmentally friendly solution for integrated water-energy-food systems. Ultimately, the voltage generation trends observed in this study reflect a dynamic interplay of biological activity, substrate composition, plant physiology, and environmental conditions. By harnessing these synergies, future PMFC designs can be tailored to maximize resilience, resource efficiency, and scalability, turning waste from an environmental burden into a cornerstone of sustainable development.

6.3.2 Wastewater Treatment (WWT) Efficiency in PMFCs

The effectiveness of PMFCs as decentralized treatment systems lies not only in their bioelectricity generation capabilities but also in their ability to remove contaminants from wastewater. Effluent samples were collected during the early operational phase on January 23 and February 1, 2018, representing the system's microbial acclimation stage. Parameters analyzed included BOD, COD, TSS, turbidity, nutrients (phosphate and nitrate), ionic content (chloride, calcium, magnesium), and overall hardness (see Figs. 6.6 and 6.7 for percentage reduction across PMFC configurations).

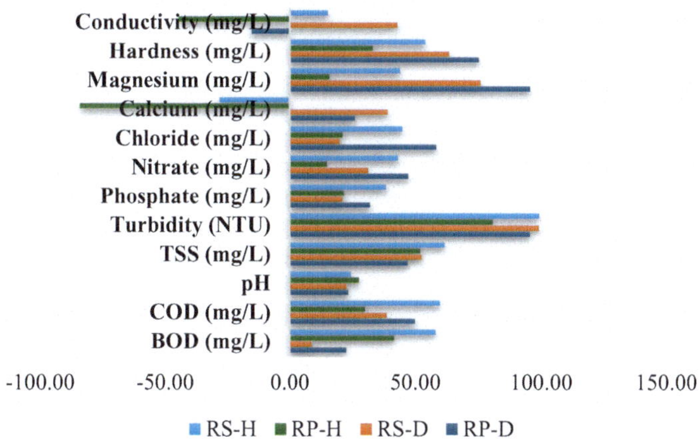

Fig. 6.6 Percentage reduction of water quality parameters in PMFCs for Effluent 1 (23 January 2018)

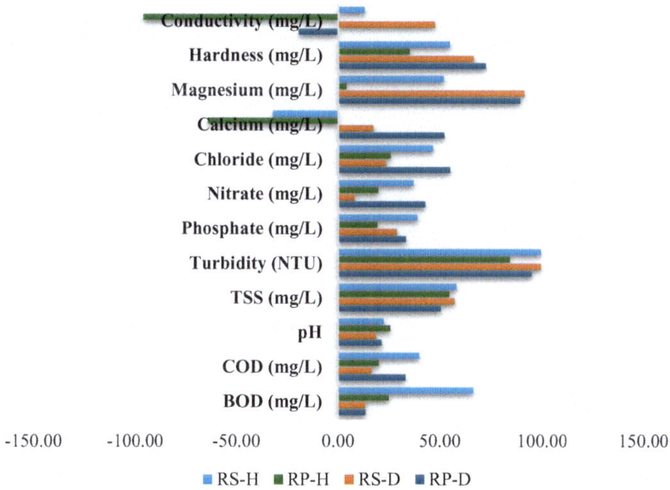

Fig. 6.7 Percentage reduction of water quality parameters in PMFCs for Effluent 2 (1 February 2018)

6.3.2.1 Organic Pollutant Removal

BOD and COD serve as primary indicators of organic load and microbial degradation activity. Across the experimental setups, RS-H consistently outperformed other configurations in BOD removal, achieving 58.33 and 66.67% reductions. This was followed by Rice Paddy–domestic wastewater (RP-H) at 41.67 and 25.00%. The superior BOD removal in domestic wastewater systems reflects the presence of readily biodegradable substrates such as peptone and meat extract, which rapidly stimulate microbial respiration and electrogenic activity in PMFCs. Conversely, dairy wastewater systems, despite slower initial degradation, achieved higher COD removal due to their complex organic composition (milk proteins, fats, carbohydrates). COD removal was highest in RP-D (50.00 and 33.33%) and RS-D (38.89 and 16.67%), suggesting prolonged microbial activity and biofilm maturation supported long-term degradation of complex organics. These findings align with earlier research emphasizing the relationship between substrate complexity and COD removal efficiency (Logan et al. 2006; Rabaey et al. 2005). The results for BOD and COD reduction across systems are visualized in Figs. 6.6 and 6.7, clearly showing superior performance in RS-H and RP-D depending on effluent type.

6.3.2.2 Nutrient Removal

Efficient phosphate and nitrate removal is essential for preventing eutrophication. RS-H showed the highest phosphate reductions (38.49 and 38.97%) and substantial nitrate removal (43.27 and 37.09%), indicating strong microbial denitrification and phosphorus uptake. RP-D also performed notably well, achieving phosphate reductions of 32.02 and 33.35%, and nitrate reductions of 47.33 and 42.91%. These trends suggest a synergistic effect between nutrient-rich environments and active microbial consortia. Dairy wastewater promoted the growth of phosphate-solubilizing and denitrifying bacteria such as *Pseudomonas* and *Paracoccus*, which contributed to enhanced nutrient transformation. Furthermore, the rhizosphere activity of Red Spinach and Rice Paddy enhanced nutrient uptake and cycling through root exudation and microbial stimulation, showcasing PMFCs as biomimetic analogs to natural wetlands.

6.3.2.3 Physical Parameters

Effluent clarity, assessed through TSS and turbidity, improved significantly across all systems. RS-H showed the highest TSS reductions (61.99 and 58.24%), while RS-D and RP-D delivered superior turbidity removal (above 94%). These results reflect differences in particle size and the nature of solids in wastewater types. Domestic wastewater, containing coarser suspended solids, facilitated greater TSS reduction due to natural sedimentation and filtration by plant roots. In contrast, dairy wastewater, with finer particulates, required biofilm-mediated trapping for turbidity reduction. SEM images confirmed that denser biofilms developed in RS-D and RP-D contributed to improved micro-particle entrapment and water clarity, further affirming the systems' capacity to emulate natural filtration processes.

6.3.2.4 Ionic and Mineral Content

Reductions in calcium, magnesium, and total hardness varied by configuration. RS-D and RP-D showed consistent calcium removal (39.13 and 26.09% on Jan 23; 17.39 and 52.17% on Feb 1), while RS-H and RP-H displayed increases in calcium concentration, likely due to ion leaching from soil or plant tissue under domestic wastewater conditions. Magnesium removal was most effective in RS-D (76.09 and 91.31%) and RP-D (95.65 and 89.13%), which also corresponded to the highest hardness reductions (63.77 and 66.67% in RS-D; 75.36 and 72.46% in RP-D). These systems demonstrated efficient ion exchange and bioavailability through sustained microbial activity and soil–root–electrode interactions. Conductivity data supported these observations, with RS-D and RS-H maintaining positive ion reductions, whereas RP-H showed substantial conductivity drops, reflecting leaching and lower retention efficiency. As depicted in Figs. 6.6 and 6.7, calcium and conductivity trends vary significantly, with negative reductions seen in RS-H and RP-H, reinforcing the ion leaching patterns discussed.

Overall, RS-H was most effective in rapid removal of organic pollutants and nutrients, attributed to the high biodegradability of domestic wastewater. RS-D and RP-D were more effective in mineral removal and long-term COD and turbidity reduction due to the complexity of dairy wastewater, which supported sustained microbial activity and robust biofilm formation. These outcomes emphasize the significance of substrate composition, microbial adaptation, and plant–microbe synergy in determining PMFC treatment performance. As decentralized, low-energy systems, PMFCs offer a multifunctional platform for wastewater purification, energy generation, and biomass production. Their effectiveness in achieving water quality improvements while closing material and energy loops aligns with global sustainability agendas, including SDG 6, SDG 7, and SDG 12. PMFCs thus represent a viable circular economy solution in areas lacking access to centralized WWT infrastructure.

6.3.3 Crop Growth and Yield in PMFC Systems

Beyond WWT and electricity generation, PMFCs demonstrate significant potential for advancing sustainable agriculture by supporting the simultaneous cultivation of food crops. Within the circular economy paradigm, PMFCs enable a multifunctional platform where waste is transformed into resources, such as, yielding clean energy, treated water, and agricultural output in a single, integrated system. Growth indicators such as plant height and fresh biomass were used to assess how substrate composition and microbial activity influenced productivity.

6.3.3.1 Red Spinach Growth in PMFC Systems

Red Spinach, a fast-growing leafy vegetable valued for its nutritional content, responded variably to wastewater type in PMFC environments. Data revealed that the RS-H configuration, supplied with domestic wastewater, produced significantly greater plant length and biomass compared to RS-D. The average plant height in RS-H was 66.5 cm, whereas RS-D plants averaged only 25.25 cm. Similarly, fresh biomass was markedly higher in RS-H, averaging 81.25 g, compared to just 19.75 g in RS-D (Fig. 6.8). These results suggest that the domestic wastewater in RS-H, rich in easily biodegradable organic compounds like peptone and meat extract, facilitated rapid microbial metabolism and efficient nutrient mineralization. Consequently, enhanced microbial activity likely increased the availability of plant-usable nutrients, thereby promoting growth. In contrast, RS-D, which utilized dairy wastewater characterized by more complex organic matter, may have supported slower microbial processing, resulting in delayed nutrient availability and reduced biomass accumulation.

Interestingly, despite the lower quantitative yield in RS-D, qualitative attributes of spinach plants, such as deeper coloration, thicker stems, and denser foliage, appeared more robust. This observation implies that while RS-H promoted faster and more voluminous growth, RS-D may have contributed to improved structural integrity, possibly due to the gradual release of micronutrients and slower microbial turnover. Importantly, electricity generation within both spinach-based PMFCs had no adverse effect on plant development, confirming that energy recovery and crop cultivation can be synergistically achieved. This dual functionality sets PMFCs apart from traditional bioenergy strategies, which often compete with food systems for land and inputs. In contrast, PMFCs support sustainable bioeconomy models by deriving both energy and edible produce from wastewater inputs, transforming liabilities into multifaceted resources.

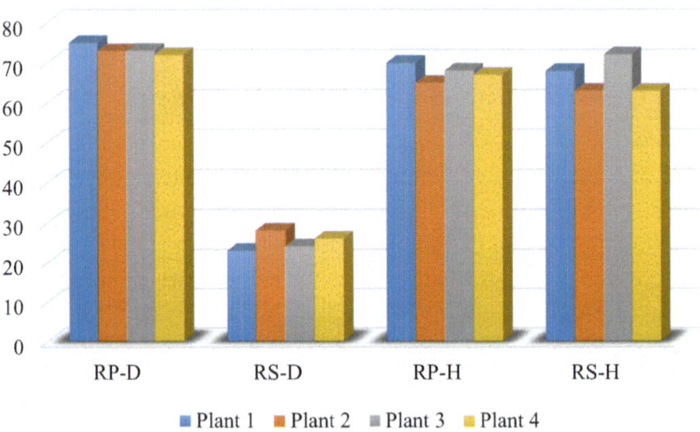

Fig. 6.8 Plant height variations in different PMFC configurations

6.3.3.2 Rice Paddy Growth in PMFC Systems

Rice Paddy systems followed a similar pattern in which wastewater composition strongly influenced plant growth metrics. The RP-D configuration (dairy wastewater) yielded superior results, with an average plant height of 73.25 cm and average fresh biomass of 55 g. In contrast, RP-H (domestic wastewater) plants were shorter (67.5 cm) and had less than half the biomass of RP-D, averaging only 23.75 g (Fig. 6.9). The better performance of RP-D can be linked to the rich nutrient content and complex organic substrates in dairy wastewater, including fats, proteins, and carbohydrates. These components likely sustained a more diverse and active microbial community over time, as confirmed by SEM observations that revealed thick, mature biofilms on RP-D electrodes. These dense biofilms enhance nutrient breakdown and facilitate greater ion exchange between the rhizosphere and plant roots.

Although RP-H systems exhibited higher initial voltage generation (discussed in Sect. 3.1), the relatively lower nutrient content and simpler substrate composition may not have supported long-term microbial activity or nutrient availability, leading to poorer crop performance. These findings underline the decoupling of electricity output and plant growth, a phenomenon where microbial electrochemical activity alone is insufficient to support optimal biomass accumulation without adequate nutrient support. It is also important to consider the experimental duration as a limiting factor. With the PMFCs operated for only 109 days, the rice plants did not complete their full growth cycle or reach the grain-bearing reproductive stage, which typically requires 120–150 days. Thus, productivity assessments were confined to vegetative metrics, such as height and biomass. Nevertheless, the observed differences in growth provide meaningful insights into how wastewater composition and microbial colonization shape plant development in PMFC-integrated systems. Additionally, the mildly acidic pH (5.5–5.7) of treated effluents

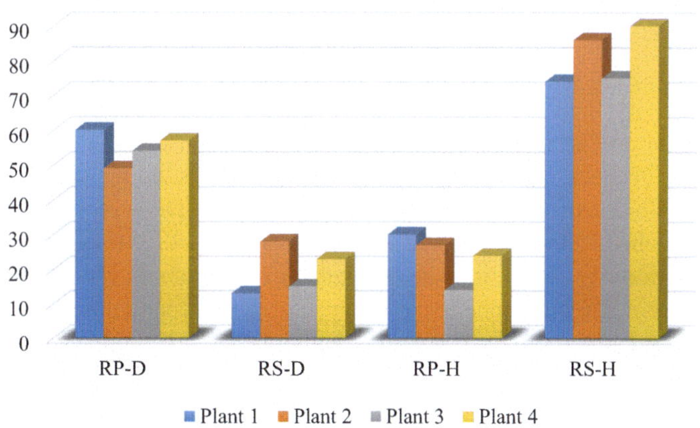

Fig. 6.9 Plant biomass (fresh weight) across PMFC configurations

across all systems likely enhanced nutrient solubility and uptake, contributing to improved overall plant health. This is particularly relevant in PMFC systems, where effluent composition is both an output and an input for sustaining plant and microbial dynamics.

The outcomes of crop cultivation across all four PMFC systems confirm the technical feasibility of integrating WWT, energy generation, and edible plant production in a single ecological platform. The RS-H system excelled in leafy vegetable productivity, while RP-D showed the most promise for cereal biomass accumulation. These findings support the idea of configuration-specific applications: spinach-PMFCs (especially RS-H) may be more suitable for rapid-cycle, high-yield vegetable production, while rice-PMFCs (particularly RP-D) are better suited for longer-cycle staple crops where nutrient richness is critical. This flexibility in design and function aligns with the goals of circular agriculture, where residual resources, such as treated effluent rich in plant nutrients, can be reused to reduce synthetic fertilizer dependency and minimize freshwater extraction. The residual nutrients in the effluent can also enhance soil fertility when reused for irrigation, adding another layer of environmental benefit.

From a systems perspective, PMFCs represent a biomimetic solution that mirrors natural ecological cycles. Microbes decompose organic matter, plants absorb nutrients, and the system as a whole outputs usable energy and food, without relying on fossil fuels or centralized infrastructure. This approach is especially relevant for resource-constrained or off-grid regions, where decentralized solutions are critical for achieving sustainability. As such, they contribute meaningfully to broader sustainability objectives, including SDG 2 (Zero Hunger) through food crop cultivation, SDG 6 via WWT, SDG 7 by producing renewable bioelectricity, and SDG 12 through nutrient recycling and waste valorization.

6.3.4 Microbial Activity, SEM Observations, and Electrode-Root Interface in PMFCs

Microbial activity is at the heart of PMFC performance, as it governs the twin processes of bioelectricity generation and WWT. Central to these mechanisms is the formation of electroactive biofilms on anode surfaces, composed primarily of bacteria capable of extracellular electron transfer (EET). These biofilms enable the conversion of chemical energy stored in organic compounds into electrical energy, while simultaneously catalyzing the breakdown of pollutants. The present study used SEM at 2.5 KX magnification to analyze the microbial colonization, biofilm morphology, and structural integration of plant roots with electrode materials across different PMFC configurations.

6.3.4.1 Electrode–Root Interface in PMFC Systems

A healthy, interactive interface between plant roots and the anode electrode is essential for successful PMFC functioning. Root exudates (such as organic acids, amino acids, and sugars) serve as primary carbon sources for the rhizospheric microbes, and the spatial proximity of roots to electrodes greatly influences microbial colonization, redox activity, and voltage output. Visual inspection and physical monitoring of the plant-root systems in all four configurations (RS-D, RS-H, RP-D, RP-H) confirmed robust root development with no visible signs of deformation, discoloration, or stress. This indicates that untreated carbon cloth electrodes used in the experiment were biocompatible and posed no toxicity risks to either crop. In Red Spinach systems (RS-D and RS-H), roots formed dense, fibrous mats that closely intertwined with the anode surface. This intimate contact ensured high rhizodeposition in the vicinity of the electrode, providing readily bioavailable organics that sustained microbial colonization. Despite RS-D showing lower plant biomass than RS-H, root health and physical integration were consistently positive in both systems. This suggests that the carbon cloth provided not just a neutral but potentially beneficial environment for root-microbe-electrode interaction.

In Rice Paddy systems (RP-D and RP-H), roots penetrated deeply into the submerged soil and made frequent contact with the anode surface. The fibrous, more spread-out architecture of rice roots, coupled with their radial oxygen loss (ROL) capacity, likely created micro-aerobic zones around the anode. This, in turn, may have supported microbial diversity and facilitated layered microbial stratification, essential for both aerobic and anaerobic metabolisms. No symptoms of root tip necrosis or other growth abnormalities were observed in rice systems either. These results align with the notion that the physical and chemical compatibility of electrodes in PMFCs is crucial. Untreated carbon cloth, due to its inertness and conductivity, emerged as a viable electrode material, supporting root development and microbial activity without needing expensive pretreatment. This feature is particularly valuable in resource-constrained settings, where material simplicity and plant-electrode harmony are essential to long-term PMFC sustainability.

6.3.4.2 SEM Analysis of Anode Biofilms

SEM imaging revealed striking differences in biofilm morphology and microbial density across different PMFC configurations. The control electrode (plain carbon cloth) showed smooth, clean fibers, with no signs of microbial adhesion or EPS, validating its pre-experiment sterility. In contrast, RS-D electrodes exhibited scattered microbial colonies, including coccoid and rod-shaped bacteria, embedded within modest EPS layers (Fig. 6.10). Although less dense than the RP-D system, these images indicated the beginning of structured biofilm formation, reflecting moderate microbial diversity and nutrient availability. RP-D electrodes stood out with the most pronounced microbial colonization (Fig. 6.10). Thick, continuous biofilms covered nearly the entire surface of the carbon filaments. The SEM imagery revealed a dense

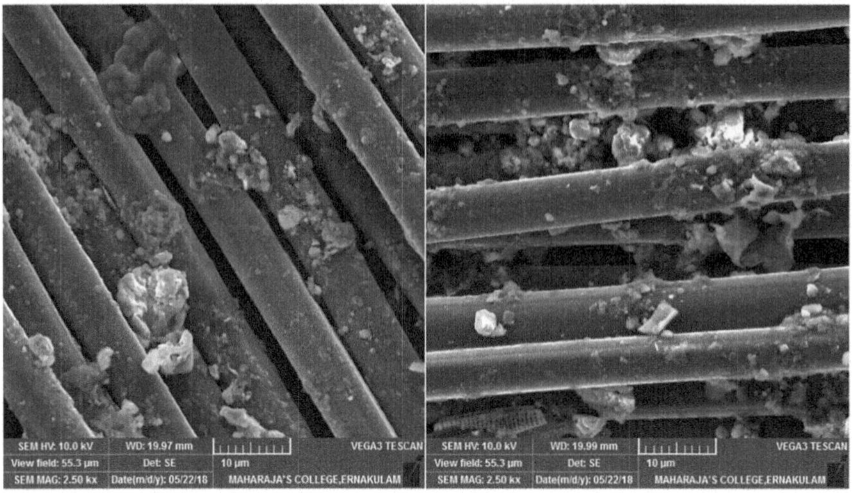

Fig. 6.10 SEM images (2500 × magnification) of anode surfaces from PMFCs fed with dairy wastewater (Left: RP-D, Right: RS-D)

matrix of bacterial communities entangled in copious EPS, signifying mature and active microbial ecosystems. The variety of shapes and sizes observed (from filamentous structures to cocci and bacilli) suggests a robust microbial consortium capable of supporting advanced EET mechanisms.

In comparison, RS-H electrodes showed thinner and patchier biofilms (Fig. 6.11). Microbial cells were scattered, and EPS layers were less visible, suggesting either slower colonization or a simpler microbial community. RP-H displayed a similar trend with relatively sparse microbial clusters and minimal EPS (Fig. 6.11). These observations align with measured system performances, where RS-D and RP-D consistently outperformed RS-H and RP-H in both voltage output and pollutant removal. Overall, the SEM analysis confirms that microbial colonization in PMFCs is sensitive to wastewater composition. Nutrient-dense dairy wastewater facilitates richer and more diverse microbial ecosystems, leading to denser biofilm growth, while the simpler organic profiles of domestic wastewater appear to constrain microbial complexity and biofilm architecture.

6.3.4.3 Role of Electroactive Bacteria and Biofilm Composition

EAB, such as *Geobacter spp.* and *Shewanella spp.,* are crucial to PMFCs due to their ability to transfer electrons extracellularly to the anode. The structural richness observed in RS-D and RP-D electrodes suggests the presence of these bacteria, supported by the availability of proteins, carbohydrates, and lipids found in dairy wastewater. The observed EPS formations provide further evidence of an active and structured microbial network, as EPS is known to serve as both a protective

Fig. 6.11 SEM images (2500 × magnification) of anode surfaces from PMFCs fed with domestic wastewater. (Left: RP-H, Right: RS-H)

matrix and a medium for electron shuttling. In domestic wastewater-fed systems (RS-H and RP-H), the microbial limitations seen under SEM likely stem from the low molecular diversity and absence of complex organics. Simple substrates, while rapidly biodegradable, may not support the development of multi-species electroactive communities required for long-term EET. Additionally, fewer nutrients may result in weaker EPS synthesis, reducing microbial adherence and biofilm longevity. These observations underscore a critical aspect of PMFC design: substrate complexity directly influences microbial ecology and, consequently, system efficiency. This emphasizes the need for tailored microbial enrichment or substrate supplementation strategies when working with low-strength wastewaters.

6.3.4.4 Electrode Material Considerations and Enhancement Strategies

Although untreated carbon cloth proved adequate for biofilm formation in nutrient-rich environments, future research should explore material enhancements for broader applications. Literature indicates that thermally or chemically treated carbon cloth exhibits improved surface properties, such as increased roughness, hydrophilicity, and conductivity, all of which promote microbial adhesion and EET efficiency (Zhou et al. 2012; Wilberforce et al. 2022; ul Haque et al. 2023). Moreover, activated carbon cloth and nanostructured composites (e.g., carbon nanotubes, graphene layers) can offer higher surface areas and more active sites for biofilm formation (Zhao et al. 2006). These advanced materials could be particularly valuable in RS-H and RP-H systems, where microbial colonization lagged due to limited substrate availability.

Another promising strategy is the pre-inoculation of anodes with enriched electroactive microbial consortia. This approach can jumpstart biofilm development, reduce lag phases, and stabilize voltage output in systems using simple or low-strength wastewaters. Such strategies would be ideal for accelerating performance in domestic wastewater-fed systems. In addition, adopting layered electrode configurations or modular stacking could facilitate higher current densities and voltage scalability without compromising plant health or soil structure. These design innovations will be pivotal for translating laboratory-scale PMFCs into real-world decentralized sanitation and energy solutions.

SEM imaging at 2.5 KX magnification revealed strong contrasts in microbial colonization between dairy and domestic wastewater systems, affirming that substrate complexity plays a decisive role in biofilm architecture and electrochemical efficiency. The findings confirm the suitability of untreated carbon cloth as an effective and plant-compatible electrode material. Future enhancement strategies, such as electrode surface modification and microbial pre-enrichment, can further amplify PMFC performance, particularly in nutrient-limited conditions. In essence, PMFCs demonstrate the viability of a bio-circular-green infrastructure, integrating food production, wastewater remediation, and clean energy recovery into a unified system. This approach closes three critical loops: energy (via microbial oxidation and EET), nutrients (via plant uptake), and water (via treated effluent reuse). As such, PMFCs serve as a living model of circular economy principles and hold immense potential for achieving SDGs 2, 6, 7, and 12. With continued innovation in microbial-electrode integration, PMFCs can evolve into a transformative solution for regenerative development in both rural and urban contexts.

6.3.5 PMFCs as a Circular Economy Solution Bridging the Circularity Gap

The findings from this study strongly support the potential of PMFCs as a circular economy solution that bridges the gap between energy generation, WWT, and agricultural productivity. The peak voltage outputs recorded is 258.31 mV for the Red Spinach with domestic wastewater (RS-H) and 171.01 mV for the Rice Paddy with domestic wastewater (RP-H) that demonstrate the viability of bioelectricity generation through natural plant–microbe interactions within the rhizosphere. The highest power density observed, approximately 2.11 mW/cm^2 in RS-H, underscores the capability of even untreated carbon cloth electrodes to facilitate effective electron transfer and support electrogenic microbial communities in soil-based systems. These results align with earlier research (Helder et al. 2010; Strik et al. 2008) and highlight the influence of diurnal solar radiation on root exudation and microbial metabolism, emphasizing the nature-synchronized operation of PMFCs. Beyond energy recovery, PMFCs in this experiment effectively treated synthetic wastewater by significantly improving effluent quality. The RS-H system achieved the highest BOD reduction

(66.67%) and substantial COD removal (40%), while RS-D and RP-D exhibited superior COD reductions (38.89 and 50%, respectively) due to the higher complexity of dairy wastewater. Turbidity reduction exceeded 99% across most systems, particularly in RS-D and RS-H, indicating dense biofilm formation and efficient particulate entrapment. Nutrient removal was also significant: nitrate and phosphate reductions reached 47.33 and 38.97%, respectively, suggesting the presence of denitrifying and phosphate-solubilizing microbial communities supported by both root exudates and wastewater composition.

The differences in pollutant removal between systems are linked to substrate complexity and microbial diversity. Dairy wastewater systems (RP-D and RS-D) supported denser and more diverse biofilms, as confirmed through SEM analysis, which revealed thick EPS matrices and complex microbial morphology. Domestic wastewater systems (RS-H and RP-H), while supporting rapid early voltage generation, exhibited patchier microbial colonization over time, possibly due to quicker substrate depletion. This highlights a critical trade-off: more biodegradable substrates enhance early bioelectricity but may compromise long-term treatment performance, whereas complex substrates support sustained microbial activity and better long-term pollutant removal. Effluent mineral content analysis further supports the reuse potential of PMFC-treated water. Hardness and chloride levels were significantly reduced in RP-D and RS-D systems, while the pH stabilized in a mildly acidic range (5.5–5.7), favorable for nutrient solubility and plant uptake. These properties make PMFC effluent suitable for non-potable applications such as agricultural irrigation, contributing to water reuse strategies in line with global sustainability goals.

Plant growth data reinforce the multifunctionality of PMFCs. RS-H produced the highest fresh biomass (average 81.25 g) and plant height (66.5–72 cm), indicating synergistic effects between nutrient-rich domestic wastewater and high electrogenic activity. RP-D, while exhibiting a slightly lower peak voltage, yielded the greatest rice biomass (55 g), suggesting that dairy wastewater fosters long-term plant productivity through consistent nutrient availability. These outcomes illustrate how PMFCs can produce edible crops while treating wastewater and generating energy, fulfilling the vision of integrated resource recovery. SEM imaging provided critical insights into the electrode-root-microbe interface. Untreated carbon cloth demonstrated excellent biocompatibility with both spinach and rice root systems, which showed no signs of necrosis or inhibition. Electrodes in RP-D and RS-D revealed dense, mature biofilms populated by diverse electroactive microbial morphologies, supporting efficient EET. In contrast, RS-H and RP-H electrodes showed sparser biofilm development, possibly due to lower substrate complexity. These results validate the importance of electrode selection and microbial compatibility in optimizing PMFC performance.

PMFCs offer unique advantages within circular economy frameworks, particularly in decentralized contexts. Compared to conventional technologies like solar photovoltaics or anaerobic digesters, PMFCs provide lower energy yields but deliver integrated benefits, including decentralized wastewater treatment, electricity generation, and food production. They operate passively, require minimal external inputs, and adapt to existing ecosystems, making them especially valuable in rural or peri-urban areas of the Global South. Furthermore, they align with SDG 2, SDG 6, SDG

7, and SDG 12. As a nature-based, multifunctional system, PMFCs exemplify regenerative infrastructure that closes resource loops. The synergy between plant growth, microbial activity, and energy production reinforces their role as a viable circular economy technology. Future advancements may include using surface-modified electrodes, inoculating with enriched EAB, and customizing configurations based on local wastewater profiles. With such innovations, PMFCs can evolve from experimental prototypes to robust, scalable solutions for sustainable development, turning waste into water, energy, and food in one integrated cycle.

6.3.5.1 Scalability and Real-World Implementation Challenges

The performance outcomes and microstructural observations in this study provide critical guidance for scaling PMFCs from lab-scale prototypes to real-world sustainable infrastructure. PMFC systems, especially those using dairy wastewater such as RS-D and RP-D, exhibited denser and well-structured microbial biofilms as visualized through SEM, correlating directly with improved electricity generation, nutrient removal efficiency, and plant growth. This supports the conclusion that substrate richness is a primary driver of system performance. Dairy effluents, with their high content of complex organic matter, promote diverse microbial communities capable of sustained electrogenic activity. In contrast, domestic wastewater systems (RS-H and RP-H), while yielding higher initial voltage due to readily biodegradable organics, demonstrated thinner, patchier biofilms and faster substrate depletion. Additionally, all four PMFC configurations confirmed strong electrode-root compatibility. No signs of phytotoxicity, necrosis, or physical deformation were observed at the electrode–root interface, affirming the biocompatibility of untreated carbon cloth. The fibrous root networks of spinach and the elongated roots of rice intertwined with the anodes, enabling effective rhizodeposition and enhancing microbe–root–electrode interactions. This validates the feasibility of integrating food crop cultivation within PMFC systems, a major advantage over traditional bioenergy systems that often compete with food production.

Importantly, microbial density (as revealed in SEM images) was a strong predictor of system performance, reinforcing the role of microbial ecology in PMFC design. For future implementation, microbial management through pre-inoculation with EAB, optimization of substrate inputs, or bioaugmentation may be vital strategies for enhancing reliability and scalability. These techniques can particularly benefit systems using low-strength domestic wastewater. The schematic in Fig. 6.12 visually encapsulates the circular nature of PMFCs, integrating WWT, energy recovery, resource cycling, and plant growth in a closed-loop system.

While the PMFC model aligns closely with the goals of the circular economy and the food-energy-water nexus, certain limitations must be addressed for full-scale deployment. The average power density achieved is 2.11 mW/cm^2 in the RS-H system, is promising for low-energy applications such as biosensors, environmental monitoring, and off-grid data loggers. However, it remains insufficient for large-scale power supply. Thus, future integration with other renewable technologies, like

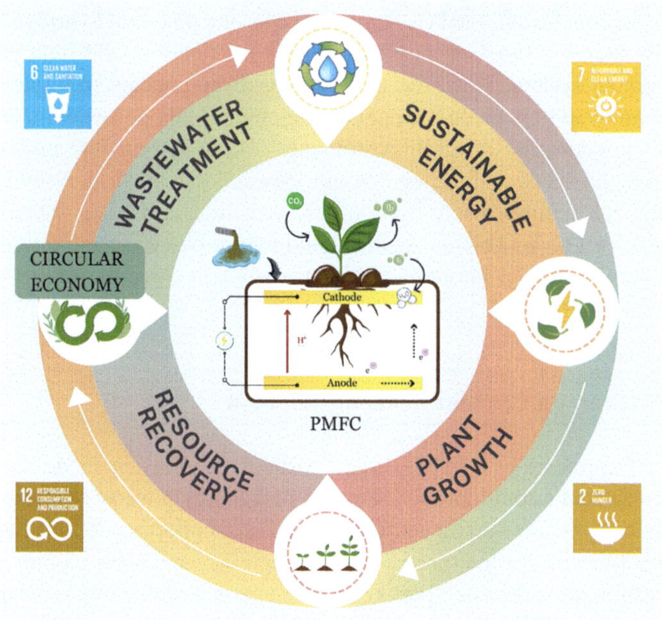

Fig. 6.12 Schematic representation of PMFCs within a circular economy framework, illustrating integrated WWT, sustainable energy generation, plant biomass production, and resource recovery. The system aligns with SDG 2, 6, 7, and 12

solar microgrids or anaerobic digesters, may be essential to meet broader energy demands. Material considerations also present a constraint. Though untreated carbon cloth electrodes demonstrated good electrochemical compatibility and biofilm adherence, their cost at scale could be a barrier. Enhancing the anode's surface properties through cost-effective techniques, such as thermal treatment, biochar infusion, or low-cost nanomaterial coatings, could improve conductivity and microbial colonization while reducing costs. Furthermore, field trials in varied environmental conditions (salinity, pH, temperature) are necessary to validate long-term performance and system resilience.

Maintenance remains a key practical challenge. Over time, biofouling, soil clogging, and electrode degradation may reduce current output and treatment efficiency. Developing simple, community-friendly cleaning and monitoring protocols, possibly with IoT-based remote alerts, could enhance system longevity in rural or decentralized contexts. Advancements in synthetic biology and plant science may unlock new performance horizons. High root-exudate crops or genetically engineered species optimized for microbial cooperation and conductivity can dramatically boost productivity. The results in this study highlight spinach and rice as promising candidates; however, region-specific alternatives (e.g., vetiver grass, Napier grass, or aquatic species like water hyacinth) should be explored for greater environmental and cultural adaptability. In conclusion, PMFCs demonstrate strong potential to bridge the circularity gap by merging bioelectricity, wastewater purification, and food production in one integrated system. However, transitioning from concept to implementation will require innovation in electrode engineering, microbial optimization, crop selection, and socio-economic integration strategies. These systems are uniquely positioned to serve the needs of the Global South, especially in off-grid, resource-scarce, and decentralized settings where clean water, sanitation, and energy remain critical unmet needs. PMFCs represent a scalable and nature-based infrastructure solution for a regenerative, circular future by modeling natural cycles and turning waste into a resource.

6.3.5.2 PMFCs as a Circular Economy Solution

The potential applications of PMFCs are particularly relevant in Asian contexts, where decentralized sanitation, food security, and energy poverty co-exist. In India, PMFCs could serve peri-urban agriculture zones, supporting treated wastewater reuse, off-grid electricity for sensors or LED lighting, and even mobile sanitation units. Integration into urban eco-parks or rooftop gardens could align with the Smart Cities Mission and AMRUT schemes. In Southeast Asian nations like Indonesia and Vietnam, PMFCs could address agricultural runoff and aquaculture wastewater while contributing to energy needs in remote villages. The modularity of PMFCs makes them ideal for community-scale deployment, particularly where land availability and water pollution are pressing concerns. PMFCs epitomize the ethos of the circular economy by bridging waste-to-resource transitions. They transform wastewater into three valuable outputs: clean water, biomass, and electricity. This multi-output nature ensures that the system maximizes resource efficiency with minimal waste generation. The integration of plants and microbes reflects biomimetic design, where ecosystems inspire solutions that are self-sustaining and regenerative. By decentralizing both energy and treatment services, PMFCs reduce reliance on centralized infrastructure, lower operational costs, and foster community ownership. This is especially relevant in the global South, where circular economy implementation requires low-tech, low-cost solutions with high adaptability.

6.3.5.3 Policy Implications and Future Research

The integration of PMFCs into mainstream sustainability strategies requires a supportive policy environment that recognizes the multi-sectoral benefits of energy–water–food nexus technologies. Current regulatory frameworks often overlook decentralized, nature-based solutions in favor of large-scale, centralized infrastructure, which may not be viable or appropriate in rural and peri-urban contexts. Therefore, policy reforms are needed to incorporate PMFCs into national and sub-national sanitation guidelines, renewable energy initiatives, and agricultural water reuse frameworks. For instance, introducing incentives for wastewater-to-energy systems, such as feed-in tariffs for bioelectricity, tax credits for sustainable agriculture practices, or capital subsidies for bio-based electrodes, could catalyze early adoption and commercial interest. Inclusion of PMFCs in flagship government programs, such as India's Smart Cities Mission, AMRUT, or Swachh Bharat Abhiyan, would provide visibility and legitimacy to the technology while encouraging innovation partnerships between academic institutions, start-ups, and local municipalities.

In addition to policy integration, capacity-building initiatives and public awareness campaigns are crucial for scaling PMFC applications. These efforts can empower communities to manage and maintain decentralized treatment-energy systems, fostering local ownership and resilience. To ensure equitable access, development agencies and public–private partnerships should prioritize demonstration projects in low-income and off-grid settings, particularly where water pollution and energy poverty intersect. From a research perspective, several avenues remain unexplored or underdeveloped. Future studies should prioritize pilot-scale implementations of PMFCs across diverse agro-climatic regions to evaluate real-world performance under varying soil types, crop varieties, and climatic conditions. Comparative life-cycle assessments and cost–benefit analyses against other decentralized technologies (e.g., anaerobic digesters, wetlands, or solar microgrids) would provide the evidence base necessary for policymakers and investors. Material innovation remains a key area, developing biodegradable or recycled electrode materials that reduce environmental impact while maintaining electrochemical performance is essential for sustainable scale-up.

Additionally, microbial engineering and synthetic biology present promising frontiers for enhancing system output. Developing tailored microbial consortia, optimized for specific wastewater types or climate zones, could increase robustness and functional stability. Genetic enhancement of plant traits, such as higher root exudation or increased conductivity, could further strengthen plant–microbe–electrode interactions. Moreover, integrating PMFCs with Internet of Things (IoT) platforms for real-time monitoring and automation can improve system efficiency, diagnostics, and user engagement. Ultimately, PMFCs represent more than just a technological innovation, they embody a systems-thinking approach to sustainable development. By aligning biological intelligence with engineering and policy innovation, they offer a pathway to rethink infrastructure through a regenerative, decentralized, and circular economy lens. Future research and policies must work synergistically to unlock this potential and embed PMFCs within the broader transition to resilient and equitable water–energy–food systems.

6.4 Conclusions

The study establishes PMFCs as an effective, nature-based technology for decentralized WWT, bioelectricity generation, and sustainable agriculture. Utilizing two common crops, Red Spinach and Rice Paddy, with synthetic dairy and domestic wastewater, PMFCs demonstrated their capacity to integrate environmental remediation and renewable energy production within a single system. Among all configurations, the Red Spinach–Domestic Wastewater (RS-H) system achieved the highest peak voltage (258.31 mV) and power density (2.11 mW/cm^2), indicating robust bioelectrochemical activity. Concurrently, significant reductions in BOD (up to 58.33%), COD (up to 60%), turbidity (>99%), and nutrients such as nitrate and phosphate were observed, aligning with water quality standards for secondary treatment. The study highlights that PMFC performance is fundamentally driven by synergistic plant–microbe–electrode interactions in the rhizosphere. Dense microbial colonization, especially in RS-D and RP-D systems, contributed to sustained energy recovery and pollutant removal, supported by SEM imaging of biofilm morphology. These interactions mirror natural wetland processes and reflect circular economy principles by converting wastewater into clean water, electricity, and edible biomass. PMFCs thus close the energy, nutrient, and water loops, offering a multifunctional, low-impact infrastructure for sustainable development. However, several challenges persist. Power output remains limited for high-demand applications, and long-term stability under field conditions requires further validation. Material costs, especially for electrode components, and risks of biofouling also constrain scalability. Nevertheless, PMFCs' compatibility with local crops, low energy demands, and passive operation make them well-suited for resource-constrained contexts in the Global South. To realize their full potential, future work should prioritize material innovation (e.g., nanostructured electrodes), microbial engineering, and inclusive policy frameworks to support scale-up. PMFCs not only represent a technical innovation but also a shift toward regenerative infrastructure that redefines wastewater as a resource within the bio-circular-green economy model.

References

Agarwal S, Jain H (2024) Investigating the potential of floral waste as a vermicompost and dual-functional biosorbent for sustainable environmental management. Water Air Soil Pollut 235(5):322

Agarwal S, Jain H, Dhupper R, Mathur A (2025) Optimizing methyl orange dye removal from aqueous solutions using white chrysanthemum floral waste-derived bioadsorbent: a study of kinetics, thermodynamics, isotherms, and RSM optimization. Int J Environ Res 19(3):104

American Public Health Association (APHA) (1995) Standard methods for the examination of water and wastewater, 19, edn. American Public Health Association Inc., New York

Benyus JM (1997) Biomimicry: innovation inspired by nature, vol 688136915. Morrow, New York

Bhardwaj LK, et al (2025) Resilience in the digital age: emerging technologies for climate adaptation. In: Choudhury M, Dixit G, Majumdar S (eds) Sustainable synergy: harnessing ecosystems for climate resilience. Climate change management. Springer, Cham. https://doi.org/10.1007/978-3-031-77957-2_11

Dakora FD, Phillips DA (2002) Root exudates as mediators of mineral acquisition in low-nutrient environments. In: Food security in nutrient-stressed environments: exploiting plants' genetic capabilities. pp 201–213. https://doi.org/10.1007/978-94-017-1570-6_23

Feng Y, Li D, Liu J, He W (2019) Carbon-based materials in microbial fuel cells. In Microbial electrochemical technology. Elsevier, pp 49–74. https://doi.org/10.1016/B978-0-444-64052-9.00003-0

Helder M, Strik DPBTB, Hamelers B, Buisman CJN (2010) The flat-plate plant-microbial fuel cell: the effect of a new design on internal resistances. Biores Technol 101(10):3541–3547. https://doi.org/10.1186/1754-6834-5-70

International Energy Agency (IEA) (2021) World energy outlook 2021. https://www.iea.org/reports/world-energy-outlook-2021

Jain H (2025) Synergizing carbon reduction technologies and nature-based solutions for a climate-neutral future. Mitig Adapt Strateg Glob Change 30:42. https://doi.org/10.1007/s11027-025-10237-6

Jain H, Dhupper R, Shrivastava A, Kumar D, Kumari M (2023) AI-enabled strategies for climate change adaptation: protecting communities, infrastructure, and businesses from the impacts of climate change. Comput Urban Sci 3(1):25. https://doi.org/10.1007/s44212-023-00044-6

Kaku N, Yonezawa N, Kodama Y, Watanabe K (2008) Plant/microbe cooperation for electricity generation in a rice paddy field. Appl Microbiol Biotechnol 79:43–49. https://doi.org/10.1007/s00253-008-1410-9

Kumar M, Kumar R, Sharma S, Mukherjee S, Bolan S, Shrivastava A, Bolan N (2025) Microplastics pollution modulating soil biological health–a review. Soil Use Manag 41(1):e70009

Logan BE, Hamelers B, Rozendal R, Schröder U, Keller J, Freguia S, et al (2006) Microbial fuel cells: methodology and technology. Environ Sci Technol 40(17):5181–5192. https://doi.org/10.1021/es0605016

Lovley DR (2008) The microbe electric: conversion of organic matter to electricity. Curr Opin Biotechnol 19(6):564–571. https://doi.org/10.1016/j.copbio.2008.10.005

MacArthur E (2013) Towards the circular economy. J Ind Ecol 2(1):23–44

Metcalf E (2003) Wastewater engineering: treatment and reuse, 4th edn. McGraw-Hill, New York

Nair NV, Sreedharan S (2023) Plant microbial fuel cell for the effective utilization of wastewater and bioenergy production. In: AIP conference proceedings, vol. 2856(1). AIP Publishing. https://doi.org/10.1063/5.0165310

Prasad PN, Kalla S (2021) Plant-microbial fuel cells-a bibliometric analysis. Process Biochem 111:250–260. https://doi.org/10.1016/j.procbio.2021.10.001

Rabaey K, Boon N, Höfte M, Verstraete W (2005) Microbial phenazine production enhances electron transfer in biofuel cells. Environ Sci Technol 39(9):3401–3408. https://doi.org/10.1021/es048563o

Saini J, Gupta AK, Shrivastava A, Dhupper R (2024) Study of urban warming in Chandigarh union territory, India: geospatial approach: J. Saini et al. Theoret Appl Climatol 155(8):8245–8261

Schröder U (2007) Anodic electron transfer mechanisms in microbial fuel cells and their energy efficiency. Phys Chem Chem Phys 9(21):2619–2629. https://doi.org/10.1039/B703627M

Strik DP, Terlouw H, Hamelers HV, Buisman CJ (2008) Renewable sustainable biocatalyzed electricity production in a photosynthetic algal microbial fuel cell (PAMFC). Appl Microbiol Biotechnol 81:659–668. https://doi.org/10.1007/s00253-008-1679-8

Tchobanoglus G, Burton F, Stensel HD (2003) Wastewater engineering: treatment and reuse. Am Water Works Assoc J 95(5):201

ul Haque S, Nasar A, Duteanu N, Pandey S (2023) Carbon based-nanomaterials used in biofuel cells–a review. Fuel 331:125634. https://doi.org/10.1016/j.fuel.2022.125634

United Nations World Water Assessment Programme (UN WWAP) (2023) United Nations world water development report 2023: partnerships and cooperation for water. UNESCO. https://unesdoc.unesco.org/ark:/48223/pf0000384655

United States Environmental Protection Agency (US EPA) (2006) Wastewater management fact sheet, energy conservation. EPA 832-F-06-024. EPA, Washington DC, USA

Wilberforce T, Abdelkareem MA, Elsaid K, Olabi AG, Sayed ET (2022) Role of carbon-based nanomaterials in improving the performance of microbial fuel cells. Energy 240:122478. https://doi.org/10.1016/j.energy.2021.122478

Yoshida S (1981) Fundamentals of rice crop science. Int Rice Res Inst

Zhao F, Harnisch F, Schröder U, Scholz F, Bogdanoff P, Herrmann I (2006) Challenges and constraints of using oxygen cathodes in microbial fuel cells. Environ Sci Technol 40(17):5193–5199. https://doi.org/10.1021/es060332p

Zhou M, Chi M, Wang H, Jin T (2012) Anode modification by electrochemical oxidation: a new practical method to improve the performance of microbial fuel cells. Biochem Eng J 60:151–155. https://doi.org/10.1016/j.bej.2011.10.014

Chapter 7
Strong Waste Management Infrastructure, Revamped Logistics and Supply Chain, and Legislative Incentives for Circular Economy

Sakshi Garg, Harshita Jain, and Anamika Shrivastava

7.1 Introduction

7.1.1 Circular Economy Overview

The need to reconsider how we engage with the material world is becoming more and more prevalent as a result of the recent environmental, sociopolitical, health, and economic challenges. The growing evidence of these crises and their impacts (from droughts to floods, pandemics, wars and the widening gap between the world's rich and poor) is prompting the gradual development of new paradigm shifts. Among these shifts, we can identify circular economy (CE) and circularity inspired ambitions as a systemic shift aimed at rethinking the world as we know it. The circular economy seeks to reduce waste and increase resource efficiency by creating closed-loop systems in which materials and products are recycled, mended, and reused. Implementing Circular Economy (CE) strategies has recently become one of the essential strategies for sustainable development and corporate social responsibility (De Jesus Pacheco et al. 2024).

Circular economy (CE) business models have emerged widely, introducing new ways of designing, producing, and consuming products that enable shifting from the so-called linear economy (LE) to an economy that can reintroduce used products/components/materials at end-of-life (EoL) back into the value chains (Panza and Peron 2025). In line with sustainability objectives and corporate social responsibility pledges, the shift can lessen its negative effects on the environment, maximize the use of resources, and improve economic resilience. The increasing use of CE business models emphasizes how important it is to have reverse logistics, waste

S. Garg · H. Jain (✉) · A. Shrivastava
Amity Institute of Environmental Sciences, Amity University, Noida, Uttar Pradesh, India
e-mail: hjain@amity.edu

H. Jain et al. (eds.), *The Circular Fallacy: Closing the Loop in a Consumerist World*, Circular Economy and Sustainability, https://doi.org/10.1007/978-3-032-22687-7_7

management infrastructure, and legislative incentives in order to move away from the linear economic paradigm effectively.

From the strategic point of view, the benefits of the CE approach arise from the reduced risk of supply disruption and price volatility as well as from the huge potential for innovation related to new technologies (needed to increase resource productivity, material substitution, waste management, and recycling), improvements of the forward and reverse cycles (optimization of the supply chain and logistics), and new and creative business models (Megevand et al. 2022). Eco-design for long-lasting and recyclable products, Extended Producer Responsibility (EPR) to make sure manufacturers manage end-of-life recycling, and reverse logistics for effective material recovery are important tactics for the adoption of the Circular Economy (CE). While blockchain, AI, and automation improve waste monitoring and recycling, waste-to-resource innovation encourages the use of secondary raw materials. CE implementation is also supported by financial and legislative incentives including landfill regulations and tax breaks. These tactics lessen reliance on resources, cut down on waste, and stimulate innovation to boost the economy.

7.1.2 *Importance of Waste Management Infrastructure*

Waste generation is the result of resource use and historically has increased with population, However, in the past century as the world's population has grown, it has also become more urban and affluent, with resource use and the resulting waste generation linked to economic activity (Kirkman and Voulvoulis 2016). The shift towards a circular economy has become imperative due to the increasing global waste crisis and the limitations of the traditional linear economy model, Sustainable waste management lies at the core of the circular economy, emphasizing waste reduction, resource efficiency, and the continual reuse of materials (Kotyal 2023).

In order to facilitate material recovery, minimize contamination, and reduce landfill usage, a robust waste management infrastructure is essential to a circular economy (CE). Effective systems for gathering, classifying, and recycling materials reduce dependency on virgin resources by reintegrating them into production. Sustainable development is fuelled by improved waste efficiency brought about by advanced technologies and legislative support. The Indore GOBARdhan Bio-CNG Plant is a great illustration of this in India; it converts organic waste into compost and Bio-CNG, minimizing landfill waste and producing renewable energy.

7.2 Waste Management Infrastructure

7.2.1 Key Components of Effective Systems

The advancement of a circular economy and sustainability depend heavily on efficient waste management systems. Important parts of these systems consist of.

7.2.1.1 Waste Collection Networks

Door-to-Door Collection: Sorting waste fractions at home for a door-to-door collection system proves to positively affect the environmental impacts of waste management strategies both by reducing the amounts of the waste landfilled and by originating new circular economies (Laurieri et al. 2020; Kumar et al. 2025).

Community Waste Bins: In metropolitan settings, strategically positioned trash cans improve resident convenience and decrease littering. Proper placement guarantees accessibility and promotes public involvement in garbage management.

Smart Waste Sensors: An efficient garbage collection model using IoT, which will relate to the inclusion of large-capacity garbage trucks as mobile warehouses, as well as to develop a model that allows to optimize the location of garbage bins for the maximum comfort of citizens (Sosunova and Porras 2022). Bins with sensors enabled by the Internet of Things (IoT) may identify fill levels and alert collection providers when they are full, improving collection routes and cutting expenses.

7.2.1.2 Sorting and Segregation Facilities

Manual vs. Automated Sorting: Nowadays, both manual and automated sorting methods, especially optical sorting, are used worldwide for sorting mixed waste (Procházka et al. 2024; Bhasin et al. 2024). Automated sorting, on the other hand, uses cutting-edge technologies like artificial intelligence (AI) and robotics to quickly and precisely identify goods. Automated technology is key for sorting plastic objects in medium- to high-volume plants (Procházka et al. 2024).

Material Recovery Facilities (MRFs): Nowadays, central sorting and separation systems, in the form of Materials Recovery Facilities (MRFs), Packaging Sorting Plants and MBT, play a pivotal role in our waste management and material recovery systems (Cimpan et al. 2015). In developed countries, MSW is usually sent to materials recovery facilities (MRFs), which use mechanical and manual sorting units to extract valuable components (Testa 2015). MRFs are essential to waste management plans that support the circular economy and lessen the usage of landfills. They promote sustainable production methods, lessen environmental pollution, and conserve natural resources by effectively recovering recyclable materials (Fig. 7.1).

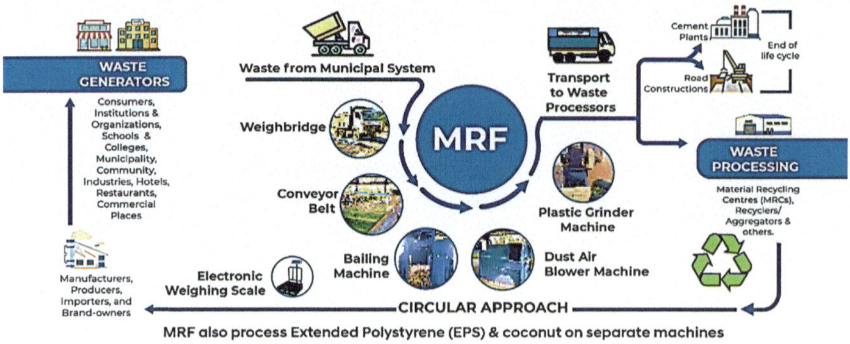

Fig. 7.1 Systems approach to waste management at material recovery facilities (MRF)

Reusing and converting waste into energy Because they can recover valuable materials and produce energy from non-recyclable garbage, facilities are essential to sustainable waste management. Reliance on virgin resources is decreased via recycling facilities such as Material Recovery Center, which use mechanical and chemical recycling to recycle waste materials such as plastics, metals, paper, and electronics.

7.2.2 *Technological Innovations in Waste Collection*

Waste management technologies have evolved in recent years, motivated by the desire to minimize the environmental impact of waste and enhance the effectiveness of waste management procedures, there is a wide range of approaches, ranging from conventional techniques like composting or landfilling, to modern, more advanced solutions using IoT technologies (Czekała et al. 2023). Machine learning (ML), artificial intelligence (AI), and image recognition can be used to automate the segregation of waste, reducing the risk of exposing labour workers to harmful waste, Radio Frequency Identification (RFID) and wireless communications enable the traceability in materials to better understand the opportunities in circular economy (Cheah et al. 2022). These technologies lower expenses, increase overall efficiency, and simplify waste management procedures. By reducing landfill trash, raising recycling rates, and guaranteeing safer handling of hazardous materials, they also advance sustainability. They also facilitate improved resource recovery, data-driven decision-making, and improved monitoring, all of which aid in the shift to a circular economy.

7.2.3 *Waste Segregation and Disposal Challenges*

India faces major environmental challenges associated with waste generation and inadequate waste collection, transport, treatment and disposal (Kumar et al. 2017). Degradation of the environment, health risks, and the loss of precious resources that could be recovered and utilized again are all results of this inefficiency. Waste management is a problem due to the ineffective management of waste and lack of knowledge of waste management (Anshuman and Kumar 2019). The recycling industry in India has experienced significant growth in recent years, However, challenges such as inadequate collection systems, poor quality segregation, and lack of awareness among the public still persist (Sk et al. 2023). The inadequate segregation of garbage at the source is a major problem in India's waste management system. The issue of unmanaged and mismanaged plastic waste is especially acute in developing countries, where the infrastructure for collection, reuse and recycling is often insufficient or lacking (Browning et al. 2021). Although the informal trash sector is essential to recycling, its effectiveness is decreased by a lack of adequate safety precautions, support, and access to contemporary sorting technologies. Furthermore, many homes and companies do not adopt source segregation, which results in polluted recyclable materials and an increased landfill burden (Agarwal et al. 2025). This makes public awareness and behavioural change a significant task. In order to address global waste concerns and guarantee long-term environmental and economic advantages, a deliberate move towards integrated waste management and sustainable disposal techniques will be essential (Bhardwaj et al. 2025; Agarwal and Jain 2024; Jain 2025).

7.3 Revamping Logistics and Supply Chain

7.3.1 *Transition to Circular Supply Chains*

The linear supply chain is predicated on the time-honoured take, make, and dispose paradigm. However, the 9Rs—Refuse, Rethink, Reduce, Reuse, Repair, Refurnish, Remanufacture, Repurpose, Recycle, and Recover—are what make the circular supply chain concept possible. Companies and governments are faced with challenges in redesigning existing linear supply chains (SCs) into circular supply chains (CSCs) (Roy et al. 2022) (Table 7.1). However, the most promising approach to achieving environmental and economic sustainability is transition. Flexibility in supply chains not only ensures quick deliveries but also helps in achieving sustainability (Dwivedi et al. 2021). Government directives and stakeholders' pressures are forcing organisations to redesign their global supply chain network to fit sustainability and circularity aims (Echefaj et al. 2024).

The circular supply chain's ultimate objective is to securely return organic materials to nature for the biological loop and to preserve the maximum value at the

Table 7.1 Comparison of linear and circular economic models in waste management

Components	Linear model	Circular model
Flow of material	One-way	Closed-loop
Use of resource	High	Optimized
Waste production	Significant	Minimal
Environmental impact	Elevated carbon emissions	Decreased carbon footprint
Economic model	Profit-driven	Value retention and cost-saving
Product lifecycle	Short and disposable	Extended and regenerative
Dependency on raw materials	High	Lower due to recycling and reuse
Role of consumers	Passive (end-user)	Active (collaborative in reuse and recycle)

product, component, and material levels for technical cycles. A circular supply chain has obvious advantages. Businesses can lessen their environmental effect and become more sustainable by cutting waste and making the best use of their resources.

7.3.2 *Role of Reverse Logistics and Material Recovery*

Reverse logistics is the continuous logistic process through which shipped products move from the consumer back to the producer or recycling enterprises for possible reuse, recycling, remanufacturing, or disposal (Burnard et al. 2015). Reverse logistics operations are one of the more interesting and significant trends in supply chain management (Elmas and Erdoğmuş 2011). It is essential to creating a circular economy because it recovers value from things that have reached the end of their useful lives (Fig. 7.2). The main operations of a reverse logistics system consist of the collection of EOL products from customers and end-users, the appropriate inspection, sorting, disassembling and/or pre-processing, the distribution of different products, parts and components to respective facilities for proper treatment, and the planning and scheduling of facility operations and transportation (Sun et al. 2022). A major challenge will be to integrate each stakeholder into a single process, considering that the current system is very decentralized (Santos and Ogunseitan 2022). Material Recovery Facilities (MRFs) are specialized facilities that sort, clean, and process collected garbage for recycling, hence they play a significant role in this for making the process more efficient and centralized. Additionally, the recyclers purchase the recyclable and sorted goods from these establishments and use them as needed. In general, the experts perceive that reverse logistics practices significantly reduce the sourcing cost of materials compared with new materials. In general, the experts perceive that reverse logistics practices significantly reduce the sourcing cost of materials compared with new materials (Pushpamali et al. 2020).

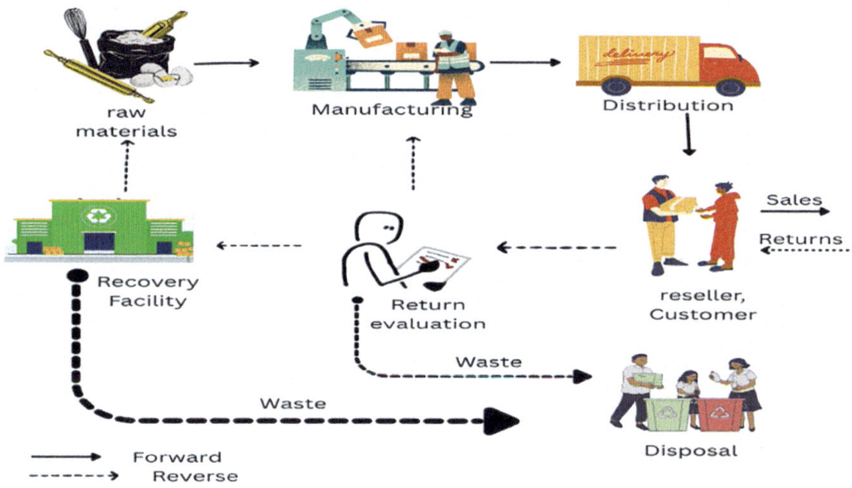

Fig. 7.2 A generic form of forward/reverse logistics

7.3.3 Sustainable Supply Chain Technologies

A sustainable supply chain incorporates social, economic, and environmental factors into every step of the supply chain, from locating raw materials to shipping completed goods and handling end-of-life disposal. For example, apparel distributors such as Nike, Disney, Levi Strauss, Benetton, Adidas or C&A have been blamed in recent years for problems occurring during the production of their clothing. Inhumane working conditions or contaminations of the (local) environment were frequently mentioned as problems (Seuring and Müller 2008). These triggers have heightened interest in sustainable supply chain management and technology that enhances sustainability.

Blockchain technology (BT) is considered one of the leading disruptive technologies that radically transform businesses and supply chains as it can improve supply chain visibility, integration, and sustainability (Khan et al. 2022). Blockchain gives businesses real-time supply chain information, enabling them to monitor, validate, and streamline procedures from the procurement of raw materials to the delivery of completed goods. This capacity is essential to the advancement of sustainability since it enables businesses to guarantee moral behavior, lessen their impact on the environment, and increase resource efficiency.

The connections among a sustainable CE, IoT capabilities, and supply chain performance are crucial elements for the future of sustainable and efficient contemporary business practices (Jum'a et al. 2024). he main goal of the IIoT-based green supply chain is to reduce the product's cost and environmental pollution rate (Qian et al. 2023). By providing real-time data and connection across the supply chain, covering both supply and demand components, the Internet of Things played a critical role.

By leveraging AI-driven technologies, businesses can significantly improve the efficiency of resource use, streamline waste management processes, and support the creation of sustainable production systems (Bashynska and Prokopenko 2024). AI can help businesses reduce material waste, manage energy consumption, evaluate big datasets in real-time, and predict maintenance needs to avoid equipment breakdowns. In supply chains, artificial intelligence (AI) can enhance demand forecasting, guaranteeing that production accurately reflects consumption while lowering overproduction and surplus inventory.

7.4 Legislative Incentives

7.4.1 Role of Legislation in Circular Economy

The adoption and application of the circular economy depend heavily on legal frameworks. These frameworks provide the structure, guidelines, and regulations needed to support the transition from a linear "take, make, dispose" model to a circular model that emphasizes resource efficiency, waste reduction, and sustainable production and consumption (Dubey 2023). Governments can provide guidance for businesses and consumers by enacting legal frameworks that clearly define goals for recycling, resource management, and waste reduction. Another critical aspect necessitating the production of a variety of environmental legislation is the strengthening of global trade forces companies for them to comply with environmental regulations and subjected to continuous modifications (Kyriakopoulos 2021).

7.4.2 Key Legislative Tools and Financial Incentives

To achieve economic, social, and environmental objectives, legislative tools are essential instruments designed to regulate, monitor, and enforce compliance in a number of domains. The tools are-

a. *Extended Producer Responsibility*

   Extended producer responsibility (EPR) is a policy tool that holds producers financially responsible for the post-use collection, recycling, and disposal of their products (Gui et al. 2015). Therefore, careful product design and material selection are anticipated to facilitate compliance and implementation of EPR systems to achieve a circular economy (Tumu et al. 2023). By supporting sustainable production methods, this strategy not only lessens its negative effects on the environment but also fosters innovation and resource efficiency.

b. *Product Design and Eco-Design Regulations*

Environmental design (eco-design) has been identified in both academic and policy circles as a point of intervention in the product lifecycle to promote environmental performance (Deutz et al. 2012). Laws can support or require eco-design concepts that encourage the use of recyclable or renewable materials, longer product lifecycles, and ease of disassembly.

c. *Intellectual Property Rights (IPR)*

   IPR can impact the circular economy by influencing access to innovative solutions, reuse, and repair; Legal frameworks should strike a balance between protecting intellectual property (Dubey 2023).

d. *Eco-Labelling*

   Eco-labelling was identified in Agenda 21 as a way of encouraging consumers to alter their consumption patterns and to make wiser use of resources and energy in the drive for sustainable development into the next century (Erskine and Collins 1997). It assists customers in identifying goods with a lower lifetime environmental impact.

Financial incentives are effective tools that stimulate investment, innovation, and compliance in a variety of industries, thereby advancing sustainable development and good behavioural change.

a. *Taxation and Incentives*

   Reduced taxes on recycled materials, tax credits for environmentally friendly practices, and subsidies for circular innovations can stimulate businesses to adopt circular economy models (Dubey 2023).

b. *Deposit-Refund System*

   A deposit-refund system combines a tax on product consumption with a rebate when the product or its packaging is returned for recycling (Walls 2011). By encouraging customers to take part in recycling initiatives, this approach lowers litter and encourages the recovery of valuable materials.

7.4.3 Global Success Stories in Circular Economy Legislation

The increasing global consumption of resources, in combination with a predominantly linear economy, is generating a debt-based unsustainable growth (Cantú et al. 2021). The implementation of strong circular economy laws in a number of nations has been spurred by this trend towards sustainable development, proving the efficacy of policy-driven initiatives in tackling waste reduction and resource efficiency. At present, organisations such as Price Waterhouse Coopers and KPMG are providing consultancy services on public private ventures in Europe and indeed world-wide (Ryan 2004).

With its Circular Economy Action Plan (CEAP), introduced as part of the European Green Deal (2020), the European Union (EU) is a global leader in circular economy policy. This framework places a strong emphasis on resource efficiency, waste reduction, and sustainable product design in important industries including textiles, electronics, and plastics. Europe's overall landfill rate has decreased from 23 to 17% between 2010 and 2022, However, the total amount of waste generated has continued to increase, Landfilled waste quantities in 2022 were 21% lower in the same period, equivalent to 306 kg of waste per year by each EU citizen in 2022 (European Environment Agency 2024). Landfill levies and prohibitions, as well as incentives for recycling and recycling infrastructure, are examples of policies that have been shown to be effective in lowering landfilling.

In line with the EU Landfill Directive, Member States must reduce the amount of municipal waste sent to landfill to 10% or less of the total amount of municipal waste generated by 2035 (European Environment Agency 2024).

7.5 Overcoming Barriers

7.5.1 Economic, Technological, and Regulatory Barriers

Circular economy has emerged as a way to achieve sustainability, Although interest in the subject is growing rapidly, barriers to its implementation are still in place (Galvão et al. 2018). There are a large number of barriers and they are most often connected to each other and clearly illustrate the complexity of CE and what is required for a transition: it is both multi-dimensional and multi-domain (Ritzén and Sandström 2017). These barriers can be broadly classified into economic, technological, and regulatory dimensions, each playing a critical role in shaping the transition.

High initial investment costs present a significant economic barrier, deterring businesses from transitioning to circular economy practices. Financial incentives and subsidies can mitigate this issue by offsetting upfront expenses, as highlighted in the International Chamber of Commerce Report.

Four major technological barriers categories have been identified: product-design, quality of recycled goods, know-how of technical solutions and IT systems for monitoring progress (Grafström and Aasma 2021). Research and development (R&D) expenditure is necessary to promote sophisticated recycling techniques.

Regulatory/Institutional/Market barriers are classified as obstacles to a circular economy adoption due to non-existent or ill-functioning of markets (Grafström and Aasma 2021). The implementation of circular economy methods can be accelerated by harmonizing national and international policies (European Circular Economy Stakeholder Platform). Additionally, classifying by-products as waste might hinder the efficient use of resources.

7.5.2 Public Awareness and Behavioural Change Strategies

The circular economy heavily relies on consumer activity. Prior research on the impact of gender on attitudes and awareness of CE techniques produced conflicting findings. For example, a survey of users of electronic and electrical appliances in the EU found that, compared with the rest of the population, respondents that are females, older than 50 years, and with medium and high education levels are more willing to purchase appliances with more environmentally friendly end-of-life scenarios (Almulhim and Abubakar 2021). We firmly believe strategic public education will be more advantageous in influencing consumers' perception of the sustainable alternative by providing facts, clarifying its performance and benefits, and justifying its green premium (Tan et al. 2022). Offering incentives like discounts for environmentally friendly purchases, maintaining transparency by clearly labelling the impact of products, involving communities through workshops and events, utilizing influencer partnerships for wider outreach, and putting in place feedback mechanisms to demonstrate the observable effects of customer decisions are all important tactics for promoting behavioural change. Stressing long-term savings and offering reliable proof of environmental benefits are key to overcoming obstacles including cost, convenience, and scepticism. All things considered, promoting a circular economy and cultivating sustainability for future generations requires behavioural change and public awareness.

7.6 Best Practices and Case Studies

7.6.1 Circular Economy Practices in Europe and Asia

The roots of the topic are in Europe; the concept stems from the 1976 report to the European Commission by Stahel and Reday (1976), with another important contribution coming from the two British environmental economists Pearce and Turner (1990) (Bassi and Dias 2019). Within the framework of the European Union (EU), all EU countries have launched a series of policies to support circularity (De Pascale et al. 2023). Countries like Japan and China emphasize industrial symbiosis and waste-to-resource systems. Successful enforcement of a CE can be seen as a way for China to tackle its urgent problem of environmental degradation and source scarcity (Su et al. 2012). The practices listed in the table underscore how historical roots and pressing environmental needs have shaped circular economy implementation in Europe and Asia. Europe's policy-driven approach and Asia's focus on resource recovery illustrate diverse pathways to sustainability (Table 7.2).

Table 7.2 Regional approaches to circular economy and waste management

Region	Practice	Description
Europe	EU circular economy action plan	Comprehensive strategy for resource efficiency
Europe	Industrial symbiosis networks	Industries sharing resources and waste streams
Asia	Waste-to-resource systems	Converting industrial waste into reusable materials
Asia	E-waste recycling programs	Large-scale electronic waste recycling initiatives

7.6.2 Corporate and Government Case Studies

Due to the environment's growing complexity and volatility, businesses must create competitive management models that strive to meet stakeholders' and society's long-term expectations while simultaneously generating profits. IKEA's initiatives to go all-green and to assist its suppliers in decreasing their negative impacts on the environment have had positive results on the retailer's overall operations (Bouhia 2022). IKEA is a prime example of how businesses may use the circular economy to balance sustainability and profitability. Among its measures are investments in renewable materials, a used furniture buy-back and resale business, and efforts to lower its carbon footprint throughout supply chains. Additionally, IKEA encourages its suppliers to use environmentally friendly production techniques, which reduces their influence on the environment and increases resource efficiency.

ESG implications showed that companies like Patagonia went through profit shrinking to safeguard societal and environmental welfare, with supply chain implications that involved factory workers and partners (Dezi et al. 2022). Local municipalities are seen to have the capacity to encourage sustainable development of the built environment (Bolger and Doyon 2019). Cities like Copenhagen and Amsterdam have incorporated circular techniques into their urban planning, placing a strong emphasis on sustainable building, resource efficiency, and waste reduction. These programs show how governance at all levels contributes to the advancement of circular economy objectives and are in line with larger government attempts to develop sustainable urban environments.

7.7 Future Directions

The circular economy's future will be influenced by developments in international collaboration, politics, and technology.

7.7.1 *Emerging Technologies Shaping the Future*

Biomimicry: Biomimicry is the study of and inspiration from biological components and natural processes to solve problems, understanding how they survive, function and evolve in a self-sustaining way (Beermann and Austin 2021).

Urban mining: the process of extracting precious resources from abandoned infrastructure and electronic waste.

Blockchain and AI in Waste Management: Technologies that improve supply chain transparency and optimize resource allocation.

Decentralized Recycling: Small-scale, localized recycling units that reduce transportation costs and emissions.

7.7.2 *The Role of Businesses and Consumers*

Circular Business Models: Shifting from ownership to models based on services, like product leasing or rental.

Consumer Awareness Campaigns: Initiatives aimed at educating consumers to adopt more environmentally friendly buying practices.

Product-as-a-Service (PaaS): a method of business innovation for traditional product companies to make their offerings more attractive, move up the value chain and find more customers (Kesavapanikkar et al. 2022).

It will need sustained investment in research, policy alignment, and consumer participation to create a genuinely global circular economy. To solve environmental issues and promote sustainable economic growth, future-oriented circular economy strategies will be essential.

7.8 Conclusion

The CE, which aims for resource efficiency, waste reduction, and environmental resilience, represents a significant departure from the conventional linear paradigm. Even if there are still technological, economic, and regulatory obstacles to overcome, advancement is being fuelled by solutions including financial incentives, policy alignment, and technological innovation. A significant part is played by behavioural change and public awareness, with tactics including eco-labelling, rewards, and educational initiatives promoting sustainable consumer choices. The success of CE-driven policies and business models is illustrated by international case studies from Europe, Asia, and companies such as Patagonia and IKEA. Cities such as

Copenhagen and Amsterdam serve as examples of how circular ideas for sustainability are included into urban development. Innovations like biomimicry, urban mining, blockchain, artificial intelligence, and decentralized recycling, which maximize resource recovery and supply chain efficiency, will influence CE in the future. Governments, corporations, and consumers must work together to make the transition to CE successful. Societies may transition to a regenerative, waste-free economy by encouraging innovation, fortifying regulations, and encouraging active engagement, guaranteeing long-term sustainability and financial success for coming generations.

References

Agarwal S, Jain H (2024) Investigating the potential of floral waste as a vermicompost and dual-functional biosorbent for sustainable environmental management. Water Air Soil Pollut 235:322. https://doi.org/10.1007/s11270-024-07144-y

Agarwal S, Jain H, Dhupper R, Mathur A (2025) Optimizing methyl orange dye removal from aqueous solutions using white chrysanthemum floral waste-derived bioadsorbent: a study of kinetics, thermodynamics, isotherms, and RSM optimization. Int J Environ Res 19(3):104

Almulhim AI, Abubakar IR (2021) Understanding public environmental awareness and attitudes toward circular economy transition in Saudi Arabia. Sustainability 13(18):10157. https://doi.org/10.3390/su131810157

Anshuman P, Kumar RP (2019) Status and challenges of waste segregation in residential communities. DOAJ (DOAJ: Directory of Open Access Journals). https://doi.org/10.22034/ijhcum.2019.04.06

Bashynska I, Prokopenko O (2024) Leveraging artificial intelligence for circular economy: transforming resource management, supply chains, and manufacturing practices. Scientific J Bielsko-Biala School Financ Law. Bielsko-Biała PL 28(2):85–91. https://doi.org/10.19192/wsfip.sj2.2024.13

Bassi F, Dias JG (2019) The use of circular economy practices in SMEs across the EU. Resour Conserv Recycl 146:523–533. https://doi.org/10.1016/j.resconrec.2019.03.019

Bhasin A, Ghosal S, Rai ID, Raina P, Shrivastava A, Hore U (2024) Wild ungulates employ spatio-temporal niche partitioning to coexist with livestock in resource-limited Trans-Himalayan rangelands. India. J Nat Conserv 82:126743

Beermann K, Austin MC (2021) An Inspection of the life cycle of sustainable construction projects: towards a biomimicry-based road map integrating circular economy. Biomimetics 6(4):67. https://doi.org/10.3390/biomimetics6040067

Bhardwaj LK et al (2025) Resilience in the digital age: emerging technologies for climate adaptation. In: Choudhury M, Dixit G, Majumdar S (eds) Sustainable synergy: harnessing ecosystems for climate resilience. Climate Change Management. Springer, Cham. https://doi.org/10.1007/978-3-031-77957-2_11

Bolger K, Doyon A (2019) Circular cities: exploring local government strategies to facilitate a circular economy. Eur Plan Stud 27(11):2184–2205. https://doi.org/10.1080/09654313.2019.1642854

Bouhia G (2022) Applying the circular economy concept to a sustainable business model for large retailers: the case of IKEA. Int J Compet 2(3):212. https://doi.org/10.1504/ijc.2022.127400

Browning S, Beymer-Farris B, Seay JR (2021) Addressing the challenges associated with plastic waste disposal and management in developing countries. Curr Opin Chem Eng 32:100682. https://doi.org/10.1016/j.coche.2021.100682

Burnard M, Tavzes Č, Tošić A, Brodnik A, Kutnar A (2015) The role of reverse logistics in recycling of wood products. In: Environmental footprints and eco-design of products and processes, pp 1–30. https://doi.org/10.1007/978-981-287-643-0_1
Cantú A, Aguiñaga E, Scheel C (2021) Learning from failure and success: the challenges for circular economy implementation in SMEs in an emerging economy. Sustainability 13(3):1529. https://doi.org/10.3390/su13031529
Cheah CG, Chia WY, Lai SF, Chew KW, Chia SR, Show PL (2022) Innovation designs of industry 4.0 based solid waste management: machinery and digital circular economy. Environ Res 213:113619. https://doi.org/10.1016/j.envres.2022.113619
Cimpan C, Maul A, Jansen M, Pretz T, Wenzel H (2015) Central sorting and recovery of MSW recyclable materials: a review of technological state-of-the-art, cases, practice and implications for materials recycling. J Environ Manage 156:181–199. https://doi.org/10.1016/j.jenvman.2015.03.025
Czekała W, Drozdowski J, Łabiak P (2023) Modern technologies for waste management: a review. Appl Sci 13(15):8847. https://doi.org/10.3390/app13158847
De Jesus Pacheco DA, Rampasso IS, Michels GS, Ali SM, Hunt JD (2024) From linear to circular economy: the role of BS 8001:2017 for green transition in small business in developing economies. J Clean Prod 439:140787. https://doi.org/10.1016/j.jclepro.2024.140787
De Pascale A, Di Vita G, Giannetto C, Ioppolo G, Lanfranchi M, Limosani M, Szopik-Depczyńska K (2023) The circular economy implementation at the European Union level. Past, present and future. J Cleaner Prod 423:138658. https://doi.org/10.1016/j.jclepro.2023.138658
Deutz P, McGuire M, Neighbour G (2012) Eco-design practice in the context of a structured design process: an interdisciplinary empirical study of UK manufacturers. J Clean Prod 39:117–128. https://doi.org/10.1016/j.jclepro.2012.08.035
Dezi L, Hysa X, Calabrese M, Mercuri F (2022) Open total quality management in the circular economy age: a social enterprise perspective through the case of patagonia. Total Qual Manag Bus Excell 1–15. https://doi.org/10.1080/14783363.2022.2051698
Diversion of waste from landfill in Europe (2024) European environment agency's home page
Dubey D (2023) Legal challenges and opportunities in implementing circular economy practices 3:1–7
Dwivedi A, Agrawal D, Jha A, Gastaldi M, Paul SK, D'Adamo I (2021) Addressing the challenges to sustainable initiatives in value chain flexibility: implications for sustainable development goals. Glob J Flex Syst Manag 22(S2):179–197. https://doi.org/10.1007/s40171-021-00288-4
Echefaj K, Charkaoui A, Cherrafi A, Tiwari S, Sharma P, Jabbour CJC (2024) From linear to circular sustainable supply chain network optimisation: towards a conceptual framework. Prod Planning Control 1–25. https://doi.org/10.1080/09537287.2024.2302479
Elmas G, Erdoğmuş F (2011) The importance of reverse logistics. Int J Bus Manag Studies 3(1):161–171
Erskine CC, Collins L (1997) Eco-labelling: success or failure? Environmentalist 17(2):125–133. https://doi.org/10.1023/a:1018552000651
Galvão GDA, De Nadae J, Clemente DH, Chinen G, De Carvalho MM (2018) Circular economy: overview of barriers. Procedia CIRP 73:79–85. https://doi.org/10.1016/j.procir.2018.04.011
Grafström J, Aasma S (2021) Breaking circular economy barriers. J Clean Prod 292:126002. https://doi.org/10.1016/j.jclepro.2021.126002
Gui L, Atasu A, Ergun Ö, Toktay LB (2015) Efficient implementation of collective extended producer responsibility legislation. Manage Sci 62(4):1098–1123. https://doi.org/10.1287/mnsc.2015.2163
Introduction to Circular Economy (2022) In World scientific eBooks (pp. 25–46). https://doi.org/10.1142/9789811241949_0002
Jain H (2025) Synergizing carbon reduction technologies and nature-based solutions for a climate-neutral future. Mitig Adapt Strateg Glob Change 30:42. https://doi.org/10.1007/s11027-025-10237-6

Jum'a L, Ikram M, Jabbour CJC (2024) Towards circular economy: a IoT enabled framework for circular supply chain integration. Comput Indus Eng 192:110194. https://doi.org/10.1016/j.cie.2024.110194

Kesavapanikkar P, Amit R, Ramu P (2022) Product as a service (PaaS) for traditional product companies: an automotive lease practice evaluation. J Indian Bus Res 15(1):40–54. https://doi.org/10.1108/jibr-04-2022-0107

Khan SA, Mubarik MS, Kusi-Sarpong S, Gupta H, Zaman SI, Mubarik M (2022) Blockchain technologies as enablers of supply chain mapping for sustainable supply chains. Bus Strateg Environ 31(8):3742–3756. https://doi.org/10.1002/bse.3029

Kirkman R, Voulvoulis N (2016) The role of public communication in decision making for waste management infrastructure. J Environ Manage 203:640–647. https://doi.org/10.1016/j.jenvman.2016.06.002

Kotyal K (2023) Sustainable waste management in the circular economy: challenges and opportunities. Environ Reports 5(2):1–5. https://doi.org/10.51470/er.2023.5.2.01

Kumar S, Smith SR, Fowler G, Velis C, Kumar SJ, Arya S, Rena N, Kumar R, Cheeseman C (2017) Challenges and opportunities associated with waste management in India. Royal Soc Open Sci 4(3):160764. https://doi.org/10.1098/rsos.160764

Kumar M, Kumar R, Sharma S, Mukherjee S, Bolan S, Shrivastava A, … Bolan N (2025) Microplastics pollution modulating soil biological health–a review. Soil Use Manag 41(1):e70009

Kyriakopoulos G (2021) Environmental legislation in European and international contexts: legal practices and social planning toward the circular economy. Laws 10(1):3. https://doi.org/10.3390/laws10010003

Laurieri N, Lucchese A, Marino A, Digiesi S (2020) A door-to-door waste collection system case study: a survey on its sustainability and effectiveness. Sustainability 12(14):5520. https://doi.org/10.3390/su12145520

Megevand B, Cao W, Di Maio F, Rem P (2022) Circularity in practice: Review of main current approaches and strategic propositions for an efficient circular economy of materials. Sustainability 14(2):962. https://doi.org/10.3390/su14020962

Panza L, Peron M (2025) The role of carbon tax in the transition from a linear economy to a circular economy business model in manufacturing. J Clean Prod 144873. https://doi.org/10.1016/j.jclepro.2025.144873

Procházka R, Valíček J, Harničárová M, Kušnerová M, Tozan H, Borzan C, Kadnár M, Palková Z, Gálik R, Slamová K (2024) Collection of plastic packaging of various types: sorting of fractions of plastic waste using both automated and manual modes. IEEE Access 12:44244–44261. https://doi.org/10.1109/access.2024.3376230

Pushpamali N, Agdas D, Rose TM, Yigitcanlar T (2020) Stakeholder perception of reverse logistics practices on supply chain performance. Bus Strateg Environ 30(1):60–70. https://doi.org/10.1002/bse.2609

Qian C, Gao Y, Chen L (2023) Green supply chain circular economy evaluation system based on industrial internet of things and blockchain technology under ESG concept. Processes 11(7):1999. https://doi.org/10.3390/pr11071999

Ritzén S, Sandström GÖ (2017) Barriers to the circular economy—integration of perspectives and domains. Procedia CIRP 64:7–12. https://doi.org/10.1016/j.procir.2017.03.005

Roy T, Garza-Reyes JA, Kumar V, Kumar A, Agrawal R (2022) Redesigning traditional linear supply chains into circular supply chains–a study into its challenges. Sustain Prod Consum 31:113–126. https://doi.org/10.1016/j.spc.2022.02.004

Ryan B (2004) Public private partnerships and sustainability principles guiding legislation and current practice

Santos SM, Ogunseitan OA (2022) E-waste management in Brazil: challenges and opportunities of a reverse logistics model. Environ Technol Innov 28:102671. https://doi.org/10.1016/j.eti.2022.102671

Seuring S, Müller M (2008) From a literature review to a conceptual framework for sustainable supply chain management. J Clean Prod 16(15):1699–1710. https://doi.org/10.1016/j.jclepro.2008.04.020

Sk M, Qamar S, Sethy T (2023) Solid waste management in Indian perspectives: a comprehensive review 12:35–45

Sosunova I, Porras J (2022) IoT-enabled smart waste management systems for smart cities: a systematic review. IEEE Access 10:73326–73363. https://doi.org/10.1109/access.2022.3188308

Su B, Heshmati A, Geng Y, Yu X (2012) A review of the circular economy in China: moving from rhetoric to implementation. J Clean Prod 42:215–227. https://doi.org/10.1016/j.jclepro.2012.11.020

Sun X, Yu H, Solvang WD (2022) Towards the smart and sustainable transformation of Reverse Logistics 4.0: a conceptualization and research agenda. Environ Sci Pollut Res 29(46):69275–69293. https://doi.org/10.1007/s11356-022-22473-3

Tan J, Tan FJ, Ramakrishna S (2022) Transitioning to a circular economy: a systematic review of its drivers and barriers. Sustainability 14(3):1757. https://doi.org/10.3390/su14031757

Testa M (2015) Modeling and design of material recovery facilities: genetic algorithm approach. Springer

Tumu K, Vorst K, Curtzwiler G (2023) Global plastic waste recycling and extended producer responsibility laws. J Environ Manage 348:119242. https://doi.org/10.1016/j.jenvman.2023.119242

Walls M (2011) Deposit-refund systems in practice and theory. SSRN Electron J.https://doi.org/10.2139/ssrn.1980142

Chapter 8
Reuse of Demolition Waste Recycled On-Site in a Residential Development Project in Perth, Australia

Salman Shooshtarian

8.1 Introduction

As cities expand rapidly worldwide, construction activities have grown at an unprecedented pace. Consequently, the construction sector now accounts for roughly 35% of all waste directed to landfills globally (Zheng et al. 2017; Solís-Guzmán et al. 2009). In the UK alone, 49 million tonnes of construction and demolition (C&D) waste were produced in 2014, with over half ending up in landfills (Menegaki and Damigos 2018). China generates approximately 2.36 billion tonnes of C&D waste annually (Zheng et al. 2017), with low recovery rates (Deng 2022), while in the United States, 516 million tonnes were recorded in 2017 (US EPA, 2019). Despite the existing disparity in C&D waste recovery rates among the In Organisation for Economic Co-operation and Development (OECD) countries, multiple forecasts suggest that significant shifts are emerging, poised to transform the landscape of C&D waste management. Recent studies estimate that the global market for C&D waste recovery could grow substantially, reaching between 149 and 300 billion US dollars by 2027–2028 (Allied Analytics LLP 2023; IMARC Group 2023; Report Ocean 2021).

Using products with recycled content (PwRC) is increasingly recognised as a sustainable approach to managing construction resources. When effectively planned, this strategy can significantly reduce waste sent to landfills, retain the value of materials within the economic cycle for longer, and lessen reliance on extracting virgin resources. Concepts such as the circular economy (CE), the waste hierarchy model, and net-zero carbon emission targets all emphasise and support the adoption of PwRC in construction projects. In addition, there is strong support from local, national, and international waste policies to reuse PwRC in construction projects (Li et al. 2020).

S. Shooshtarian (✉)
School of Property, Construction and Project Management, RMIT University, Melbourne, Australia
e-mail: Salman.shooshtarian@rmit.edu.au

H. Jain et al. (eds.), *The Circular Fallacy: Closing the Loop in a Consumerist World*, Circular Economy and Sustainability, https://doi.org/10.1007/978-3-032-22687-7_8

The adoption of PwRC is considered a key enabler for governments striving to achieve waste management targets (Active Sustainability 2020). This approach is intended to tackle a broad range of environmental, social, and economic challenges associated with waste generation and resource use.

Building on this understanding, several studies have investigated the factors influencing the use of PwRC in construction projects. Findings indicate that adopting PwRC is far from straightforward, as it is shaped by a range of internal and external factors that must be fully understood (Park and Tucker 2017; Shooshtarian et al. 2020). Currently, PwRC use in the built environment sector is largely confined to low-value applications (Di Maria et al. 2018) often described as downcycling in waste management literature. Previous research has highlighted barriers such as high energy and transport costs, limited awareness, technological constraints, poor material quality and contamination, weak market demand, ineffective waste regulations, and negative stakeholder perceptions (Shooshtarian et al. 2020).

To date, much of the research has focused on capturing broad stakeholder experiences or has been heavily technical in nature, offering limited understanding of the contextual factors that influence PwRC application. To address this gap, the present study adopts a case study approach to deliver practical, current insights into how CE principles are applied within Australia's built environment sector at the project level.

8.1.1 Waste Regulatory Framework in Australia

In Australia, the management of C&D waste is governed by three levels of government: federal, state or territory, and local. The federal government typically has no direct role in regulating C&D waste, intervening only when state or territory regulations conflict with Australia's international obligations (such as Agenda 21, the Basel Convention, or the Stockholm Convention) or when environmental issues of national significance arise. The majority of legislation and regulatory oversight sits with state and territory governments, each of which operates under its own distinct regulatory framework. This decentralised approach has resulted in inconsistencies between jurisdictions, creating barriers that hinder the efficient and unified management of C&D waste across the country (Hyder Consulting Pty Ltd 2011; Senate Environment and Communications References Committee 2018).

8.2 Research Approach

The case study approach was utilised to explore the experience of the Australian key stakeholders in using PwRC in the case project. In consultation with the waste experts, the criteria for the selection of case studies were set. Case studies had to meet the following criteria:

1. *Significant use of PwRC; and*
2. *Completed recently or due to be completed.*

8.2.1 Case Study Description

The selected case project included a residential project situated in Perth, Western Australia (WA). The project was government-owned projects. This project which was managed by WA Development aimed to redevelop an old high school and the neighbour lands (11.9 ha) into a residential estate to provide residential opportunities for WA communities. The old school was closed in 2017 after 55 years of operation according to the WA Department of Education. The project budget was $2.7 m and completed over 12 months in 2019. In this project, through the extraction of demolition resources from the old senior high school around 86% of waste diversion was achieved (Fig. 8.1).

Recognition of Environmental Sustainability

OneOneFive Hamilton Hill, part of DevelopmentWA's Innovation Through Demonstration initiative, was designed to achieve exceptional sustainability standards. The project focused on delivering an environmentally conscious development by significantly reducing water and energy usage while preserving the site's natural landform, existing trees, and biodiversity.

At the time of the case study, the project had been awarded six petals of the Urban Development Institute of Australia (UDIA) under its benchmark EnviroDevelopment certification program. The petals include Ecosystems, Waste, Materials Energy, Water and Community. Furthermore, the project has been recognised by Water Corporation WA as a Waterwise Platinum development, the highest Waterwise Development status, and also won the Waste Sorted Award offered by Waste Authority WA in 2021. Figure 8.2 illustrates the main project certifications and awards achieved to date.

The project has delivered several economic, social, and environmental benefits. The estimated cost saving was around $100,000 which was realised in reduced waste levies and transportation costs. This cost saving could offset the increased cost to complete the crushing process and keep recovered materials suitably stored and organised.

This vibrant new estate is planned to include around 232 lots, offering a variety of housing choices such as single residential homesites, townhouses, compact lots, and four designated grouped housing areas. Furthermore, the project reserved the site heritage value by incorporating the extracted materials into the new construction.

Lastly, with the minimisation of the need for material movements, the project had a minimum impact on the local road networks. There was 1000 less truck movements, 9000 L of diesel saved and therefore 663 tonnes of CO_2 were avoided. This carbon emission reductions were achieved in three keys areas: onsite, through the

Fig. 8.1 An overview of the construction of OneOneFive Hamilton Hill project. *Source* TABEC (2022)

replacement of raw materials with recovered demolition products; offsite, through recycling; and transport, through reduced truck movements.

8.2.2 Data Collection and Process

Once the case project was selected, input was gathered from four key stakeholder groups directly involved in the use of PwRC: the designer, client, head contractor, and supplier (Fig. 8.3). Each stakeholder received an email invitation to schedule a convenient interview time, and for every chosen project, four individual online interviews were conducted.

Fig. 8.2 The project sustainability accreditations and awards achieved to date

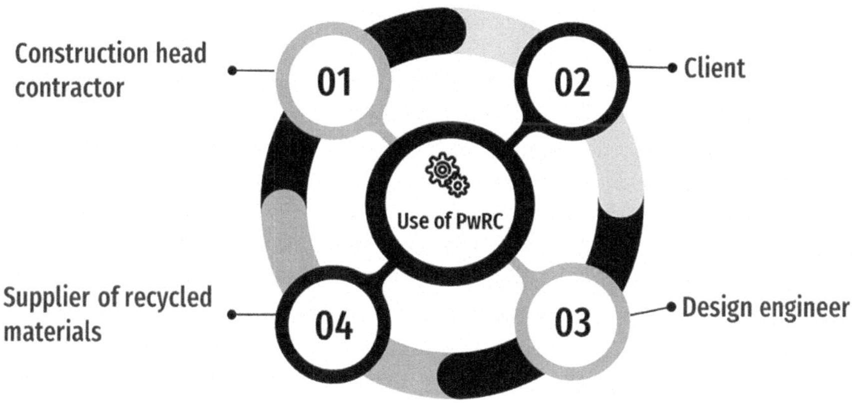

Fig. 8.3 Stakeholders who were interviewed in each case study

Between April and July 2022, interviews were carried out virtually via Microsoft Teams. Alongside gathering demographic information, professional background, and expertise in using PwRC within construction projects, participants were invited to discuss key drivers, barriers, and potential opportunities for expanding the use of these materials in the examined case studies. The recorded audio was transcribed verbatim by a professional service, after which the research team verified the accuracy of the transcripts. Subsequently, the qualitative data were processed and coded using NVivo Pro 12 software to support systematic thematic analysis.

8.3 Case Study: One-One-Five Hamilton Residential Project

8.3.1 Use of On-Site Recycled Demolition Waste

About 10,000 tonnes of demolition waste resources were extracted from the old school using the methodical deconstruction method. Most of the processed demolition waste resources—PwRC—applications were in the civil and landscape construction works including drainage system, local access roads, landscapes, and footpaths. About 86% of demolition resources were crushed onsite for reuse in the project. Through the reuse of these materials, carbon emissions were achieved by reducing the need for material transportation to and from recycling facilities. The following represent the PWRC quantity and type that were used in this project:

- *Incorporation of salvaged timber for landscaping elements such as shaded structures and seating areas.*
- *Repurposing 40,000 clay bricks and roof tiles as aggregate material for subsurface drainage works.*
- *Reusing aged bricks to construct new brick walls and a brick-built toilet facility.*
- *Recycling crushed bricks, tiles, concrete, and similar materials for use in road sub-base layers, reducing reliance on quarried resources like crushed limestone.*
- *Application of 2,425 m^3 of recycled concrete in building retaining walls, eliminating the need for newly quarried limestone.*
- *Deployment of 400 tonnes of PwRC for multiple purposes, including the construction of temporary truck access roads.*

Furthermore, the head contractor has imported 1,050 tonnes of recycled roabdase from other demolition sites to use in this project. The project used IPWEA/WALGA Specification for the Supply of Recycled Road Base (IPWEA 2016) to assess PwRC quality and performance.

8.3.2 Stakeholders' Interaction Modelling in the Case Project

Figure 8.4 shows the interaction model between the four study stakeholders in this project. The modelling was done based on the responses captured during the interviews.

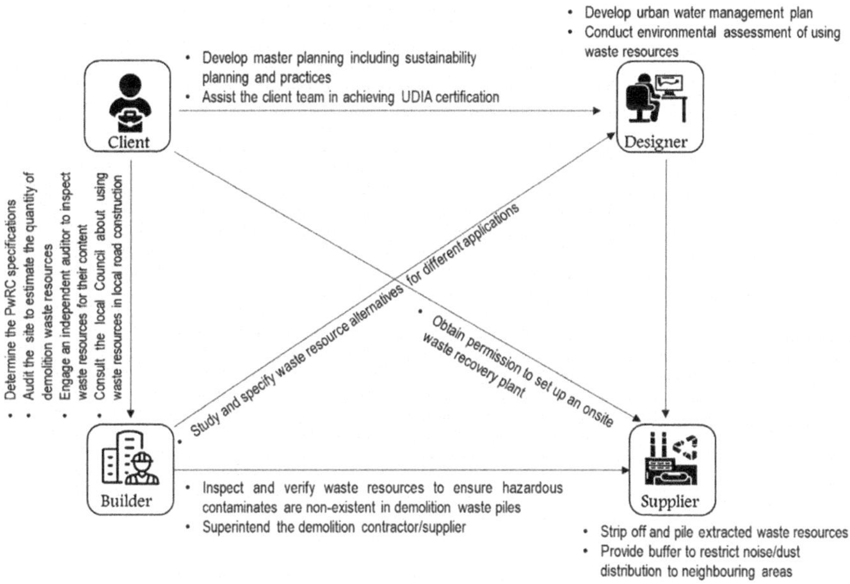

Fig. 8.4 Stakeholders' interaction model in this project

8.3.3 *What Were the Motivations for Reusing Demolition Waste Resources?*

As illustrated in Fig. 8.5, the participants in this case study highlighted various reasons for using PwRC on the construction site. The most frequently mentioned factor was to reduce waste disposal or otherwise increase the waste recovery rate.

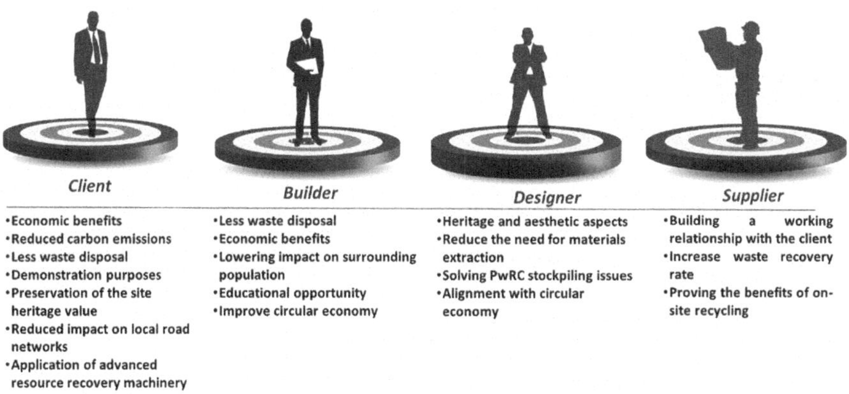

Fig. 8.5 The main motives for using PwRC in OneOneFive Hamilton Hill project

C_4P_1 gave various justifications for the utilisation of the PWRC in this construction project. The reasons included economic benefits, reduced carbon emissions, less waste resource disposal, demonstration purposes, preservation of the land's heritage characteristics, reduced impact on the local road network and the application of modern waste recovery technologies. As previously mentioned, the project utilised a substantial amount of waste resources on-site, resulting in cost and carbon savings. This participant stated that their unique waste management solution was deemed superior to landfill disposal and obtaining PwRC from recyclers. Furthermore, the use of PwRC reduced the need for material transportation leading to minimum impact on the local road networks.

> So they actually realised that they could save some money by not tracking all that material away. They realised that the cost to crush it on-site was less than them tracking the material off-site… Because this purely worked out to be the most economical, most advantageous in terms of carbon footprint and disruption to the residence [C_4P_1]
> We put recycled aggregates from the school under the drainage infrastructure, so we didn't have to buy and import those new materials. And that way, we're basically cutting down on what new material needs to be mined [C_4P_1]
> The truck movements and the impact on the road networks, the impact on the suburban residential streets, the impact to people, through truck movements and the dangers, you know that is far worse than allowing a temporary facility that's going to crush their own material on-site [C_4P_1]

Furthermore, this project was the opportunity for the client's organisation, which is a state government agency, to demonstrate the successful utilisation of PwRC in residential construction to other public and private stakeholders.

> Certainly, the purpose of these projects for us is to be able to demonstrate industry. Whereas as state government, obviously, we want to be showing that we're pushing the boundaries and trying to, I suppose provide demonstration projects that show what can be done and within a normal commercial environment [C_4P_1]

Lastly, the land on which this project was being built had some heritage values in two periods the European heritage and the indigenous heritage. In the first period, the land was created, farmed and then become a school. Prior to that, the indigenous population used the land as scrubland and bushland where they have also used it as an important place to survey their surroundings as it sits on the hill. According to C_4P_1, the use of PwRC derived from the old school could preserve the heritage spirit of the residential development.

> In terms of the school heritage, we actually retained a lot of material that we didn't crush that we actually reused in landscaping elements and for different features within the state that then show that sort of that old school being reconstituted into the new development…we're also providing that opportunity for people that perhaps went to this school to come back and look at the new residential development and still recognise certain elements out of the school that they might have gone to landfill [C_4P_1]

The head-contractor representative, C_4P_2, highlighted the six motives for using PwRC in this case project. The diversion of waste resources from landfills and the cost savings associated with it were the first two main reasons why their organisation was motivated to use PwRC.

> We then had huge stockpiles all over the site of crushed materials and ordinarily, they would have been put into trucks and disposed of quite expensive to a landfill facility that would have been ordinary. Probably outcome we thought we'll let's assist the opportunity of reusing this as recycled Rd material and there was probably that opportunity to promote the circular economy [C_4P_2]

Furthermore, C_4P_2 indicated that the use of PwRC primarily extracted from the demolition resource reduced the impact of construction activities on the surrounding population. By minimising the truck movements due to keeping demolition resources on the site and subsequent applications, and not having to ship new materials to the site, the usage of PwRC provided positive social and environmental impacts. The educational opportunity in using PwRC was the last motivation indicated by C_4P_2. The participant stated that PwRC was an opportunity to educate their organisation, design consultants and the wider industry about the possibilities of using PwRC in construction projects. Lastly, C_4P_2 mentioned that the client's strong support facilitated the usage of PwRC in this project.

> Educate consultants and the wider industry on what can be done, how you can do it and try and, you know educate ourselves that we can take that kind of projects and slowly sort of do our little bit in our pushing the industry forward and sharing that knowledge you know throughout [C_4P_2]
>
> [the client was] very supportive in that respect that you don't just sort of limit you know education to one project or one that a group of people try and do your part in nudging the whole industry forward in that respect so. And there was actually consistent support from [the client] the whole along there [C_4P_2]

In response to the reasons why the design consultancy organisation was involved in this project and pursue the utilisation of PwRC, C_4P_3, indicated that environmental impacts, heritage value and aesthetic aspects of PwRC as the two main drivers. In terms of environmental sustainability impacts, the participant noted that the utilisation of PwRC reduces the need for new extractions, solving the ongoing issue of stockpiling in the state and alignment with the circular economy aspiration.

> In line with a circular economy aspiration, recognising that there is value in these in these materials, you know, reduce the need for the extraction and processing of new materials, and avoiding the issue of seeing materials coming out of the demolition of this project stockpile which we know is an ongoing [C_4P_3]

The two last motivating factors were related to the heritage value of the construction site which should be reflected in the new design. Furthermore, the PWRC produced from the demolition resource had strong visual impacts that will be beneficial from a design perspective.

> It was very clear from the early community consultation with the public, that people wanted a sense of connection with the old school. So, they didn't want to be erased… you know, reusing the old bricks, reusing salvaged timbers reusing some of the other school elements in public art, and so forth, is a way of, of keeping that character [C_4P_3]

According to C_4P_3, for the demolition and recycling contractor, the three motives for being engaged in this project were building a working relationship with the client, increasing the waste recovery rate and proving the benefits of on-site recycling.

For me. It was purely and simply about getting maximum recycling and reuse on-site and working with the client WA Development who was sensational and working with them. So it was really clear that we had to maximise reuse and recycling. That was really my goal [C_4P_4]

8.3.4 What Hindered the Broad Application of Demolition Waste Resources?

The participants in this case study mentioned 15 unique challenges that hindered the application of PwRC in OneOneFive Hamilton Hill project. As illustrated in Fig. 8.6, the most frequently mentioned challenge was the delays that occurred in getting permission to use the temporary waste recovery plan on the demolition site.

The client's respondent, C_4P_1, noted four major barriers to using PwRC in this project. First, the unsupportive state waste regulations made the permission for temporary setup of waste recovery (e.g., crushing) machinery unreasonably long. The participant mentioned that they had to go through the same process and meet the same requirements as they would in the case of establishing a permanent recycling facility.

I think at least three months, maybe four months for approval, we had to meet all the same requirements as a permanent facility on the noise side of things, which actually doesn't make sense…It effectively meant we had to go through a process if we were setting up a permanent crushing facility that was going to be operating for, you know, 5 10 years or something. So that process is something we tried to challenge, and we were unsuccessful because it is legislated [C_4P_1]

The second challenge reported by C_4P_1 was to ensure that the final PwRC can meet the standard requirements for the intended applications. To this occur, the participants had issues fully separating the waste resources on the site.

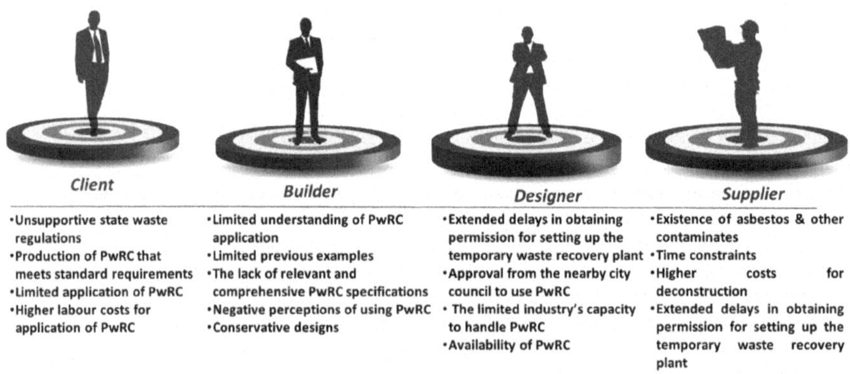

Fig. 8.6 The major issues around using PwRC in OneOneFive Hamilton Hill project

> In that [Engineers Australia] specifications, there's certain percentages that are allowed. So that was a big challenge for us how the demolition contractor was separating bricks from concrete. Sometimes they're attached, obviously. So, if you're talking [about a] big high school, where there's concrete and columns, bricks built in between the columns, then typically the material just all comes down together [C_4P_1]

The two last challenges were related to the application of PwRC. The participant observed that PwRC cannot be applied everywhere due to regulatory restrictions or lower performance. Moreover, there was a higher labour cost for using salvaged materials compared to brand-new ones as it needs different handling and application method.

> So really, what it became a challenge there just to try and find what at every opportunity, where could we put recycled material versus new material. I guess that's what we've tried to do. It's not always possible [C_4P_1]

> There's probably a cost involved because buying new bricks is cheaper than the labour required to say chip off the old mortar off an old brick [C_4P_1]

The second participant, C_4P_2, who represented the organisation overseeing the civil works named five barriers. These barriers were the limited understanding of PwRC applications, limited previous examples, the lack of relevant and comprehensive PwRC specifications, negative perceptions of using PwRC, and conservative designs. C_4P_2 noted that the application of PwRC was not common and there were very limited examples of this application that could be used as a reference. Furthermore, the participant mentioned that the only available specification (IPWEA 2016) could not help the design team to understand the best applications for the available PwRC in the project.

> There was not a lot of sort of strong understanding because there had been in sort of to my knowledge only limited examples of recycled C&D waste being used in road construction in Perth [C_4P_2]

> There was no common specification that we could go and refer to or point to and say oh that's appropriate to use on this project. There's a WALGA specifications … So that's what I guess sort of specifications with a no clear reference that we consider turn to and so what we did is essentially developed our own sort of bespoke specifications [C_4P_2]

C_4P_2, highlighted the negative perceptions from the local council regarding the use of PwRC in road construction projects. The participant indicated that a previous application did not perform well which made negative perceptions among stakeholders. This in turn was a challenge for design works resulting in producing conservative designs.

> There was some negative perception with the local council who sort of take some ownership of the roads, be due to sort of, I guess, historic trials I say of roads that had been constructed using recycled materials, and the asphalt had done performed poorly over time, and so they wanted to know how that was sort of being addressed [C_4P_2]

The design organisation's representative, C_4P_3, noted five primary challenges in using PwRC in this project. These include extended delays in receiving permission for the temporary waste recovery plant operation on the site, approval from the nearby city council to use PwRC, the industry capacity to handle PwRC and the availability of PwRC. The latter factor caused several stages of design refinements to address the issue of PwRC availability.

> In terms of the design and detail for the landscape architects, again, no problems with the contractors being happy to use the materials. But just at times, you know, when designing a large project, but only having a certain amount of materials, that it just means you have to be more nimble in how you're designing [C_4P_3]

C_4P_4 noted several challenges in using demolition resources. The existence of asbestos and other contaminants in the waste resources was the biggest challenge.

> So, the biggest challenge was asbestos. We found asbestos in all sorts of places randomly in the buildings behind whiteboards behind blackboards behind cupboards in window frames underneath and packing them out…and so that was the biggest challenge was making sure the material we crushed was clean [C_4P_4]

The other challenges included time, cost and getting permission for waste recovery on-site. In terms of time, the client sought to complete the project as soon as possible. According to C_4P_4, there were extra costs for additional work to ensure that the waste resources extracted from the old school were free of contaminants. The extra work also impacted their organisation's operation on site.

> The time on the asbestos removal and I think in the future you would just change your methodology. Because obviously, we were finding it ad hoc. So, we'd start another building...So while we're doing that, we have to go find something else to do. So, it was a bit all over the place because of that issue [C_4P_4]

> Then cost, because of the extra work that we had to go through to make sure that all asbestos was collected and removed. Obviously, this cost impacts the job in doing that [C_4P_4]

> We're in the middle of a residential area and we're talking about crushing. You're basically in Perth. It never happened in a residential area before. So that took, I think, almost six months to get approvals to be allowed to crush [C_4P_4]

8.3.5 How to Overcome the Barriers?

The participants in this case study provided 12 strategies that could potentially overcome the issues faced in this project. The main strategy that was indicated by three participants was to set up demonstration projects in which PwRC are successfully used. Figure 8.7 summarises the key findings in this section.

The participant representing the client in this project provided four main strategies to increase the utilisation of PwRC and also remove the barriers mentioned. According to C_4P_1 the local councils should come on board at the early stage of the project development, as some of PwRC can be used in their infrastructures.

> So when you're constructing those roads, you have to construct them in a way that's acceptable to the council because they have the long-term liability to maintain those roads... So, it's important that the council comes on board because councils also manage footpaths [C_4P_1]

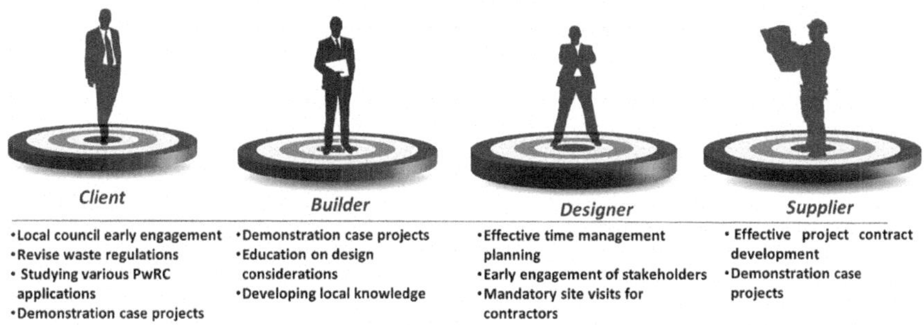

Fig. 8.7 The major strategies to remove barriers to using demolition waste resources in OneOneFive Hamilton Hill project

The second was to revise the regulations determining the permission for establishing waste recovery facilities. The participant suggested that permission should be given on a project-by-project basis and revised in the favour of on-site waste recycling and production of PwRC due to the massive benefits associated.

> I think put into the minds of the regulators that they need to probably consider a different process for short term operations to affect that we could crush the whole school and around about three weeks…But it has to be done on a site-by-site and project-by-project basis, it's not going to suit every demolition… I suspect that if we hadn't crushed this material on site, the amount of recycled content in our project would be lower than it is now [C_4P_1]

Furthermore, studying the variety of applications for PwRC can increase the uptake of these materials. In this project, according to the client participant, the project team workshopped the opportunities for PwRC applications. Even though they cannot be used in all circumstances, the project team was successful in finding some innovative applications for these materials.

> What we ended up doing was just doing a blanket instruction to all consultants that in all their design work, whether it's the civil engineer or the landscape architect, or anyone that's involved in the land development phase of the project, is that where there's a recycled material option to specify that…we basically said, look, brief to us that if [it] can be built in recycled material at the same cost [C_4P_1].

The last strategy was to facilitate the demonstration aspects of construction projects in which PwRC are used. The client organisation used this strategy to promote this idea among the major stakeholders in WA.

> When we do the projects, we're now looking at ways to demonstrate. So we had a tour of the project…and we talked through one of Perth, sort of bigger private developers, their project managers came through so we're able to point out to them, the recycled material, the fact that we'd had these stockpiles of crushed material, showed them our retaining walls and asked them if they recognise them and they didn't recognise them [C_4P_1]

The builder representative, C_4P_2, reported that having demonstration projects, education on design considerations and local knowledge are the main strategies to promote the use of PwRC.

> Probably [there] needs to be an educational piece around what are the design considerations that you need to be sort of thought through…before you might sort of charge off and say I wanna do a project with recycled base. OK let's step through what you need to do there [C_4P_2]

C_4P_3 provided three strategies to reduce the impact of challenges towards using PwRC in construction projects. The strategies include effective time management planning, early engagement of stakeholders, and mandatory site visits for contractors. Time management can reduce the impacts of different risk factors and will provide insight into the time implications. Likewise, the early engagement of stakeholders will contribute to managing risks and ensure that all parties involved understand the project goals.

> It's about good planning, understanding the timeline implications. Certainly, [through time management planning] you would hope that early engagement with regulators helps identify risk and ensuring that all parties understand what the intention of the project is. So for example, once we designed the landscape and packaged it up, we had a section in the landscape specification that actually articulated the intent of waste avoidance and resource recovery for the project [C_4P_3].

According to C_4P_3, *a* mandatory site inspection is a rather common practice in the industry to make sure that the pricing of the job is done with a high level of confidence and that the project goals are clearly understood by the contractors.

> [we] had a mandatory site visit inspection for contractors who are going to be tendering for the landscape construction works. I'm pretty sure the civil engineer did the same for the civil contractors to come and actually see the material to understand what was there and before they did their pricing… So that would improve our confidence that they would be pricing appropriately [C_4P_3].

C_4P_4 suggested that having a clear contract that is informed by negotiating sessions would diminish a lot of issues concerning time and money. This participant noted that for instance, in the negotiations it had to be clear that it is a full strip-out kind of job with the potential for having asbestos.

> So it's probably the only way to overcome that is if you're going to do a major crushing and reuse on-site, Is that in the initial contract negotiation that you need to make it clear that it's a full strip out with the potential for asbestos, that would probably save a bit of time and money, potentially [C_4P_4].

The other suggestion was to showcase the project to set an example of such activities to ensure that future projects will run more smoothly. This will aid in getting the necessary approvals in a timely manner and also to determine the best methodology for handling the waste resources.

8.4 Challenges and Limitations

One notable limitation of this study lies in the difficulty of generalising its findings due to the case study approach. Concerns about generalisability are common in qualitative research and require deliberate strategies to address them. In this study, several measures were taken to mitigate this issue. First, the research process, including data collection methods and analytical procedures, was transparently described to enable readers to evaluate the study's rigour and applicability. Second, pilot interviews with waste management professionals were carried out to refine the interview guide, improving the internal reliability of the findings. Finally, purposive sampling was employed to ensure that the selected case study offered insights relevant to the wider context.

8.5 Future Directions

In the current climate, it is anticipated that the use of PwRC will soon become a standard practice in construction projects worldwide. As a result, the industry must prepare to adapt to this emerging norm. Such adaptation may involve revising business models to more fully embrace circular economy principles, contributing to the development of circular supply chains, and collaborating closely with key stakeholders to ensure that the risks associated with these resources are reasonably and equitably shared. Furthermore, as recycling technologies advance, a wider variety of PwRC is expected to enter the market. This underscores the importance of increased investment in research and development, ensuring that the building and construction sector can safely and successfully integrate these innovative materials into future projects.

References

Active Sustainability (2020) Expanding reuse opportunities for recycled construction materials. Waste Authority, Perth, Australia

Allied Analytics LLP (2023) Construction and demolition waste recycling market is booming worldwide to reach $149.2 billion by 2027. Newswires

Deng D (2022) An insight into China's construction and demolition waste recycling industry. Invest Northern Ireland

Di Maria A, Eyckmans J, Van Acker K (2018) Downcycling versus recycling of construction and demolition waste: combining LCA and LCC to support sustainable policy making. Waste Manag 75:3–21

Hyder Consulting Pty Ltd (2011) Construction and demolition waste status report: management of construction and demolition waste in Australia. Department of Energy and Environment (Department of Sustainability, Environment, Water, Population and Communities), Australia

IMARC Group (2023) Construction and demolition waste management market: global industry trends, share, size, growth, opportunity and forecast 2023–2028. SR112023A1230. The International Market Analysis Research and Consulting Group

IPWEA (2016) IPWEA/WALGA specification for the supply of recycled road base

Li J, Yao Y, Zuo J, Li J (2020) Key policies to the development of construction and demolition waste recycling industry in China. Waste Manag 108:137–143

Menegaki M, Damigos D (2018) A review on current situation and challenges of construction and demolition waste management. Curr Opin Green Sustain Chem 13:8–15

Park J, Tucker R (2017) Overcoming barriers to the reuse of construction waste material in Australia: a review of the literature. Int J Constr Manag 17:228–237

Report Ocean (2021) Construction and demolition waste recycling market research report. Report Ocean, Chicago, USA

Senate Environment and Communications References Committee (2018) Never waste a crisis: the waste and recycling industry in Australia. Parliament of Australia, Canberra, Australia

Shooshtarian S, Caldera S, Maqsood T, Ryley T (2020) Using recycled construction and demolition waste products: a review of stakeholders' perceptions, decisions, and motivations. Recycling 5(4):31

Solís-Guzmán J, Marrero M, Montes-Delgado MV, Ramírez-De-Arellano A (2009) A Spanish model for quantification and management of construction waste. Waste Manag 29(9):2542–2548

TABEC (2022) 115 Hamilton hill stage 1. TABEC Civil Engineering Consultants, Perth, Australia

US EPA (2019) Advancing sustainable materials management: EPA 530-F-19-007. The US Environmental Protection Agency, Washington, US

Zheng L, Wu H, Zhang H, Duan H, Wang J, Jiang W, Dong B, Liu G, Zuo J, Song Q (2017) Characterizing the generation and flows of construction and demolition waste in China. Constr Build Mater 136:405–413

Chapter 9
Reimagining Sustainable Tourism Through Circular Economy Principles for Long-Term Environmental and Economic Sustainability

Harshita Jain and Anamika Shrivastava

9.1 Introduction

The tourism industry has emerged as one of the most dynamic, rapidly expanding, and economically significant sectors in recent years, operating at global, national, regional, and local scales. To maximize its economic and social benefits, substantial policy attention and strategic planning initiatives have been undertaken (Singh and Alam 2024). As one of the largest and fastest-growing industries worldwide, tourism plays a crucial role in driving economic growth, generating employment, and fostering intercultural exchange (Wani et al. 2024). In 2019 alone, the sector generated over $1.7 trillion in export revenues, supported by the movement of approximately 1.5 billion international tourists.

However, the environmental footprint of tourism is considerable. Owing to its heavy reliance on transportation, accommodation, and recreational activities, the sector accounts for nearly 8% of global greenhouse gas emissions. Unregulated tourism growth can lead to severe environmental consequences, including pollution, deforestation, depletion of natural resources, biodiversity loss, and ecological degradation, particularly in regions already facing resource constraints (Agarwal and Jain 2024; Agarwal et al. 2025). Consequently, the rapid expansion of tourism may come at a significant environmental cost (Lenzen et al. 2018).

In response to these challenges, there has been a paradigm shift toward more environmentally and socially responsible forms of tourism. Sustainable tourism aims to minimize negative environmental impacts while preserving the cultural, social, and

H. Jain (✉) · A. Shrivastava
Amity Institute of Environmental Sciences, Amity University Uttar Pradesh, Noida, India
e-mail: hjain@amity.edu

A. Shrivastava
e-mail: ashrivastava1@amity.edu

H. Jain et al. (eds.), *The Circular Fallacy: Closing the Loop in a Consumerist World*, Circular Economy and Sustainability, https://doi.org/10.1007/978-3-032-22687-7_9

ecological integrity of destinations. Recognizing its importance, the United Nations General Assembly declared 2017 as the International Year of Sustainable Tourism for Development. Similarly, the 2016 Beijing Declaration highlights tourism as a "driver of development and peace," promoting harmonious coexistence among nations (Dłużewska 2019).

Parallel to this shift, the concept of the circular economy (CE) has gained prominence as a transformative framework for addressing climate change and advancing sustainability (Velenturf and Purnell 2021). The circular economy seeks to replace the traditional linear "take–make–dispose" model with a regenerative system focused on resource efficiency, waste minimization, and material recovery (Kirchherr et al. 2023a). Within this framework, products and materials are maintained in use for as long as possible, waste is minimized through reuse and recycling, and natural systems are restored rather than depleted (Blomsma and Tennant 2020).

The integration of circular economy principles into the tourism sector offers significant potential to enhance sustainability outcomes. By improving resource efficiency, reducing waste generation, and promoting regenerative practices, tourism can lower its environmental impact while strengthening long-term resilience (Kirchherr et al. 2023a). Moreover, circular approaches can unlock new economic opportunities in areas such as eco-design, green infrastructure, and renewable energy. These strategies not only contribute to environmental conservation but also enhance the overall quality of tourist experiences by offering more authentic, sustainable, and meaningful travel options (Tiwari et al. 2024).

This chapter aims to explore the intersection of circular economy principles and sustainable tourism, focusing on their combined potential to promote environmental sustainability, economic resilience, and social well-being. It begins by defining key concepts, followed by an analysis of the environmental impacts of tourism and the application of circular economy strategies within the sector. The chapter further examines how circular practices can transform the tourism industry, identifies key barriers to implementation, and proposes future pathways for advancing sustainable tourism.

Figure 9.1 provides a conceptual overview of the interlinkages between tourism activities, environmental impacts, and the role of circular economy principles in promoting sustainability. While tourism remains a vital economic driver, its environmental implications necessitate a transition toward more sustainable and circular practices that prioritize resource conservation and ecological balance.

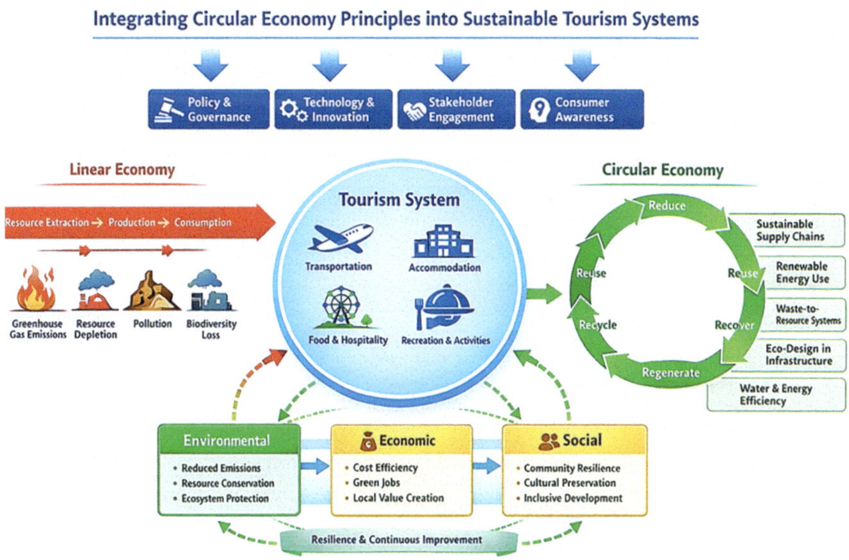

Fig. 9.1 Integrating circular economy principles into sustainable tourism systems

9.2 The Imperative for Sustainable Transformation in Tourism

The relationship between the tourism industry and the environment is complex and multifaceted. On one hand, ecotourism and nature-based tourism provide valuable opportunities for economic development, environmental conservation, and cross-cultural exchange (Zhao et al. 2024). On the other hand, mass tourism exerts significant pressure on natural ecosystems, particularly in popular destinations where high tourist inflows intensify pollution, habitat degradation, and resource depletion. These challenges are further compounded by climate change. Rising global temperatures, shifting weather patterns, and sea-level rise increasingly threaten many tourist destinations, especially those located in coastal, island, and mountainous regions (Arabadzhyan et al. 2020). Such environmental stressors not only degrade ecosystems but also jeopardize the long-term viability of tourism-dependent economies.

As awareness of these impacts continues to grow among tourists, policymakers, and industry stakeholders, there is an increasing demand for more sustainable and environmentally responsible tourism practices. This shift reflects a broader recognition that tourism must transition toward models that minimize environmental harm while enhancing socio-economic benefits for local communities. Achieving sustainability in tourism requires a fundamental structural transformation in how destinations manage resources and how tourists engage with services and experiences. In this context, the circular economy provides a robust and practical framework for

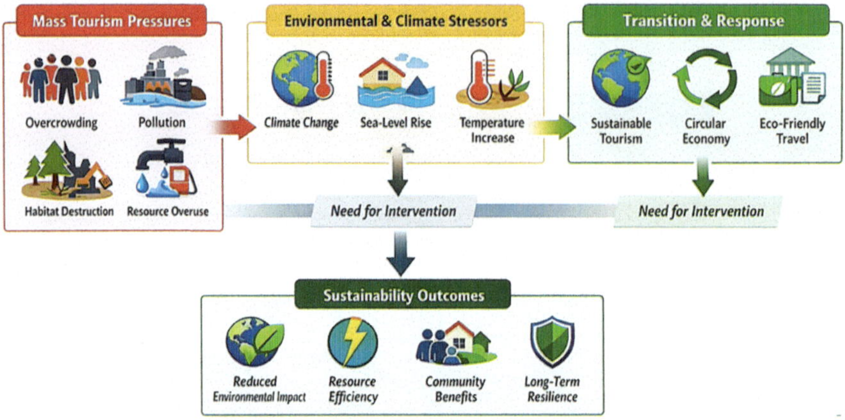

Fig. 9.2 Drivers for transition toward sustainable and circular tourism

reform by emphasizing resource efficiency, waste reduction, reuse, and regeneration (Strippoli et al. 2024).

Figure 9.2 illustrates the urgent need for reform in the tourism sector by highlighting the environmental consequences of mass tourism, including pollution, habitat loss, and climate-related vulnerabilities. It also emphasizes the growing demand for greener travel alternatives driven by increased environmental awareness and sustainability concerns.

9.3 Foundations of the Circular Economy Framework

The circular economy is founded on three core principles that collectively aim to create a sustainable and regenerative system (Manickam and Duraisamy 2018). Figure 9.3 illustrates these foundational pillars as applied to the tourism sector: (1) elimination of waste and pollution, (2) preservation and optimization of resources, and (3) regeneration of natural systems. The framework begins with the design of tourism processes and services that proactively minimize waste generation and environmental pollution. It then advances toward extending the lifecycle of materials, products, and infrastructure through strategies such as reuse, repair, refurbishment, and efficient resource management. Finally, it emphasizes regenerative practices that restore ecosystems, enhance biodiversity, and replenish natural resources. These interlinked principles form a continuous cycle that supports the development of a sustainable and resilient tourism model. By integrating environmental protection with long-term economic viability and social well-being, the circular economy provides a comprehensive pathway for transforming the tourism sector toward sustainability.

Fig. 9.3 Core principles of the circular economy in tourism

The circular economy challenges the conventional perception of waste as an unavoidable outcome of production and consumption by promoting its elimination at the design stage. This approach emphasizes the development of systems, services, and products that minimize waste generation, pollution, and resource use from the outset (Kirchherr et al. 2023; Barros et al. 2020). In the tourism sector, such strategies can be implemented through sustainable design and operational practices in hotels, resorts, and attractions, including the use of eco-friendly materials, energy-efficient technologies, and optimized water management systems (Ioannidis et al. 2021; Habibullah et al. 2023). These measures not only reduce environmental impact but also enhance operational efficiency and long-term sustainability.

Another central principle of the circular economy is the preservation of goods and materials by extending their lifecycle through reuse, repair, refurbishment, remanufacturing, and recycling (Blomsma and Tennant 2020). Within the tourism industry, this can be achieved by encouraging the use of reusable products such as cutlery, water bottles, and toiletries, implementing efficient waste management systems, and maintaining infrastructure through regular maintenance and upgrades (Juvan et al. 2023; Bittner et al. 2024). Such practices contribute to reduced resource consumption, improved cost efficiency, and enhanced sustainability performance.

Beyond minimizing negative impacts, the circular economy also prioritizes the regeneration of natural systems. This involves restoring ecosystems, enhancing biodiversity, and promoting the use of renewable resources (Dagilienė et al. 2021). In tourism, regenerative practices may include afforestation, eco-friendly landscaping, and the integration of renewable energy sources into tourism facilities. Furthermore, tourism enterprises can actively support conservation initiatives and sustainable land-use practices, thereby contributing to the restoration and protection of local ecosystems (Jain 2025a, b).

By integrating these interconnected principles, the tourism sector can transition toward a more sustainable and regenerative model that not only minimizes environmental degradation but also enhances resource efficiency and supports positive socio-economic outcomes for local communities. This transformation aligns with broader global efforts to address climate change and promote sustainable development within resource-intensive industries (Arabadzhyan et al. 2020; Khatib 2023).

9.4 The Relevance of Circular Economy Approaches in the Tourism Industry

The application of circular economy principles offers significant potential for transforming the tourism industry toward greater sustainability. Given its strong dependence on natural resources such as land, water, and energy, and its vulnerability to environmental degradation, the tourism sector must adopt more responsible and resource-efficient practices. At the same time, evolving tourist preferences increasingly favor destinations and experiences that align with environmental values, positioning tourism as a powerful platform for advancing sustainability (Li et al. 2023).

Integrating circular economy strategies provides practical solutions to several pressing challenges within the sector. Waste generation, particularly in high-density tourist destinations with limited waste management infrastructure, remains a critical concern. Circular approaches promote waste minimization through practices such as recycling, composting, and the reduction or elimination of single-use plastics. For example, hotels can implement food waste reduction strategies through portion control and redistribution initiatives, while resorts may adopt closed-loop water systems that enable wastewater reuse for non-potable applications like landscaping (Juvan et al. 2023).

Energy consumption is another major issue, as tourism-related activities—including accommodation, transportation, and recreation—are energy-intensive. Circular economy practices encourage the adoption of energy-efficient technologies, such as LED lighting and smart climate control systems, alongside the integration of renewable energy sources like solar and wind power to reduce overall energy demand (Miera et al. 2024).

Furthermore, the use of sustainable, locally sourced, and recycled materials represents an important dimension of circular implementation in tourism. By prioritizing such materials, tourism enterprises can reduce the environmental impacts associated with production and transportation while simultaneously supporting local economies and communities (Barros et al. 2020).

In addition, the growing demand for authentic and environmentally responsible travel experiences provides further impetus for adopting circular practices. Destinations that incorporate circular economy principles can enhance visitor engagement by offering meaningful experiences that foster a connection with local culture and the natural environment. This not only improves tourist satisfaction but also strengthens the appeal of destinations among environmentally conscious travelers (Zhang and Deng 2024) (Table 9.1).

9.4.1 Environmental Impacts of Tourism

9.4.1.1 Resource Consumption in Tourism

The tourism industry is highly resource-intensive, relying extensively on natural resources such as water, energy, and raw materials. Hotels, resorts, and recreational facilities require substantial water for daily operations, including guest services, maintenance, landscaping, and amenities such as swimming pools and golf courses. Energy demand is also considerable, driven by transportation systems (air travel, road transport), accommodation services (heating, cooling, and electricity), and tourism-related activities such as guided tours and entertainment.

In addition, the construction and maintenance of tourism infrastructure depend heavily on materials such as wood, concrete, and metals, leading to resource extraction and, in many cases, deforestation. This high level of resource consumption places significant pressure on local ecosystems, particularly in environmentally sensitive regions such as small islands, coastal zones, and developing countries where resource availability is often limited (Zhu et al. 2024).

9.4.1.2 Waste Generation in Tourism

Waste generation represents another major environmental challenge associated with tourism activities. The sector produces waste from multiple sources, both direct and indirect. Hospitality establishments such as hotels, restaurants, and resorts generate substantial quantities of food waste, particularly in buffet-style dining systems where overproduction is common.

The widespread use of single-use plastics—including water bottles, packaging materials, and disposable cutlery—further exacerbates the problem, often leading to accumulation in landfills and marine environments. Inadequate sewage treatment infrastructure in many tourist destinations, especially in developing regions, results in the discharge of untreated or partially treated wastewater into natural water bodies, causing pollution and posing risks to aquatic ecosystems.

Additionally, the construction and renovation of tourism facilities contribute to significant volumes of solid waste, including concrete, metals, and other non-biodegradable materials. Improper waste management practices can severely

Table 9.1 Circular economy relevance and environmental implications in the tourism sector

Dimension	Key issues in tourism	Circular economy strategies	Expected benefits	Key references
Resource consumption	High demand for water, energy, and raw materials in hotels, transport, and infrastructure; pressure on fragile ecosystems (coastal, islands)	Efficient resource use; water recycling systems; renewable energy integration; sustainable construction materials	Reduced resource depletion; improved efficiency; lower ecological pressure	Zhu et al. (2024)
Waste generation	Large volumes of food waste, plastic waste, construction debris; inadequate waste management systems; marine and land pollution	Recycling, composting, elimination of single-use plastics; food waste reduction strategies; closed-loop systems	Reduced landfill burden; minimized pollution; improved waste management efficiency	Nayanarangani et al. (2022)
Energy consumption	High energy demand from accommodation, transport, and tourism activities; reliance on fossil fuels	Energy-efficient technologies (LEDs, smart systems); adoption of renewable energy (solar, wind)	Lower energy consumption; reduced operational costs; decreased emissions	Miera et al. (2024)
Carbon emissions	Significant CO_2 emissions from aviation, cruise tourism, and hospitality sector operations	Low-carbon strategies; energy transition; sustainable transport systems	Reduced carbon footprint; contribution to climate mitigation goals	Khatib (2023), Jain (2025a), Yadav and Jain (2024), Jain (2025b)
Material use	Extensive use of non-renewable and high-impact materials (wood, concrete, metals); deforestation and extraction impacts	Use of recycled, locally sourced, and eco-friendly materials; circular design in infrastructure	Reduced environmental footprint; support to local economies	Barros et al. (2020)
Tourist behavior & demand	Growing demand for environmentally responsible and authentic travel experiences	Promotion of eco-tourism, sustainable destinations, and circular tourism models	Increased tourist satisfaction; enhanced destination competitiveness	Zhang and Deng (2024), Li et al. (2023)

(continued)

Table 9.1 (continued)

Dimension	Key issues in tourism	Circular economy strategies	Expected benefits	Key references
Water & wastewater management	High water consumption; discharge of untreated wastewater in many destinations	Closed-loop water systems; wastewater reuse for landscaping and non-potable uses	Water conservation; reduced pollution of water bodies	Juvan et al. (2023)
Overall circular transition	Linear consumption patterns; inefficient resource utilization; environmental degradation	Integration of circular economy principles across tourism value chain	Sustainable tourism development; resilience; long-term environmental and socio-economic benefits	Li et al. (2023), Juvan et al. (2023)

impact fragile ecosystems such as beaches, forests, and marine habitats, ultimately threatening biodiversity and ecological stability (Nayanarangani et al. 2022).

9.4.1.3 Carbon Emissions from Tourism Activities

The tourism sector is a notable contributor to global greenhouse gas emissions, particularly through transportation. Long-haul flights, cruise ships, and road travel are major sources of carbon dioxide (CO_2) emissions. The accommodation sector also contributes significantly, as many hotels and resorts rely on non-renewable energy sources for electricity, heating, and cooling.

The cumulative effect of these emissions has intensified concerns regarding the role of tourism in accelerating climate change. Addressing the carbon footprint of tourism has therefore become a critical priority in global sustainability efforts. Reports from the Intergovernmental Panel on Climate Change emphasize the urgent need to reduce emissions across sectòrs, including tourism, to meet international climate targets (Khatib 2023). Supporting studies further highlight the importance of integrating low-carbon strategies and sustainable practices within the tourism industry to mitigate its environmental impact (Jain 2025a; Yadav and Jain 2024; Jain 2025b).

9.5 Circular Economy in Tourism

The adoption of circular economy principles within the tourism sector provides a comprehensive pathway to enhance sustainability by addressing key operational challenges such as waste generation, resource consumption, and environmental degradation. By integrating strategies focused on waste minimization, resource efficiency, renewable energy utilization, and sustainable supply chains, tourism systems can transition toward more regenerative and resilient models.

One of the most critical areas of intervention is waste reduction. The tourism industry generates substantial quantities of food waste, packaging materials, and single-use plastics, particularly in hospitality operations. Circular approaches emphasize prevention at the source through measures such as portion control, composting organic waste, and redistributing surplus food to local communities. The replacement of single-use plastics with reusable or biodegradable alternatives further contributes to minimizing environmental burdens. Additionally, well-structured recycling systems in hotels, resorts, and tourist destinations ensure that materials such as glass, paper, and plastics are reintegrated into production cycles, thereby reducing landfill dependency (Ioannidis et al. 2021).

Energy efficiency and the transition to renewable energy sources represent another vital dimension of circular tourism. Tourism-related infrastructure, including hotels, recreational facilities, and transport systems, is highly energy-intensive. The incorporation of energy-efficient technologies such as LED lighting, smart energy management systems, and improved insulation can significantly reduce energy demand. Furthermore, the adoption of renewable energy sources such as solar, wind, and hydropower can help decrease reliance on fossil fuels and lower carbon emissions. Sustainable mobility solutions, including electric vehicles, cycling infrastructure, and public transportation systems, also contribute to reducing the environmental footprint of tourist activities (AlKheder 2021).

Water resource management is equally essential, particularly in regions where water scarcity is already a concern. Circular strategies promote efficient water use through the installation of low-flow fixtures and water-saving technologies in accommodation facilities. Wastewater recycling systems enable the reuse of greywater for non-potable purposes such as irrigation and sanitation. Rainwater harvesting further supplements water availability and reduces dependence on freshwater resources, supporting sustainable water cycles within tourism operations (Habibullah et al. 2023).

The implementation of sustainable and circular supply chains also plays a crucial role in enhancing the environmental performance of the tourism sector. By sourcing goods and materials from local and environmentally responsible suppliers, tourism enterprises can reduce transportation-related emissions while supporting regional economies. Preference for products made from recycled or renewable materials further strengthens circularity and encourages sustainable production practices (Yang et al. 2022).

In addition, extending the lifespan of products and infrastructure is a fundamental principle of the circular economy. In tourism, this can be achieved through the repair, refurbishment, and reuse of furniture, appliances, and construction materials, thereby reducing the need for new resource extraction. Maintenance and repair of recreational equipment used in activities such as hiking, skiing, and water sports can also prolong usability and reduce waste generation (Pröbstl-Haider et al. 2020).

Collectively, these strategies demonstrate how circular economy principles can be operationalized within the tourism industry to reduce environmental impacts, optimize resource utilization, and promote long-term sustainability (Table 9.2).

Table 9.2 Circular economy strategies in the tourism sector

Strategy area	Key practices	Application in tourism	Sustainability outcomes	References
Waste reduction	Composting, recycling, reducing single-use plastics, food redistribution	Hotels managing food waste, replacing disposable items, recycling systems in resorts	Reduced landfill waste, lower pollution, improved resource recovery	Ioannidis et al. (2021)
Energy efficiency and renewable energy	LED lighting, smart systems, solar/wind energy, sustainable transport	Energy-efficient hotels, renewable-powered resorts, promotion of EVs and public transport	Reduced carbon emissions, energy savings, climate mitigation	AlKheder (2021)
Water management	Low-flow fixtures, wastewater recycling, rainwater harvesting	Greywater reuse in hotels, irrigation systems, water-efficient infrastructure	Reduced water consumption, improved water security	Habibullah et al. (2023)
Sustainable supply chains	Local sourcing, eco-friendly materials, reduced transport	Procurement from local producers, use of recycled/renewable materials	Lower carbon footprint, support for local economies	Yang et al. (2022)
Product life extension	Repair, refurbishment, reuse, maintenance	Reusing furniture, maintaining tourism equipment, infrastructure upgrades	Reduced resource extraction, cost efficiency, waste minimization	Pröbstl-Haider et al. (2020)

9.6 Barriers to Implementing Circular Economy Practices in the Tourism Sector

The transition toward a circular economy within the tourism industry is accompanied by several structural, financial, and institutional challenges that can hinder effective implementation. Despite the long-term environmental and economic benefits, the adoption of circular practices often requires overcoming significant initial constraints. A major limitation lies in the high upfront investment required for sustainable infrastructure and advanced technologies. The installation of energy-efficient systems, renewable energy solutions such as solar panels, and water recycling technologies involves substantial capital expenditure. This financial burden is particularly challenging for small and medium-sized tourism enterprises, which may lack access to sufficient funding or face longer payback periods, thereby discouraging adoption (Bittner et al. 2024).

In addition to financial barriers, limited awareness and technical capacity among stakeholders pose a significant challenge. Many tourism operators, employees, and even travelers may not fully understand the principles or benefits of the circular economy. This knowledge gap restricts the widespread adoption of sustainable practices and makes behavioral change difficult to achieve. Without targeted education, training programs, and awareness campaigns, the transition toward circular systems remains slow and fragmented (Bocken and Short 2021). Regulatory and policy-related constraints further complicate implementation efforts. In many regions, outdated or inadequate policy frameworks fail to support circular economy initiatives effectively. Challenges such as insufficient recycling infrastructure, lack of incentives for renewable energy adoption, and rigid or unclear waste management regulations can limit the ability of tourism businesses to transition toward circular practices. The absence of coordinated governance and supportive policy mechanisms often acts as a barrier to innovation and large-scale adoption (Dagilienė et al. 2021).

9.7 Future Pathways for Sustainable Tourism Through Circular Economy Approaches

The advancement of sustainable tourism through circular economy principles is increasingly supported by emerging opportunities in technology, governance, and stakeholder engagement. These enabling factors can accelerate the transition toward more resource-efficient and environmentally responsible tourism systems.

Technological innovation is expected to play a pivotal role in facilitating circular practices within the tourism sector. The integration of smart technologies allows for real-time monitoring and optimization of resource use, including energy, water, and waste management. For example, smart sensors and automated systems in hotels can track consumption patterns and improve operational efficiency, while digital platforms enhance data-driven decision-making. Additionally, blockchain technology offers potential for increasing transparency and traceability within tourism supply chains, ensuring that sourcing and operational practices align with sustainability goals (Tyan et al. 2021).

Supportive policy frameworks and financial incentives are equally critical in promoting the adoption of circular economy strategies. Governments can encourage sustainable transitions by offering economic instruments such as tax incentives, subsidies, and grants to tourism enterprises investing in green infrastructure and technologies. At the same time, regulatory measures—such as mandatory sustainability standards, waste reduction targets, and energy efficiency requirements—can drive accountability and ensure compliance across the sector. The alignment of policy mechanisms with sustainability objectives can significantly enhance the scalability and effectiveness of circular initiatives in tourism (Naruetharadhol et al. 2024).

Furthermore, raising awareness and fostering behavioral change among both tourists and industry stakeholders is essential for the successful implementation of circular economy practices. Educational campaigns and outreach programs can inform stakeholders about the environmental and economic benefits of sustainable tourism, encouraging more responsible consumption patterns. Tourism businesses can actively engage visitors by offering eco-friendly services and experiences, while promoting practices such as waste segregation, water conservation, and energy efficiency. Such participatory approaches not only strengthen sustainability outcomes but also enhance visitor satisfaction and long-term engagement (Santamarina-Campos et al. 2024).

9.8 Conclusions

The tourism sector possesses substantial potential to advance sustainable development; however, realizing this potential requires a fundamental transition toward circular economy principles. By prioritizing waste minimization, efficient resource utilization, and environmentally responsible practices, the industry can significantly reduce its ecological footprint while continuing to generate economic value and social benefits. Despite persistent challenges, including high initial investment costs and regulatory limitations, the long-term advantages of adopting circular approaches outweigh these constraints. The circular economy provides a transformative framework that enables the tourism industry to move beyond conventional linear models toward more regenerative and resilient systems. In essence, embedding circular economy principles within tourism operations offers a strategic pathway for reshaping the sector. Through the reduction of waste, optimization of resources, and restoration of natural ecosystems, tourism can evolve into a model that supports environmental sustainability while enhancing visitor experiences and strengthening local communities. The successful transition toward sustainable tourism will depend on the combined efforts of technological innovation, supportive policy frameworks, and increased awareness among stakeholders. By embracing these drivers, the tourism industry can build a more adaptive, inclusive, and sustainable future, ensuring its continued growth while safeguarding environmental resources for generations to come.

References

Agarwal S, Jain H (2024) Investigating the potential of floral waste as a vermicompost and Dual-Functional biosorbent for sustainable environmental management. Water Air Soil Pollut 235(5):322

Agarwal S, Jain H, Dhupper R, Mathur A (2025) Optimizing methyl orange dye removal from aqueous solutions using white chrysanthemum floral waste-derived bioadsorbent: a study of kinetics, thermodynamics, isotherms, and RSM optimization. Int J Environ Res 19(3):104

AlKheder S (2021) Promoting public transport as a strategy to reduce GHG emissions from private vehicles in Kuwait. Environmental Challenges 3:100075. https://doi.org/10.1016/j.envc.2021.100075

Arabadzhyan A, Figini P, García C, González MM, Lam-González YE, León CJ (2020) Climate change, coastal tourism, and impact chains—a literature review. Curr Issue Tour 24(16):2233–2268. https://doi.org/10.1080/13683500.2020.1825351

Barros MV, Salvador R, Prado GFD, De Francisco AC, Piekarski CM (2020) Circular economy as a driver to sustainable businesses. Clean Environ Syst 2:100006. https://doi.org/10.1016/j.cesys.2020.100006

Bittner N, Bakker N, Long TB (2024) Circular economy and the hospitality industry: A comparison of the Netherlands and Indonesia. J Clean Prod 444:141253. https://doi.org/10.1016/j.jclepro.2024.141253

Blomsma F, Tennant M (2020) Circular economy: Preserving materials or products? Introducing the Resource States framework. Resour Conserv Recycl 156:104698. https://doi.org/10.1016/j.resconrec.2020.104698

Bocken NM, Short SW (2021) Unsustainable business models—recognising and resolving institutionalised social and environmental harm. J Clean Prod 312:127828. https://doi.org/10.1016/j.jclepro.2021.127828

Dagilienė L, Varaniūtė V, Bruneckienė J (2021) Local governments' perspective on implementing the circular economy: A framework for future solutions. J Clean Prod 310:127340. https://doi.org/10.1016/j.jclepro.2021.127340

Dłużewska AM (2019) Well-being versus sustainable development in tourism—The host perspective. Sustain Dev 27:512–522. https://doi.org/10.1002/sd.1903

Habibullah N, Sahrir S, Ponrahono Z (2023) Integrating rainwater harvesting and greywater recycling to increase water efficiency in office buildings. Planning Malaysia 21. https://doi.org/10.21837/pm.v21i29.1369

Ioannidis A, Chalvatzis KJ, Leonidou LC, Feng Z (2021) Applying the reduce, reuse, and recycle principle in the hospitality sector: Its antecedents and performance implications. Bus Strateg Environ 30(7):3394–3410. https://doi.org/10.1002/bse.2809

Jain H (2025) Data analytics enabled by the Internet of Things and artificial intelligence for the management of Earth's resources. In: Data analytics and artificial intelligence for earth resource management, pp 19–36. Elsevier

Jain H (2025b) Synergizing carbon reduction technologies and nature-based solutions for a climate-neutral future. Mitig Adapt Strat Glob Change 30(6):42

Juvan E, Grün B, Dolnicar S (2023) Waste production patterns in hotels and restaurants: An intra-sectoral segmentation approach. Ann Tour Res Empir Insights 4(1):100090. https://doi.org/10.1016/j.annale.2023.100090

Khatib AN (2023) Climate change and travel: Harmonizing to abate impact. Curr Infect Dis Rep 25(4):77–85. https://doi.org/10.1007/s11908-023-00799-4

Kirchherr J, Yang NN, Schulze-Spüntrup F, Heerink MJ, Hartley K (2023) Conceptualizing the circular economy (revisited): An analysis of 221 definitions. Resour Conserv Recycl 194:107001. https://doi.org/10.1016/j.resconrec.2023.107001

Kirchherr J, Yang NN, Schulze-Spüntrup F, Heerink MJ, Hartley K (2023) Conceptualizing the circular economy (revisited): An analysis of 221 definitions. Resour Conserv Recycl 194:107001. https://doi.org/10.1016/j.resconrec.2023.107001

Lenzen M, Sun YY, Faturay F et al (2018) The carbon footprint of global tourism. Nature Clim Change 8:522–528. https://doi.org/10.1038/s41558-018-0141-x

Li K, Cipolletta G, Andreola C, Eusebi AL, Kulaga B, Cardinali S, Fatone F (2023) Circular economy and sustainability in the tourism industry: critical analysis of integrated solutions and good practices in European and Chinese case studies. Environ Dev Sustain 26(7):16461–16482. https://doi.org/10.1007/s10668-023-03395-7

Manickam P, Duraisamy G (2018) 3Rs and circular economy. In: Elsevier eBooks, pp 77–93. https://doi.org/10.1016/b978-0-08-102630-4.00004-2

Miera K, Botts A, Lemar P, Kamath D, Wenning T (2024) Unlocking manufacturing sustainability: energy efficiency opportunities through the US department of energy's better plants program energy treasure hunts (2023–2024). Sustainability 16(18):7918. https://doi.org/10.3390/su1618 7918

Naruetharadhol P, ConwayLenihan A, McGuirk H (2024) Assessing the role of public policy in fostering global eco-innovation. J Open Innov Technol Mark Complex 10(2):100294. https://doi.org/10.1016/j.joitmc.2024.100294

Nayanarangani NMDP, Dissanayaka NDMSH, Jayaneththi NJPHU (2022) Solid wastes generation and its management in tourist destinations of Sri Lanka. Glob J Eng Technol Adv 12(2):106–112. https://doi.org/10.30574/gjeta.2022.12.2.0144

Pröbstl-Haider U, Hödl C, Ginner K, Borgwardt F (2020) Climate change: Impacts on outdoor activities in the summer and shoulder seasons. J Outdoor Recreat Tour 34:100344. https://doi.org/10.1016/j.jort.2020.100344

Santamarina-Campos V, Mas-Gil MÁ, De-Miguel-Molina M, Catalá-Pérez D (2024) Impact culture and the circular economy in the tourism industry: An analysis of challenges and recommendations for sustainability. In: SpringerBriefs in business, pp 19–31. https://doi.org/10.1007/978-3-031-49689-9_3

Singh D, Alam Q (2024) Is tourism expansion the key to economic growth in India? An aggregate-level time series analysis. Ann Tour Res Empir Insights 5(2):100126. https://doi.org/10.1016/j.annale.2024.100126

Strippoli R, Gallucci T, Ingrao C (2024b) Circular economy and sustainable development in the tourism sector—an overview of the truly-effective strategies and related benefits. Heliyon:e36801. https://doi.org/10.1016/j.heliyon.2024.e36801.

Tiwari S, Mentel G, Mohammed KS, Rehman MZ, Lewandowska A (2024) Unveiling the role of natural resources, energy transition and environmental policy stringency for sustainable environmental development: Evidence from BRIC+1. Resour Policy 96:105204. https://doi.org/10.1016/j.resourpol.2024.105204

Tyan I, Yagüe MI, Guevara-Plaza A (2021) Blockchain technology's potential for sustainable tourism. In: Springer eBooks, pp 17–29. https://doi.org/10.1007/978-3-030-65785-7_2

Velenturf AP, Purnell P (2021) Principles for a sustainable circular economy. Sustain Prod Consum 27:1437–1457. https://doi.org/10.1016/j.spc.2021.02.018

Wani MJG, Loganathan N, Mujalli A (2024) The impact of sustainable development goals (SDGs) on tourism growth. Empirical evidence from G-7 countries. Cogent Soc Sci 10(1). https://doi.org/10.1080/23311886.2024.2397535

Yadav K, Jain H (2024) Integrating urban master planning with environmental impact assessment for sustainable and resilient cities. In: Cities of tomorrow: Urban resilience and climate change preparedness. Springer Nature Singapore, Singapore, pp 3–23

Yang M, Chen L, Wang J, Msigwa G, Osman AI, Fawzy S, Rooney DW, Yap P (2022) Circular economy strategies for combating climate change and other environmental issues. Environ Chem Lett 21(1):55–80. https://doi.org/10.1007/s10311-022-01499-6
Zhang Y, Deng B (2024) Exploring the nexus of smart technologies and sustainable ecotourism: A systematic review. Heliyon 10(11):e31996. https://doi.org/10.1016/j.heliyon.2024.e31996
Zhao X, Li T, Duan X (2024) Studying tourism development and its impact on carbon emissions. Sci Rep 14:7463. https://doi.org/10.1038/s41598-024-58262-w
Zhu F, Cao X, Zhuang D, Jin S, Gao L, Yang X (2024) Effects of tourism on local green environment in China: economy for environment? Environ Dev Sustain. https://doi.org/10.1007/s10668-024-05488-3

Chapter 10
Product Longevity, Service Models, Producer Accountability, Reuse, Disassembly, and Recycling

Jyoti Saini, Monika Tiwari, Anamika Shrivastava, and Harshita Jain

10.1 Introduction

10.1.1 Overview of Product Life Cycles in Linear Versus Circular Economies

The linear economic model of take-make-waste has dominated industry and consumer systems for over 100 years. In this model, the cycle is linear, in which natural resources are extracted, processed, consumed, and then disposed of, and through which the volume of discarded materials increases and depletes finite resources (Stahel 2016). Given the global increase in consumption and rapid obsolescence, the linear model has become unviable. In direct opposition to this is the CE, a restorative model that keeps products, components, and materials at their highest utility and value in use for as long as possible. The CE concept takes an interest in "circular economy" (CE) in a bid to tide over the" open loop" through means like product life extension, remanufacturing, reuse, and recycling (Ellen MacArthur Foundation 2015). In contrast to a linear economy, the necessity to develop intelligent designs and business model innovations in circular economy is very high, to disconnect economic activity from resource use and environmental impact (Kirchherr et al. 2018).

J. Saini (✉)
Integrated Centre for Adaptation to Climate Change, Disaster Risk Reduction and Sustainability (ICARS), Indian Institute of Technology Roorkee (IIT Roorkee), Greater Noida, Uttar Pradesh, India
e-mail: jyotee.s@gmail.com

M. Tiwari
GSL Associates Pvt Ltd, Saidulajab, New Delhi, India

A. Shrivastava · H. Jain
Amity Institute of Environmental Sciences, Amity University Uttar Pradesh, Noida, India

H. Jain et al. (eds.), *The Circular Fallacy: Closing the Loop in a Consumerist World*, Circular Economy and Sustainability, https://doi.org/10.1007/978-3-032-22687-7_10

At the heart of the two approaches, there's a key disparity in what value is and how it accrues. In the linear model, value is in volume and throughput on other side, the circular model, is where the value is in durability, reuse and service. Shifting gears from a linear to a circular economy will require systemic changes in the way products are designed, in corporate responsibility, and in consumer behaviour.

10.1.2 Importance of Product Longevity for Sustainability

Although performance here is a concept that is debatable and often difficult to accurately quantify, is the ability of a product to work with the intended function on a long-term basis without serious performance degradation in function, appearance, with reasonable safety. Extended product life results directly in a reduction of material throughput, emissions and environmental pollution, and indirectly in decreased energy consumption throughout a product's life cycle (Cooper 2005). Product longevity extends replacement cycles, lowers extraction of raw material and decrease the reliance on resource-intensive manufacturing. In the case of high-consumption societies the latter has become more important, as short-lived consumer goods (such as electronics, textiles and household appliances) have come to represent significant environmental loads (Bakker et al. 2014).

In addition, more durable products induce more sustainable consumption patterns and constitute an essential precondition for business models such as leasing, repair services, or pay-per-use. There's longevity in taking a product that has been well built and not just operating on materials longevity and design for longevity, but also accepting the fact that if you need to change the oil or a tyre on your car, that is possible, or upgrade the software on your computer, or upgrade a component in your computer, or actually repair your computer when it's broken.

10.1.3 Scope and Objectives

This chapter provides an overview of the circular product design, service-oriented innovation and producer responsibility. It offers an ordered examination of-

- Why product longevity is important and its benefits
- The rise of business models such as Product-as-a-Service (PaaS) that change ownership-based relationships to usage-based relationships, and
- The role of legislative mechanisms including 'Extensive Producer Responsibility' (EPR) in refashioning the after-markets obligations levied on manufacturers;
- Real world applications in enabling and promoting reuse, disassembly, and recycling through design thinking and system's innovation,
- Success stories of reuse, refurbishing and remanufacturing in the various industries.

Through these integrated themes, the chapter ultimately seeks to engage with the wider circular economy debate around more sustainable practices that minimise material waste, strengthen product responsibility and integrate sustainability in the production of products; from cradle to grave. It also pinpoints where policy and corporate practices are lagging and provides recommendations for future enhancement.

10.2 Product Longevity: Design and Material Strategies

In a circular economy, the design of products for longevity, adaptability, repair, and upgradeability is an important lever to prevent resource and environmental factors. In this part, the main strategies to increase product service life through choice of materials, modular design, repair-friendly design and the use of standards and certifications are discussed.

10.2.1 Materials and Modular Product Durability Concept

The use of good quality materials which are robust and last over time, is core to allowing life to the product. Materials are also expected to degrade slowly over time as a function of wear and tear, environmental exposure, and times of use. For example, stainless steel, recycled aluminium, tempered glass and engineered polymers are being increasingly applied to durable consumer goods, such as electronics, appliances and furniture (den Hollander et al. 2017). Also significant is the modular design, by which an item is made up of replaceable and independently functionable sub-items. The approach is designed not just to encourage repair and upgrading but also to facilitate dismantling for recycling at the end of the product's life. One of the most well-known is Fairphone, a modular smartphone that permits its users to replace the battery, camera and screen without specialist tools (Bocken et al. 2020).

Modularity is also aligned with the principles of design-for-disassembly, which targets easier product recovery, part reuse, and material separation. The product variety phenomenon has transmitted from services to products, and industries of electronics and office furniture have benefited from the effect of modularity that has facilitated the inclusion into the design process, and thereby enabled to lower life-cycle costs, and reducing environmental impacts (Charter and Tischner 2017)

10.2.2 Design for Repairability and Upgradeability

Design for repairability and upgradeability is a simple measure to increase the lifespan of a product, which however does not receive wide consideration in commonly used product development methods. When products are intentionally made to be repairable or upgradable, users and repair professionals can replace a bad part, update software, or upgrade the hardware without tossing the entire device in a garbage bin and reducing waste and overall consumption. Key design considerations include fastening technologies (screws rather than epoxied or proprietary fasteners), publicly available repair manuals and diagnostic software, the availability of spare parts for the life of the product, and software architectures that can be backwards-compatible and updated at regular intervals.

The Right to Repair (RtR) is an international movement that encourages the access to tools, parts, and information needed for product maintenance and repairs, especially of electronic devices. This has forced large manufacturers such as Apple and Microsoft to establish authorised self-repair programmes and to open the market for components and documentation (Watson et al. 2023). In addition, a modular design has increasingly become an important solution, most prominently in the electronics industry. Products like the Framework laptop are a proof point of modularity and user centric design by giving customers the ability to swap out or upgrade components like processors, memory modules and I/O ports and in doing so extending the useful life of such product and discouraging premature obsolescence.

10.2.3 Standards and Certifications for Longevity

Efforts to standardize longevity in design and production are global. Various standards and certifications exist today to gauge and incentivize durable, repairable, and reusable items.

A number of international standards and certification schemes are being developed to encourage design for a long product life, repairability and environmental responsibility. The EU Ecodesign Directive is a pioneer legislative plan establishing minimum requirements for products with a direct or indirect impact on energy consumption, and its latest development extended these requirements to include reparability, durability and availability of spare parts, promoting sustainable design principles in the stages of the products (European Commission 2021). TCO Certified is a global sustainability certification for IT products, including monitors, notebooks and smartphones. It rates products (by durability, repairability and recycling etc.), encouraging both producers and enterprise customers to purchase more responsibly (TCO Development 2022). Based in Germany, the Blue Angel (Blauer Engel) eco-label promotes products with durable built-in service life and eco-friendly use and disposal, thus indicating supply-chain integrated product durability (German Environment Agency 2020). AU$ Any such initiatives are best represented with examples, such as the sustainability service standard, UL 110, for mobile phones, developed by UL (Underwriters Laboratories), which examines environmental criteria in terms of recyclability, energy performance and toxic substance reduction (UL 2021). Together these standards motivate the manufacturer to consider the entire lifecycle of products and allow consumers and organizations to make an informed choice.

These labels affect both consumer decision-making and how those satisfying an increasing consumer demand lead their lives. For example, the French Reparability Index, required by law in 2021, includes a score (out of 10) on how repairable a product is. This has even resulted in design modifications of smartphones and laptops marketed in France (ADEME 2021). Additionally, voluntary standards like ISO 14006 (eco-design) and IEC 62430 (environmentally conscious design) help manufacturers incorporate elements of sustainability and product longevity into their processes for product development.

From a Circular Economy (CE) viewpoint, the transition from product ownership toward a service-oriented consumption is one of the most significant factors for innovation. New service models focus on 'access' rather than 'ownership' and promote longer product lifecycles, dematerializing, resource efficiency and waste reduction. This section presents the three principal service-led strategies of PaaS, the sharing economy and collaborative consumption. It also provides a critical appreciation of the possibilities and limitations of these models.

10.2.4 Product as a Service (PaaS) and Leasing Models

PaaS is a circular business model, in that the customer is paying for a result/resulting utility (the "function)"and not the product as such. In this mode the ownership is kept by the producer and the user pays to use the service that the product offers for example, subscription, rent or pay-per-use. PaaS encourages producers to make durable products that are also easily repaired, up-gradable, because the financial liabilities from products failure always stay with them. It also provides a closed loop feedback system for businesses having interests in both the sustainability and business benefit (Tukker 2015). An example in this context is Phillips' "Pay per lux" model where the system provider still has ownership rights on the lighting system and the client only pays for light as a service. This platform ensures energy-saving, on-time maintenance, and prolonged lifecycle. Leasing forms also have the potential to support the preservation of durable goods, particularly for high-value items such as cars, electronics and industrial machinery, by limiting premature disposal. Corporations such as Michelin and Xerox sell services (tires, printing) in a performance-outcome context backed by real-world measurable performance rather than in units (Lacy and Rutqvist 2015).

10.2.5 Sharing Economy and Collaborative Consumption

The sharing economy is an additional disruptive service-based model which allows individuals to access each other's underutilized products and resources by facilitating sharing (often through digital or online platforms) to maximize utilization and minimize redundancy. A subtype of this collaborative consumption consists of the shared usage or exchange of products among users. There are a growing number of service-based models that are promoting circularity and resource optimisation, across all industries. Car-sharing platforms (e.g., Turo and Zoomcar) in which one person shares their car with another to increase utilization (thereby decreasing the need for individual car ownership) are examples of peer-to-peer car-sharing systems (Shaheen and Cohen 2019). For example, tool libraries and equipment lending libraries provide community access to underutilized items such as power tools or gardening equipment or camping equipment, reducing the need for multiple households to purchase such items and reducing the space to store them (Fishman and Russel 2018). In the fashion sector, companies like Rent the Runway allow consumers to rent clothing for special events, mitigating the environmental toll of fast fashion by increasing the availability of clothing and decreasing demand for production (Lang et al. 2022). Similarly, shared offices, such as those offered by WeWork and Impact Hub, illustrate the efficient use of infrastructure by sharing office facilities among multiple users thereby reducing energy and material requirements per head (Frenken et al. 2020). These are examples of how P2P can facilitate life-styles that are more sustainable, however, the realisation of this depends on responsible user behaviour, reliable platforms and equal digital access.

These systems optimize product usage, prevent premature future investments and optimize the use and allocation of resources (Botsman and Rogers 2010). In the city, the sharing economy is also associated with lower emissions, less congestion and lower throughput of materials. However, the environmental benefits of these platforms depend on effective regulation and responsible user behavior. For example, if shared products are used to the extent that they are overused or not well maintained, or if usage leads to an increase in demand for related goods (rebound effect), then the overall net gains in terms of sustainability can be undermined (Frenken and Schor 2017).

10.2.6 Benefits and Challenges of Service-Based Consumption

Circulating business models based on service provision are a suite of a variety of environmental, economic and social advantages. By changing the incentive structure from one of ownership to one of access, such models can prolong the duration of the product, increase resource efficiency and encourage more sustainable consumption patterns. Producers maintain ownership of their products, engendering design for durability, repair and end-of-life recovery. Consumers, meanwhile, enjoy lower one-time costs and higher-quality products and services. Furthermore, companies receive significant information on how the product is used and can use that data to improve products and raise customer satisfaction. But many barriers stand in the way of the big adoption rush these models are promised. A significant obstacle lies in the consumer mindset that needs to change, where people still find much better value in owning and are wary of service providers. Infrastructure and logistics related to services such as maintenance, reverse logistics, digital solutions must be strong enough to technically support continuous operations. Furthermore, if not properly managed, service-oriented models may contribute to rebound effects, where savings in environmental performance are thwarted by increased overall use (either through cheaper or more convenient access), undermining the sustainability of the service. And because many of the service platforms are digital, they could also exclude people who do not have ready access to the Internet. It therefore will require cooperation in product design, consumer knowledge, infrastructure and enabling public policy systems in order to achieve its full potential of service-based consumption.

10.3 Producer Accountability and Policy Instruments

In modern principles of product responsibility, producer accountability refers to the degree in which manufacturers are held liable for the environmental and communal impacts of their goods, particularly at conclude-of-life. This approach represents a noteworthy shift away from traditional linear economic models, where obligation for waste predominantly fell to customers and municipalities, toward circular strategies prioritizing shared and transferred accountability. By embedding accountability within the design, fabrication, and post-consumer periods, producers are motivated to consider the full product lifecycle in business decisions. Such models are perceived as essential drivers of sustainable material administration and are being actively advanced through progressive policies, like Extended Producer Responsibility (EPR), that seek to transform waste from a burdensome yield into a managed, and often valuable, resource flow (Brown et al. 2023). The transition responds not only to accumulating environmental matters such as resource depletion and pollution but also reflects a developing acknowledgment that meaningful change toward circularity necessitates those who design and benefit from products to take primary duty for their ultimate destiny (Raghunandan et al. 2024).

10.3.1 Extended Producer Responsibility (EPR) Frameworks

Extended Producer Responsibility regulations are a central strategy in advancing corporate accountability while shifting both financial and logistical responsibility for post-consumer waste management from municipalities to manufacturers. This framework operates under the premise that producers—typically manufacturers, importers, or brand owners—are obligated to manage the environmental consequences of their goods throughout the entire lifecycle, especially once these items become refuse. That duty can incorporate collecting, sorting, recycling, and properly disposing of products, with the overt goal of mitigating public waste management burdens, fostering circularity, and reducing environmental damage (Monier et al. 2014).

EPR is commonly executed through various regulatory tools, like mandatory take-back schemes, advance disposal tariffs, deposit-refund systems, and mandatory recycling targets. Truly, producers may organize these duties individually or collectively through Producer Responsibility Organizations, which assist in compliance monitoring, fee collection, and system coordination. A significant advancement within EPR is the recent adoption of eco-modulated charges, where producers pay lower fees for items designed to be more easily reused or recycled, thus directly incentivizing sustainable design choices (Organisation for Economic Co-operation and Development 2001). This approach not only internalizes waste management costs but also aligns economic incentives with eco-design, encouraging product innovation for longevity and material recovery (Monier et al. 2014).

By integrating environmental expenses into product pricing and promoting accountability for end-of-use impacts, EPR frameworks represent a shift from a linear consumption model—where items are disposed of at end-use without direct producer responsibility—to a circular model in which resources are retained in the economy for as long as possible. This shift supports higher resource efficiency, greater recycling rates, and the overall goal of transitioning to sustainable production and consumption systems (Hadley 2024).

By integrating these elements, EPR operationalizes the "polluter-pays" principle, making producers key agents in the move toward a circular economy and responsible resource management (WWF International Network 2019).

10.3.2 Legal and Economic Incentives for Sustainable Production

Legal and economic requirements are key approaches for encouraging sustainable production and imposing more liability upon producers and are used to different extents by government bodies. Regulatory instruments such as Extended Producer Responsibility (EPR) and similar are also crucial; they shift the onus of post-consumption waste management from the local authorities to product manufacturers. and also pushes manufacturers to understand the environmental impact throughout the life of a product. EPR mandates producers to organize take-back and disassembly for recycling and other forms of recovery, often with mandatory recycling targets and public reporting (Mallick et al. 2024). This legal framework also integrates environmental protection with product design, innovation, and resource management, in creating a circular economy where products are to be reutilized, repaired and remanufactured, rather than wasted (Shrivastava et al. 2024).

On the economic side, governments use a variety of policy devices to support sustainable practices. These instruments range from preferential taxes and subsidies or grants to lower value added taxes (VAT) for environmentally sound services and goods, making sustainable production economically rewarding (Vence et al. 2021). Finally fines like landfill taxes, resource extraction fines and fees are used to discourage unsustainable practices. Market-based mechanisms, such as tradable permits, and the internalization of the environmental costs through the use of the polluter-pays principles, also have incentives to increase the amortization and durability of the products, repairability and the use of resources efficient. Furthermore, through eco-labeling and product certification, companies have a competitive gain, generating more income and reputation in sustainability awareness markets (Porter and Linde 1995).

And, beyond legal mandates and economic triggers, are the voluntary certification programs, supply chain sustainability targets, and public-private partnerships that are both driving innovation and adding velocity to the circular approaches to doing business (Genovese et al. 2015). Together, such incentives help to make sustainability core to compliance and a source of competitive advantage, in effect requiring companies to seek the most profitable and operationally-effective ways to protect the environment and contribute positively to society.

10.3.3 Corporate Sustainability and Accountability Practices

Corporate sustainability and accountability initiatives have an important role to play in driving producer responsibility, especially under schemes based on product durability, circular servitization models, and end-of-life treatments (reuse, disassembly and recycling) (Stabler et al. 2024). With values about environmental responsibility and social responsibility becoming more and more ingrained, this pressure will only increase as companies will be required to operate in a way that considers environmental and social impacts in their entirety (Wu and Tham 2023). But when we talk about accountability, it's not just about following regulations; it's also about transparency and action throughout the product life cycle of design, production, use and final recovery.

Sustainability sustainability reporting and disclosure, an essential activity in promoting sustainability, is globally practiced wher policies to encourage the practice is aligned. They expect firms to provide reports on their environmental, social and governance (ESG) impacts, which will make them public and open to scrutiny by all stakeholders (Martiny et al. 2024). Compulsory sustainability reporting mandated in most markets levels the playing field and provides a transparent basis for consumers, investors and regulators to judge performance on sustainability and hold businesses to account for their promises.

Industry leaders are integrating circularity into their product and service models, working to reconcile innovation, profitability and environmental integrity. This means creating durable, modular products, using sustainable materials, and making repair and breakdown as straightforward as possible. That way you can reuse the product and recover valuable resources at end-of-life—elements needed for the so-called closed-loop economy. Producers have a strong obligation to address the overall lifecycle impacts of their products across the supply chain, and in jurisdictions where they answer to EPR laws, they have a direct responsibility to manage collection, recycling and reporting—all of which further subjects them to public scrutiny (Tong et al. 2024).

That is why transparency is the bedrock of corporate sustainability. Creditable accountability is only possible through transparent, honest and timely reporting of sustainable practices, issues and progress. This is enabling businesses to build trust with stakeholders, and show progress towards global sustainability goals such as the United Nations SDGs (Sustainable Development Goals) (Shayan et al. 2022). Firms that rank highest on these practices generally have supplier codes of conduct, engage in sustainable procurement, and have cross functional teams with responsibility for managing ESG risks, and they therefore incorporate sustainability and accountability into the fabric of corporate governance (Torres et al. 2023).

And as the pressure mounts, an increasing number of forward-looking companies see sustainability and accountability not as things being foisted upon them from outside or above, but as opportunities. By incorporating these principles into innovation, brand reputation, and the supply chain, companies are establishing a competitive edge in a world where long-term value creation is guided by responsible stewardship of both resources and what society trusts in.

10.4 Reuse, Disassembly, and Recycling: Closing the Loop References Left

Reuse, Disassembly and Recycling: "Closing" the loop the creation of a real circular economy requires the closure of the loop: materials and products should not travel along a linear anything all the way from resource to the trash can but instead be kept in a loop as long a possible and at the highest value possible. This approach is well visualized in the famous "Butterfly Diagram" of the circular economy which discriminates between the technical and biological cycles. And technically, just to keep things interesting, referring to a vast majority of all manufactured products (not to mention some other products that can, if push comes to shove, be called recycled): materials rotate through various vehicles, such as reuse, repair, refurbishment, remanufacture and, as a last resort, five: r-e-c-y-c-l-i-n-g.

Reuse is the core of closing the loop. Goods or parts that are in reusable condition may be reused without further reprocessing. This also holds for the intrinsic value of the material as well as its embedded energy and labor from the time of its manufacture. Reuse is evident throughout systems like second-hand markets, and business models supporting product-as-a-service or leasing, where products are moved from one user to another, significantly extending product life.

Taking things apart, by playing a supporting role, is a key enabler that makes reuse, repair and recycle, both technically and economically feasible. Ease of disassembly: Products that are designed to be re-disassembled make it easier to recycle materials, or to replace individual parts. Disassembly-friendly design cuts down contamination, reduces the loss of materials, and lowers the demands on labor and energy consumption while improving the quality of recycled materials, creating a more resource-efficient loop.

Recycling comes into play when products can't be reused or refurbished. Here, products are reduced to their constituent raw material which are then processed and returned to manufacturing streams to produce new products. Although recycling also entails some additional loss of embedded value relative to reuse or repair, it is essential for the extraction of useful resources and minimization of dependence on primary materials. But under the hierarchy of the circular economy, recycling is the "outermost loop," to be tapped only when higher-value loops (like reusing or refurbishing) have no better to yield.

The Butterfly Diagram in Fig. 10.1 (Mboli et al. 2020), created by The Ellen MacArthur Foundation, demonstrates the flow of materials in the circular economy with an emphasis on value retention throughout the product life-cycle. In the technical cycle, the smaller loops—reuse, sharing, maintenance, and repair—are the most efficient in terms of maintaining the value of a product by saving embodied energy and resources and they maintain the product in use without changing the product in any significant way. Options continue to decrease in value even after they have become worthless, and products can be refurbished or remanufactured, which are more labour intensive but still valuable options since components or even whole products can be made functional again. Followed by deconstruction, where parts are more easily disassembled and can go on to be reintroduced into production. Recycling is the loop that sits farthest on the outer side, a last resort of bringing materials down to their rawest forms, at the expense of energy and quality. The diagram illustrates that the higher the value and the worst impact on the environment as the materials are flowing away from the center. So, the idea is to maintain materials cycling through the smaller loops for as long as possible. Through the design for durability, modularity, reparability, and end-of-life recovery, the system could be a regenerative system that is expected to generate less waste and dependence on virgin resources (Calzolari et al. 2021).

10.4.1 Strategies for Reuse, Refurbishment, and Remanufacturing

Reuse, refurbishment, and remanufacturing strategies are critical aspects of promoting a circular economy, which create societal value by allowing products to last longer, minimizing waste, and maximizing resource utility. At the basis of these strategy is the concept of keeping the highest possible value in a product, a component or a material as long as possible, with this'mindset it can be limit environmental concern however it also leads to economic advantages.

Reuse The act of reusing products, materials, or parts that can be used again, without processing them. Successful strategies for reusing will depend on the quality

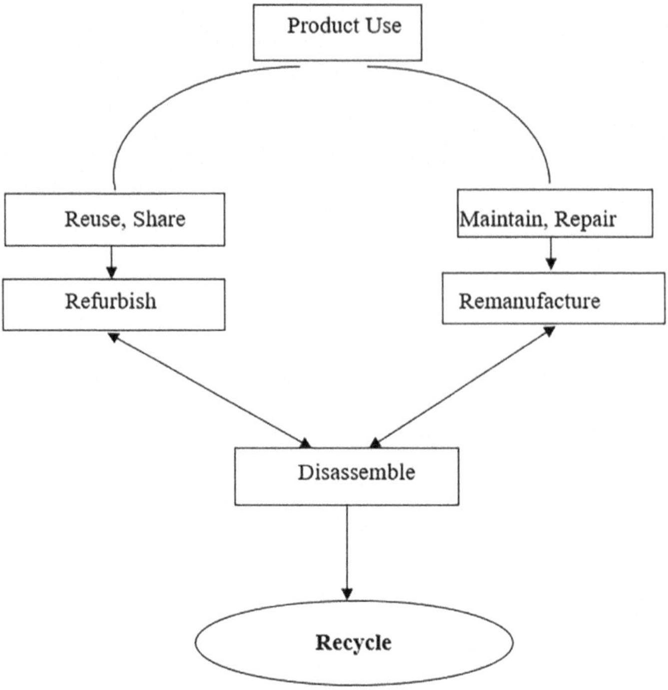

Fig. 10.1 Simplified conceptual structure adapted from the Ellen MacArthur Foundation's Butterfly Diagram (Mboli et al. 2020)

and longevity of designs—along with logistics to easily recover, clean, and redistribute it. Through product-as-a-service business models, which transfer ownership from consumer to producer, manufacturers are incentivized to produce goods for a prolonged material life and reuse. Examples include a uniform shipping pallet and various glass bottles that can hold liquids or food products, and which are used respectively in many industries and on a household scale; and transport container templates, which are sanitized and reused or disposed of.

Refurbishment goes a step further by making the used stuff work again and perform well in a second- or third-hand life. They are then disassembling, checking the condition of each component, replace or repair any parts that are worn down, reassemble, and sell the product. Refurbishing is especially popular in the electronics market for the return of devices that can be upgraded with software and new parts before receiving recertification to be put back on the market. For refurbishment at scale, the products should first be designed for modularity and repairability when designing for disassembly and component replacement to be done compliantly and safely.

Remanufacturing is an even deeper process that involves taking a product apart to the core and returning it to like-new condition, including modification and improvements with upgrades, with additional oversight into the remanufacturing process.

Remanufacturing, unlike refurbishment, is often guided by industry standards and is widespread in markets such as automotive, heavy vehicles and industrial equipment. It lets parts that are still in good shape be incorporated in new products that have the same function, and this reduces resource intake and environmental emissions factors of a million fold over making something new from virgin resources. Because remanufactured products come with the same types of warranties as new products, they are trusted by consumers and are acceptable on the market.

Successfully implementing such strategies requires strong reverse logistics, a convincing business case for producers and consumers and, crucially, an enabling regulatory environment driving companies to design for circularity. Where products are developed for ease of disassembly, repair, and upgrade, and then supply chains can collect, sort, and process used items economically, the loop is closed. These circular systems not only further reduce the requirements for virgin resource extraction and landfill but also innovate value opportunities by maintaining a flow of materials and components within the economy.

10.4.2 Design for Disassembly and Material Recovery

Design for disassembly and materials recycling is an advanced concept of sustainable product development dealing with entire product lifecycle, from production till disposal. It makes the need for materials and process choices that source-level design engineers and designers make, ways that allow for products to be taken apart easily to access defective components for repair, reuse, remanufacture or recycling (Harivardhini et al. 2017). This means using reversible fastening methods—including screws, snaps, and modular connections—instead of permanent ones like adhesives or welds, and focusing on materials that can be disassembled without contamination so that recycle streams stay pure and free from dangerous chemicals (Harivardhini et al. 2017). They are also making use of modular design methods that adds a little homogenisation to the parts sets and therefore upgrading (or repair) is made easier, stretching the life-span of a product. Further, there must be proper documentation and parts labelling to provide recyclers and repair technicians with the appropriate information to enable handling of the product (Gershenson et al. 1999). Despite that these design principles might require a bit more planning and be a bit more expensive in the development, it will save on environment impact (of transports, waste and utilization of extractive resources) and potential economical revenue from reclaiming valuable material will pay off. Crucially, designing for disassembly and remanufacture also aids in compliance with regulation, underpins brand reputation and promotes producer responsibility and extended responsibility applying pressure on manufacturers to care about the afterlife of their products and switch towards a more circular, sustainable economy (Soh et al. 2015).

10.4.3 Recycling Technologies and Their Limitations

Closed loop in product system, through reuse, disassembly, and recycling, is a key concept in a circular economy (Bocken et al. 2016), but this idea is thermodynamically and technologically limited, mainly in the context of recycling technology, but also based on product system and logistics, as the requirement of disassembly (Vikström et al. 2013). Contemporary strategies to recycle, i.e., mechanical, chemical and biological recycling, represent an important alternative to obtain raw materials from waste materials bearing in mind that each technology faces limitations. One of the most significant constraints is material quality loss: many recycled materials, particularly plastics and metals, deteriorate in quality with each processing cycle, and may eventually no longer be fit for high-quality applications, resulting in the need for virgin material feedstock's in order to maintain product specifications (Alassali et al. 2021). Furthermore, some of the materials and multi-layered products might be extremely difficult or not cost-effective to be recycled, for which state-of-the-art sorting technologies might be, in some cases, not present or hardly available yet, in particular for electronics, batteries or composites.

Recycling is also highly reliant on infrastructure, and many areas do not have facilities to collect, sort and process increasing flows of post-consumer waste, resulting in contamination and low yields of recovered material. Recycling is kept in check also by its economic viability: with its high collection, transport and processing costs and unstable market for recycled material, recycling is not an economically sustainable activity. The policy and consumer dimension also serves as an obstacle, given weak regulatory push and a continued throw-away society culture that does not promote upscaled circularity (Aldieri et al. 2025). Last but not least, recycled materials generally suffer that they are only market-wise conditionable for their quality and their performance, which is a barrier for their true implementation in the new manufacturing loops.

Together, while the importance of recycling technology in closing material loops and maintaining sustainability cannot be overstated, the limitations of existing technologies—including degradation, technology gaps, infrastructure limitations, and market factors—point to the urgent importance of continued innovation, improved end of life design considerations and systemic support across policy, industry and consumer segments (Velenturf and Purnell 2021).

10.4.4 Case Studies of Effective Circular Practices

An array of case studies, highlighting new product longevity, service models, producer responsibility, re-use, disassembly and recycling are provided to demonstrate successful implementation of circular initiatives. TerraCycle is an example of cutting edge recycling innovation, making hard to recycle materials into a raw material that is usable again and made into products, with investment from Closed

Loop Partners that makes sure these sorts of local solutions can get bigger (Rubio et al. 2018). Their model significantly reduces waste while championing producer responsibility. Likewise, Interface's "Mission Zero" initiative illustrates the difference strategies for designing for disassembly and recycling can make: a monumental decrease in virgin material hauled out of the ground and a closed loop through modular product and corporate responsibility. In fashion, Patagonia's "Worn Wear" program encourages repairs and the resales of garments, which extends product life and animates consumer involvement in a circular economy. Examples of a new type of service model focusing more on the use-phase are subscriptions, such as Mud Jeans' Lease A Jeans, to discourage disposability and challenge the ownership paradigm, as well as, to extend product lifetime through the capacity to reclaim the use of the product (Smulders et al. 2007). An example of an automotive product remanufacturing is illustrated by Renault's "Choisy-le-Roi" factory where, disassembled components are refurbished, and enhanced, extending the life cycle of component parts, and saving raw resources at the same time (Dzwigo et al. 2021). They also highlight the need for policy frameworks and cross-industry collaboration to build infrastructure for repair, refurbishment and recycling. These cases all demonstrate clearly that the seamless integration of durable product design, circular business models, producer responsibility and effective end-of-life strategies can indeed scale operationalize circularity and deliver both environmental impact and economic opportunity in a wide range of sectors. They can function as inspiring and actionable examples to advance circular practice at the level of product systems around the world.

10.5 Conclusion

10.5.1 Summary of Key Insights

This chapter highlights the importance to prolong product life and be engaged in service models towards a circular economy. Beyond the linear model of take-make-dispose, circular means regeneration and closed resource loops. Long-lasting elements, modular designs and repairability support product sustainability, bolstered by standards and certification that help to engender consumer certainty. Novel service concepts like product-as-a-service, leasing, sharing decrease ownership and maximise resource utilisation at the same time, although they have economic and logistic issues. Incentive-based policies such as Extended Producer Responsibility (EPR) motivate producers to take the end-of-life into account, and facilitate joint resource stewardship between parties involved in production and consumption. Proper reuse, refurbishment and recycling therefore depend on disassembly design and on overcoming today's technological bottlenecks, with best practice examples proving the scalability of circular approaches. Combining these efforts leads to a unified system that contributes to lowered anthropogenic pollution and to advances in wider sustainability targets.

10.5.2 Future Directions for Product Longevity and Circular Service Models

Future developments will include scaling repair, reuse, remanufacturing and refurbishment, enabled by technologies such as AI and digital product passports to enhance lifecycle traceability and recycling efficiency. Integrating biomaterials and circular design principles will drive durable, modular, and repairable products. Product as a service, and leasing models will also promote resource-efficient consumption based more on use, but pose operational dilemmas. Reinforced international policy frameworks in particular, EPR, will drive industry towards circular design and corporate sustainability. Platformed solutions and artificial intelligence will drive supply-chain transparency, resource regeneration and economy-wide cooperation, creating a dynamic ecosystem that helps to extend product life, reduce waste and support sustainable resource management after 2025.

10.5.3 Policy Recommendations and Industry Implications

Strong circular economy promoting legislation is needed to pave the way for the transition. Non-technical measures Governments should introduce well-designed EPR systems, durability and repairability requirements and digital product passports on (minimum) material trace-ability of these products. To scale circular practices, enabling policies should also finance infrastructure for repair, upcycling, recycling and reverse logistics. Industry has to go circular, working with policy makers and disruptors to embed sustainable materials, modular design and responsible end-of-life at the heart of businesses. Transparency and consumer engagement are critical in driving the demand for circular products and services. The transition must pass policy; innovation and accountability; technological, economic, and behavioral hurdles to succeed. This convergence serves the environment and creates new economic drivers, ensuring supply chain resilience, and competitive market positioning for the habitats and industries we hope to see beyond 2025, to achieve the circular economy and sustainability objectives.

References

ADEME (2021) The french reparability index: a new tool for promoting more sustainable products. French Agency for Ecological Transition. https://librairie.ademe.fr

Alassali A, Picuno C, Chong ZK, Guo J, Maletz R, Kuchta K (2021) Towards higher quality of recycled plastics: limitations from the material's perspective. Sustainability 13(23):13266. https://doi.org/10.3390/su132313266

Aldieri A, Gamage TPB, La Mattina AA, Li Y, Loewe A, Pappalardo F, Viceconti M (2025) Consensus statement on the credibility assessment of machine learning predictors. Brief Bioinform 26(2). https://doi.org/10.1093/bib/bbaf100

Bakker C, Wang F, Huisman J, den Hollander M (2014) Products that go round: exploring product life extension through design. J Clean Prod 69:10–16. https://doi.org/10.1016/j.jclepro.2014.01.028

Bocken NMP, Morales ME, Lehner M, Chirumalla K (2020) Product design and business model strategies for a circular economy. J Ind Prod Eng 37(5):308–320. https://doi.org/10.1080/21681015.2020.1812139

Bocken NMP, De Pauw I, Bakker C, Van Der Grinten B (2016) Product design and business model strategies for a circular economy. J Ind Prod Eng 33(5):308–320. https://doi.org/10.1080/21681015.2016.1172124

Botsman R, Rogers R (2010) What's mine is yours: the rise of collaborative consumption. Harper Business

Brown A, Laubinger F, Börkey P (2023) Organisation for economic co-operation and development. New aspects of EPR: extending producer responsibility to additional product groups and challenges throughout the product lifecycle. In: OECD environment working papers (Report No. 225). OECD. https://www.oecd.org/content/dam/oecd/en/publications/reports/2023/11/new-aspectsof-epr-extending-producer-responsibility-to-additional-product-groups-and-challenges-throughout-the-product-lifecycle_84483c40/cfdc1bdc-en.pdf

Calzolari T, Genovese A, Brint A (2021) The adoption of circular economy practices in supply chains–an assessment of European multi-national enterprises. J Clean Prod 312:127616. https://doi.org/10.1016/j.jclepro.2021.127616

Charter M, Tischner U (2017) Sustainable solutions: developing products and services for the future. Routledge

Cooper T (2005) Slower consumption: reflections on product life spans and the "throwaway society." J Ind Ecol 9(1–2):51–67. https://doi.org/10.1162/1088198054084671

Dzwigo H, Trushkina N, Kwilinski A (2021) The organizational and economic mechanism of implementing the concept of green logistics. Virtual Econ 4(2):41–75. https://doi.org/10.34021/ve.2021.04.02

Ellen MacArthur Foundation (2015) Towards a circular economy: business rationale for an accelerated transition. https://ellenmacarthurfoundation.org

European Commission (2021) Circular economy action plan: for a cleaner and more competitive Europe. Eur Comm. https://environment.ec.europa.eu/strategy/circular-economy_en

Extended Producer Responsibility: a guidance manual for governments. https://www.oecd.org/content/dam/oecd/en/publications/reports/2001/03/extendedproducerresponsibility_g1gh2874/9789264189867-en.pdf

Fishman T, Russel J (2018) The sharing economy and the environment: a review and synthesis of environmental implications. Environ Res Lett 13(12):123001. https://doi.org/10.1088/1748-9326/aaf3e4

Frenken K, Schor JB (2017) Putting the sharing economy into perspective. Environ Innov Soc Trans 23:3–10. https://doi.org/10.1016/j.eist.2017.01.003

Frenken K, van Waes A, Pelzer P, Smink M, van Est R (2020) Safeguarding public interests in the platform economy. Policy Internet 12(3):400–425. https://doi.org/10.1002/poi3.219

Genovese A, Acquaye AA, Figueroa A, Koh SL (2015) Sustainable supply chain management and the transition towards a circular economy: evidence and some applications. Omega 66:344–357. https://doi.org/10.1016/j.omega.2015.05.015

German Environment Agency (UBA) (2020) Federal ministry for the environment, climate action, nature conservation and nuclear safety. https://www.bundesumweltministerium.de/en/ministry/tasks-and-structure/federal-and-laender-authorities/germanenvironment-agency-uba

Gershenson JK, Prasad GJ, Allamneni S (1999) Modular product design: a life-cycle view. J Integr Des Process Sci 3(4):13–26. https://doi.org/10.3233/jid-1999-3402

Hadley S (2024) Extended producer responsibility–how can it drive global circularity and sustainability? Plastic Collective. https://www.plasticcollective.co/extendedproducer-responsibility-how-can-it-drive-global-circularity-and-sustainability/

Harivardhini S, Krishna KM, Chakrabarti A (2017) An integrated framework for supporting decision making during early design stages on end-of-life disassembly. J Clean Prod 168:558–574. https://doi.org/10.1016/j.jclepro.2017.08.102

den Hollander MC, Bakker CA, Hultink EJ (2017) Product design in a circular economy: development of a typology of key concepts and terms. J Ind Ecol 21(3):517–525. https://doi.org/10.1111/jiec.12610

Kirchherr J, Reike D, Hekkert M (2018) Conceptualizing the circular economy: an analysis of 114 definitions. Resour Conserv Recycl 127:221–232. https://doi.org/10.1016/j.resconrec.2017.09.005

Lacy P, Rutqvist J (2015) Waste to wealth: the circular economy advantage. Palgrave Macmillan

Lang C, Armstrong CM, Brannon LA (2022) Collaborative fashion consumption: understanding motivations, attitudes, and behaviors. Sustainability 14(7):4081. https://doi.org/10.3390/su14074081

Mallick PK, Salling KB, Pigosso DC, McAloone TC (2024) Designing and operationalising extended producer responsibility under the EU Green Deal. Environ Challenges 16:100977. https://doi.org/10.1016/j.envc.2024.100977

Martiny A, Taglialatela J, Testa F, Iraldo F (2024) Determinants of environmental social and governance (ESG) performance: a systematic literature review. J Clean Prod 456:142213. https://doi.org/10.1016/j.jclepro.2024.142213

Mboli JS, Thakker D, Mishra JL (2020) An Internet of Things-enabled decision support system for circular economy business model. Softw Practice Exp 52(3):772–787. https://doi.org/10.1002/spe.2825

Monier V, Hestin M, Cavé J (2014) Arcadis Ecologic, Institute for European Environmental Policy (IEEP), & Umweltbundesamt (UBA). Development of Guidance on Extended Producer Responsibility (EPR). https://ec.europa.eu/environment/pdf/waste/target_review/Guidance%20on%20EPR%20%20Final%20Report.pdf

Organisation for Economic Co-operation and Development (2001)

Porter ME, Van Der Linde C (1995) Toward a new conception of the environment-competitiveness relationship. J Econ Perspect 9(4):97–118. https://doi.org/10.1257/jep.9.4.97

Raghunandan G, Thangam D, Patil S, Murthy BMR, Juned M, Duraisamy D, Narasimhaiah R, Ponnada NR (2024) Transitioning toward a circula economy through E-waste management. In: Practice, progress, and proficiency in sustainability, pp 311340. https://doi.org/10.4018/979-8-3693-1018-2.ch020

Rubio S, Ramos TRP, Leitão MMR, Barbosa-Povoa AP (2018) Effectiveness of extended producer responsibility policies implementation: the case of Portuguese and Spanish packaging waste systems. J Clean Prod 210:217–230. https://doi.org/10.1016/j.jclepro.2018.10.299

Shaheen S, Cohen A (2019) Shared mobility: the potential of ride hailing and carsharing to reduce vehicle emissions. Transp Policy 81:86–93. https://doi.org/10.1016/j.tranpol.2019.06.009

Shayan NF, Mohabbati-Kalejahi N, Alavi S, Zahed MA (2022) Sustainable Development Goals (SDGs) as a framework for Corporate Social Responsibility (CSR). Sustainability 14(3):1222. https://doi.org/10.3390/su14031222

Shrivastava A, Abhishek K, Gupta AK, Jain H, Kumari M, Patel M, Sharma P (2024) Removal of micro-and nano-plastics from aqueous matrices using modified biochar–a review of synthesis, applications, interaction, and regeneration. J Hazard Mater Adv 16:100518

Smulders MMJ, Schenning APHJ, Meijer EW (2007) Insight into the mechanisms of cooperative self-assembly: the "Sergeants-and-Soldiers" principle of chiral and achiral C3-symmetrical discotic triamides. J Am Chem Soc 130(2):606–611. https://doi.org/10.1021/ja075987k

Soh S, Ong S, Nee A (2015) Application of design for disassembly from remanufacturing perspective. Procedia CIRP 26:577–582. https://doi.org/10.1016/j.procir.2014.07.028

Stabler D, Hakala H, Huikkola T, Mention A (2024) Aligning servitization and circularity: the role of institutional confluence in sustainable business models. J Clean Prod 462:142666. https://doi.org/10.1016/j.jclepro.2024.142666
Stahel WR (2016) The circular economy. Nature 531(7595):435–438. https://doi.org/10.1038/531435a
TCO Development (2022) TCO certified-impacts and insights: driving sustainability in IT products. TCO Development
Tong X, Wang F, Chen J, Li Y (2024) Advancing circular economy through product life extension and resource efficiency: a systems perspective. J Clean Prod 426:139123. https://doi.org/10.1016/j.jclepro.2023.139123
Torres L, Ripa D, Jain A, Herrero J, Leka S (2023) The potential of responsible business to promote sustainable work–an analysis of CSR/ESG instruments. Saf Sci 164:106151. https://doi.org/10.1016/j.ssci.2023.106151
Tukker A (2015) Product services for a resource-efficient and circular economy–a review. J Clean Prod 97:76–91. https://doi.org/10.1016/j.jclepro.2013.11.049
Underwriters Laboratories Inc (2021) Standard for safety fixed and location-dedicated electric room heaters. In UL 2021 (Fourth Edition). https://inspectapedia.com/heat/UL-2021-2021-en.pdf
Velenturf AP, Purnell P (2021) Principles for a sustainable circular economy. Sustain Prod Consumption 27:1437–1457. https://doi.org/10.1016/j.spc.2021.02.018
Vence X, De Jesus López Pérez S (2021) Taxation for a circular Economy: new instruments, reforms, and architectural changes in the fiscal system. Sustainability 13(8):4581. https://doi.org/10.3390/su13084581
Vikström H, Davidsson S, Höök M (2013) Lithium availability and future production outlooks. Appl Energy 110:252–266. https://doi.org/10.1016/j.apenergy.2013.04.005
WWF International Network (2019) Legal framework Study of extended producer responsibility
Watson D, Bauer B, Milios L, Siderius HP (2023) Unlocking the right to repair: policy frameworks and product design in the circular economy. Resour Conserv Recycl 193:106782. https://doi.org/10.1016/j.resconrec.2023.106782
Wu Y, Tham J (2023) The impact of environmental regulation, environment, social and government performance, and technological innovation on enterprise resilience under a green recovery. Heliyon 9(10):e20278. https://doi.org/10.1016/j.heliyon.2023.e20278

Chapter 11
E-Waste Reduction and Optimization; Algorithmic Waste, Massive Data Centre Energy Usage and Climate Change

Niketa Bhati, Renu Dhupper, Harshita Jain, Anamika Shrivastava, and Meenam Bhatia

11.1 Introduction

The swift digital transformation during the twenty-first century presents a dual-edged outcome that results in unmatched convenience and technological advances while simultaneously causing severe environmental harm. The environmental crisis hinges around electronic waste (e-waste) which represents discarded electronic devices from our growing dependence on technology such as smartphones, tablets, computers, household appliances and industrial electronics According to the Global E-waste Monitor 2024, the annual generation of e-waste is rising by 2.6 million tonnes annually, projected to reach 82 million tonnes by 2030—a 33% increase from 2022. E-waste encompasses discarded electronic products with plugs or batteries, many of which contain hazardous substances like mercury, posing health and environmental risks, which establishes e-waste as the most rapidly expanding domestic waste stream globally (Ankit et al. 2021a, b). The recycling rate for this waste remains worryingly low as only 17.4% of e-waste generated in 2019 underwent formal collection and recycling while the rest ended up in landfills, incinerators, or informal dismantling processes that endanger human health and the environment (Forti 2024).

Modern consumer habits combined with economic systems and technological advancements drive the increase in electronic waste. The current global technology market promotes annual device upgrades through planned obsolescence which leads consumers to treat their electronics as disposable. The shortage of repair facilities and insufficient incentives to reuse electronics makes replacement more practical and

N. Bhati (✉) · R. Dhupper · H. Jain · A. Shrivastava
Amity Institute of Environmental Sciences, Amity University Uttar Pradesh, Noida, India
e-mail: niketabhati1998@gmail.com

M. Bhatia
Department of Botany, Daulat Ram College, University of Delhi, Delhi, India

H. Jain et al. (eds.), *The Circular Fallacy: Closing the Loop in a Consumerist World*, Circular Economy and Sustainability, https://doi.org/10.1007/978-3-032-22687-7_11

financially sensible than repair (Kavil et al. 2025). The decentralization of electronics manufacturing and distribution through globalization presents significant challenges for tracking and managing electronic goods throughout their life cycles. Nations with high income levels generate the most e-waste per person with Norway, the UK, and the US at the top (Buchholz 2021). The majority of e-waste mismanagement falls upon low- and middle-income countries because they receive vast amounts of exported electronics waste for informal recycling and dismantling. Workers along with children in these environments face frequent exposure to harmful substances like lead, mercury, cadmium, and brominated flame retardants because they lack protective gear and safety measures (Bakhiyi et al. 2017).

The rapid expansion of digital infrastructure accompanies the increase in electronic devices throughout the world with data centers being the most prominent example. Data centers serve as the core physical structure for the internet and all modern digital services including cloud computing, social media platforms, artificial intelligence systems and more. Throughout the last twenty years the data center industry worldwide has seen exponential expansion due to growing needs for storage capacity, computational power and immediate information processing (Kez et al. 2022). The massive facilities that tech giants such as Amazon, Google, Microsoft, and Facebook operate represent the digital economy through hyperscale data centers. The global count of hyperscale data centers reached around 600 in 2020 and projections show that this number will surpass 1,000 by 2025 (Swinhoe 2021).

Daa centers produce substantial environmental impacts that span multiple dimensions. Their energy consumption stands as one of the primary concerns. Data centers need large amounts of electrical power to operate servers and storage units alongside networking equipment and keep climate control systems running to prevent overheating. Data centers consumed about 1–1.5% of worldwide electricity demand in 2022 but projections show this could expand to 8% by 2030 without strong efficiency measures (Thangam et al. 2024). The amount of electricity consumed in this way directly affects greenhouse gas emissions when power sources include fossil fuels. Despite certain companies pledging to renewable energy use and carbon neutrality goals many data centers still depend on traditional energy grids which leads to increased climate change impacts.

The clustering of data centers in specific areas puts stress on nearby ecosystems and power systems. The growing population of data centers in Northern Virginia has resulted in grid reliability challenges and increased water use for cooling along with land use disputes (Ciampoli 2024). Data centers consume such a large fraction of electricity in nations like Ireland and the Netherlands that regulators have enforced moratoriums and tightened planning rules (Moss 2024). The construction and operation of data centers become more complex due to the integration of rare earth elements, heavy metals, and non-renewable resources within servers and storage devices. The extraction of these materials demands substantial energy efforts and their recovery becomes challenging when devices reach end-of-life which strengthens the connection between digital growth and electronic waste issues (Goel et al. 2024).

Although hardware receives most attention when people discuss digital sustainability environmental impact discussions overlook the critical yet hidden influence of software. Algorithmic waste has become more recognized recently as it describes software inefficiencies which create excess computational workload while increasing energy consumption (Bharany et al. 2022). Software inefficiencies manifest through extensive codebases together with inadequate algorithm optimization and needless processing activities that overload hardware without enhancing user experience. High-frequency trading platforms along with social media algorithms and AI-powered recommendation engines require continuous computational power that leads to increased energy consumption and requires advanced processors which results in faster hardware wear and tear and subsequent e-waste (Aldoseri et al. 2024). The energy expenditure required to train a large language model equals the lifetime energy consumption of five cars which reveals the hidden energy demands of digital automation and convenience (Ueda et al. 2024).

Algorithmic inefficiencies affect more than just backend systems. User-facing applications frequently experience software bloat because ongoing updates add features but reduce performance. Users experience device slowdowns that lead them to consider replacements even though their hardware remains operational. Software optimization failures result in a cycle that generates increased electronic waste (Islam and Huda 2018). The widespread use of machine learning and artificial intelligence technologies has caused computational resource needs to grow faster than energy-efficient algorithm developments. Deep learning models that train neural networks on massive datasets need thousands of GPU hours to function properly (Justus et al. 2018). These models create a burden on data centers and speed up hardware becoming outdated because current components do not meet the power requirements for new applications.

To mitigate the environmental effects of the digital economy the solution must encompass system-wide interventions throughout all stages of technology development from production to end-of-life management. Implementing circular economy principles in manufacturing processes is crucial. The design process should focus on creating devices that last longer and can be repaired easily along with enabling component recycling (Bovea and Pérez-Belis 2018). The implementation of modular design structures together with standardized components and recyclable material usage enables this transition. The Fairphone project shows how sustainable smartphone design can work by providing devices with features like replaceable batteries and upgradable components made from ethical materials. The examples of sustainable device design must transition from being rare to becoming standard practice (Van Der Velden 2014).

Data centers must aggressively work towards improving their energy efficiency during operational processes. To improve data center energy efficiency organizations use liquid cooling systems while implementing AI for workload optimization and constructing facilities in cold regions to minimize cooling requirements (Zhang et al. 2021). Utilizing renewable energy sources from providers alongside battery storage and demand-response technologies delivers enhanced sustainability benefits. Google and Microsoft have committed to carbon-free operations by 2030 while making substantial investments in wind, solar, and geothermal energy sources (Joppa 2021). To realize effective outcomes, these initiatives require strong regulatory structures and international partnerships for consistent and accountable results. The government holds a critical function through the enforcement of right-to-repair legislation alongside offering tax benefits for green computing approaches and funding e-waste recycling facilities (Grandhi et al. 2024).

Sustainable digital growth requires education and public awareness to function effectively. People must understand the environmental consequences of their digital activities so they can choose technologies that support sustainability objectives. Users can contribute to sustainability by prolonging their electronic device usage duration while selecting efficient appliances and joining electronics recycling initiatives (Dwivedi et al. 2021; Shrivastava et al. 2024). Software developers and engineers must receive training in sustainable coding practices that focus on maximizing algorithmic efficiency while optimizing performance. The digital industry will evolve toward a model which values ecological responsibility equally with innovation and profitability through the integration of sustainability into technical education and professional standards.

The digital revolution has created substantial changes across the global socioeconomic landscape through its transformative effects on communication, productivity, and connectivity. The unchecked expansion of this field generates substantial risks to environmental sustainability. The world faces mounting electronic waste while data centers draw increasing amounts of global energy which, together with inefficient algorithms, quietly compromise our vision of a smart sustainable future. A shared responsibility among digital value chain stakeholders, including manufacturers, policymakers, software developers, infrastructure providers, and end-users, is essential to resolving these issues in a comprehensive manner. We should redirect innovation efforts to guide technology development toward advancing human progress while safeguarding planetary health. Our dedication to reducing the environmental impact of the growing digital economy needs to progress through strategic design approaches and systemic changes along with promoting conscious consumption.

11.2 E-Waste Reduction and Sustainable Management

11.2.1 Sources and Types of E-Waste

E-waste describes discarded electrical and electronic products which have reached their functional lifespan termination. The range of e-waste covers both tiny personal electronics and massive industrial machinery (Jain et al. 2023). Developing effective reduction and sustainable management strategies requires a comprehensive understanding of the various sources and types of e-waste because their composition, quantity, and hazardous potential differ greatly based on origin and device type.

This section examines the primary origins and classifications of e-waste that lay the groundwork for sustainable management discussions.

Electronic waste emerges from various sectors and user groups which can be broadly classified into residential, commercial, industrial, and institutional sources. Household electronics from residential sources make up a large portion of e-waste which encompasses items like smartphones, laptops, televisions, refrigerators and microwaves among other appliances (Noorakma et al. 2024). Technological progress alongside evolving consumer choices and intentional product obsolescence creates a quick replacement cycle for consumer electronics which leads to increased residential electronic waste. Smartphones dominate global markets but their brief lifespan of about 2–3 years results in millions being thrown away every year. The frequent replacement of flat-screen televisions and laptops creates large amounts of electronic waste from domestic consumers.

Business and institutional operations generate substantial e-waste primarily from office hardware and communication systems. Businesses maintain their productivity and security standards by conducting regular upgrades of computers, servers, printers, and telecommunications equipment which results in continuous electronic waste generation (Amankwah-Amoah 2016). Large-scale machinery alongside control systems medical devices and manufacturing equipment forms the specialized e-waste generated by the industrial sector. The intricate nature of these devices together with their rare and hazardous material composition requires specialized handling and recycling procedures.

E-waste sources become easier to understand when product categories are used to determine material composition and management methods. The United Nations University and the International Telecommunication Union categorize e-waste into six primary groups: The primary categories of e-waste include temperature exchange equipment as well as screens and monitors alongside lamps big and small equipment and IT and telecommunication devices (Priyashantha et al. 2022). Refrigerators, air conditioners and heat pumps fall under temperature exchange equipment because they use refrigerants recognized as strong greenhouse gases which require strict handling to avoid releasing them into the atmosphere. The category of screens and monitors includes CRT displays (cathode ray tubes), LCD displays (liquid crystal displays), as well as other display technologies. The widespread use of CRTs ended because these devices hold large amounts of lead and other heavy metals which create environmental dangers when they are not disposed of properly (Priyashantha et al. 2022).

Compact fluorescent lamps (CFLs) along with fluorescent tubes contain mercury that poses serious threats to both human health and the environment. Household appliances such as washing machines, ovens, and dryers constitute large equipment which incorporates electronic control circuits in many models. Small equipment consists of cameras and handheld tools as well as electric shavers while small IT and telecommunication equipment contains mobile phones tablets and personal computers (Priyashantha et al. 2022).

A practical classification method for e-waste considers both its material makeup and potential dangers. This classification divides e-waste into three broad groups: The

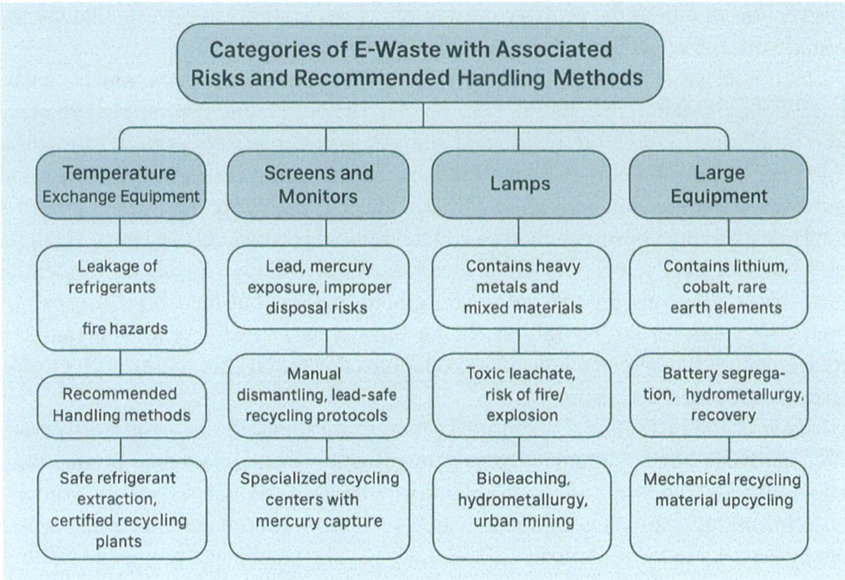

Fig. 11.1 Categorization of E-Waste with Associated Challenges and Solutions. This flowchart visually categorizes common types of e-waste—including temperature exchange equipment, screens and monitors, lamps, and large equipment

three main categories of e-waste include hazardous materials with toxic elements like lead and mercury, valuable components featuring precious metals such as gold and silver, as well as non-hazardous materials composed of plastics and non-toxic metals like aluminum and copper (Ilankoon et al. 2018). This perspective underscores the dual challenge of e-waste management: The goal of e-waste management involves both the prevention of environmental pollution by hazardous substances and the maximization of valuable material recovery to lessen resource extraction demands (Rajesh et al. 2022) (Fig. 11.1).

The two most rapidly expanding e-waste streams worldwide consist of small IT devices and mobile phones because of strong consumer demand and frequent device replacements. The fast-paced technological evolution of smartphones makes them the main contributor to this segment while causing these devices to become obsolete quickly (Annamalai and Gurumurthy 2020). The Global E-Waste Monitor 2020 reports that almost 17% of e-waste produced in 2019 consisted of discarded mobile devices. Large household appliances make up about 38% of global e-waste by weight since they take up more space than other items.

The complexity of managing e-waste increases because electronic devices consist of various different materials. The composition of modern electronics includes numerous metals, plastics, glass, and ceramics which are usually combined into multilayer structures. Printed circuit boards (PCBs) which are essential parts of almost all electronic devices consist of several metal layers such as copper and gold that are set within fiberglass and plastic substrates (Huang et al. 2014). Batteries serve as essential elements for numerous devices yet present extra difficulties because their chemical composition demands strict management of lithium-ion batteries to avoid fire risks during collection and recycling (Windisch-Kern et al. 2021).

Geographic variations in e-waste sources exist because economic development levels as well as technology adoption and regulatory frameworks differ from region to region. High-income nations produce more e-waste per person because they consume electronics more rapidly and their products have shorter lifespans. High-income countries ship their e-waste to developing areas claiming "reuse" or "donation" purposes which results in informal recycling sectors processing the materials without proper environmental supervision (Miljövetenskap 2024). The practice of exporting e-waste to poorer regions transfers environmental and health burdens onto vulnerable populations which underscores the necessity for global cooperation and enforcement of agreements like the Basel Convention on the Control of Transboundary Movements of Hazardous Wastes (Abalansa et al. 2021).

Different source streams require specifically designed e-waste reduction strategies that address their distinct characteristics. Public awareness initiatives alongside take-back schemes and right-to-repair laws help extend residential e-waste product lifecycles while advancing recycling efforts. The implementation of policies that promote leasing options and refurbishment and enforce responsible disposal methods for commercial and institutional sources will lower waste output and decrease environmental damage (Sakao et al. 2024). Specialized treatment facilities become necessary in industrial environments for the safe management of large and complex electronic equipment.

The sustainable management of e-waste requires various stages which encompass collection, sorting of components, refurbishment processes, recycling activities and final disposal methods. The effectiveness of waste collection systems plays a vital role in controlling both the amount and condition of waste that reaches recycling facilities. Material recovery rates increase and informal sector leakage decreases through formal collection programs run by governments or PROs. The sorting process plays an essential role in identifying and removing hazardous components along with materials that hold economic recycling value. Modern sorting equipment including automated optical systems and X-ray fluorescence analyzers along with eddy current separators enhance both accuracy and operational efficiency (Gaur et al. 2023).

The sustainability framework relies heavily on refurbishment and reuse which allows operational devices to extend their service life before they reach the recycling phase. The method both minimizes material usage and enables disadvantaged communities to access affordable electronics (Velenturf and Purnell 2021). Reuse programs need to evaluate data protection measures along with system functionality and adherence to global regulations.

Recycling processes vary according to material types. The mechanical recycling process consists of shredding and component separation whereas chemical recycling extracts precious metals through hydrometallurgy and pyrometallurgy. Extracting metals such as gold and silver presents economic benefits while contributing to resource conservation (Dutta et al. 2022). Certain recycling processes have the potential to emit toxic substances when they are not managed correctly. Building recycling facilities that prioritize environmental safety through rigorous regulations and worker safety measures becomes crucial (Abera 2023).

The integration of circular economy principles remains a crucial element for achieving sustainable e-waste management (Agarwal et al. 2025). Product design should focus on end-of-life recovery by promoting modularity along with easy disassembly. The process includes building supply chains that return materials to production facilities which helps reduce waste and resource extraction. Extended producer responsibility (EPR) policies ensure manufacturers take responsibility for their products throughout their entire lifecycle and promote sustainable design as well as efficient take-back systems (Sakao et al. 2024).

Technological advancements aid in reducing e-waste while promoting sustainable management practices. Innovations like solid-state batteries offer improved safety and durability for energy storage and could decrease waste from batteries (Machín et al. 2024). The creation of biodegradable electronics and printed electronics could minimize environmental effects in future developments. Blockchain technology enhances tracking and transparency in e-waste supply lines to fight illegal disposal and encourage responsible recycling practices (Faiz et al. 2024).

The multitude of sources and types of electronic waste demonstrates the complexity and extensive range of modern electronic technology that evolves quickly. The residential, commercial, institutional, and industrial sectors produce separate waste streams that present different management challenges and opportunities. The path to reducing and sustainably managing e-waste demands knowledge about its diverse sources and comprehensive strategies that integrate policy development, technological advancements, industry practices, and consumer behaviour modification. We can only overcome the environmental and health risks from e-waste while extracting economic value from electronic refuse through comprehensive strategies (Table 11.1).

11.2.2 Strategies for Repair, Refurbishment, and Recycling

Electronic waste (e-waste) is among the fastest-evolving waste streams worldwide, creating huge environmental and health problems. With an increasing number of discarded electronic items, there is a need to establish and practice efficient e-waste management techniques for sustainable practices. Some of the most effective strategies include repair, refurbishment, and recycling, which are all instrumental in prolonging product lifespan, minimizing resource extraction, and limiting hazardous waste. This part elaborates on these strategies at length, identifying their underlying principles, methodology, advantages, and drawbacks, and demonstrating their imperative role in creating a circular economy for the electronics industry.

The most direct and impactful strategy of curtailing e-waste is repair, as it extends the life of electronic items directly by restoring their functionality. Repair entails

Table 11.1 Categories of E-waste with associated risks and recommended handling methods

E-waste category	Examples	Environmental and health risks	Recommended handling method	References
Temperature exchange equipment	Refrigerators, Air conditioners	Leakage of refrigerants (greenhouse gases), Fire hazards	Safe refrigerant extraction, Certified recycling plants	Priyashantha et al. (2022)
Screens and monitors	CRT and LCD TVs, Computer monitors	Lead, Mercury exposure, Improper disposal risks	Manual dismantling, Lead-safe recycling protocols	Priyashantha et al. (2022)
Lamps	Fluorescent tubes, CFLs	Mercury contamination, Inhalation risks	Specialized recycling centers with mercury capture	Priyashantha et al. (2022)
Large equipment	Washing machines, Ovens, Industrial machinery	Contains heavy metals and mixed materials	Disassembly, Component-level material separation	Ilankoon et al. (2018)
Small IT and telecommunication	Mobile phones, Tablets, Laptops	Contains lithium, Cobalt, Rare earth elements	Urban mining, Battery-safe recycling	Annamalai and Gurumurthy (2020)
Hazardous components	PCBs, Batteries	Toxic leachate, Risk of fire/ explosion	Battery segregation, Hydrometallurgical recovery	Windisch-Kern et al. (2021)
Valuable metals	Gold, Silver, Palladium in connectors and PCBs	Economic loss if not recovered	Bioleaching, Hydrometallurgy, Urban mining	Dutta et al. (2022)
Non-hazardous materials	Plastic casings, Aluminium frames	Long degradation time, Landfill volume	Mechanical recycling, Material upcycling	Huang et al. (2014)

fault diagnosis, replacing faulty or worn-out components, and verifying that devices are meeting performance standards (Ilankoon et al. 2018). The process of repair can be implemented on a range of levels, ranging from basic battery changes in mobile phones to intricate repairs in industrial machinery. Promoting repair not only minimizes the number of devices ending up in landfills but also reduces the environmental burden linked with producing new products, such as raw material extraction and production energy (Hazelwood and Pecht 2021).

One of the key obstacles to repair has been the built-in planned obsolescence in a lot of new electronics, where products are deliberately made to be not very repairable. Companies frequently incorporate proprietary screws, glue pieces, or assemble items in a manner that prevents them from being easily disassembled or safely disassembled (Sydor and Stańczyk 2025). In addition, software updates can make older devices obsolete or incompatible, driving consumers to replace instead of repair. To reverse these trends, international advocacy for "right to repair" legislation has gained strength (George et al. 2024). "Right to repair" laws mandate manufacturers to make available spare parts, repair guides, and diagnostic software to enable consumers and independent repair professionals to repair devices at reasonable cost and minimal complexity. The United States, the European Union, and South Korea have made or are debating legislation to promote repairability and consumer choice (George et al. 2024).

Repair efforts also gain from the emergence of repair cafes and community workshop organizations, where volunteers with training help individuals repair their devices. Grassroots campaigns promote sustainable consumption awareness and establish local capacity for device repair. Online platforms that link users to certified repair professionals further enhance access to cost-effective and timely repairs. In addition, repairability scores and eco-labels, for example, the iFixit repairability score, alert consumers to how easily a given product can be repaired, encouraging producers to create more repairable devices (Dangal et al. 2025).

Refurbishment takes repair a step further by not only putting devices into operating condition but also readying them for resale or donation to other users. Refurbished electronics are thoroughly tested, worn-out parts replaced, software reapplied, and cleaned for appearance. This practice lengthens the useful life of equipment, postpones entry into waste streams, and makes technology affordable, especially in low-income areas or under-served communities. The international marketplace for used electronics has grown very quickly, facilitated by both commercial refurbishers and non-profit groups (Wallner et al. 2024).

An essential component of effective refurbishment is quality assurance. Consumers' confidence in refurbished products relies on dependability, warranties, and disclosure of device history and condition. Grading systems standardized by industry categorize refurbished products by cosmetic and functional standards, serving to manage consumer expectations and support market expansion (Belbağ and Belbağ 2023). Refurbishment also serves to discourage demand for newly made devices, easing pressure on natural resources and reducing greenhouse gas emissions from production and transportation.

Refurbishment initiatives are being more commonly used within corporate social responsibility (CSR) plans and extended producer responsibility (EPR) programs (SAP 2024). Most tech firms have instituted take-back schemes in which customers are provided with the option to return old devices to be refurbished or recycled. Apple's Renew initiative and Dell's Asset Resale and Recycling Services are notable examples of how manufacturers are integrating refurbishment into product life cycle management (Apple 2025). But the effectiveness of such initiatives is dependent on good collection systems, consumer response, and logistics with minimal environmental impacts.

Recycling is the last action on the waste hierarchy but is also vital in the management of devices that are no longer repairable or refurbishable. Recycling entails recovering valuable content from junk electronics and disposing of hazardous components in a safe manner. Electronic devices hold a vast range of materials, ranging from precious metals such as gold, silver, and palladium to base metals such as copper and aluminium, plastics, glass, and other toxic materials (Ankit et al. 2021a, b). Effective recycling saves virgin material extraction, saves energy, and avoids environmental pollution.

Most recycling processes start with collection and sorting. Collection systems can be extremely diverse, ranging from municipal curbside collection to specific e-waste drop-off points and retailer take-back initiatives. Accurate sorting is required to disassemble various types of devices and materials in order to maximize recovery and prevent contaminated mixing of hazardous wastes with non-toxic substances (National Academies Press (US) 2011). In most parts of the world, unorganized e-waste recycling industries function outside the lines of governance, using unsafe techniques like open incineration or acid baths to release metals, which are very harmful to health and cause environmental contamination. Formal e-waste recycling plants try to avoid these dangers by following strict environmental and safety protocols.

After sorting, mechanical and chemical recycling processes are used to reclaim material. Mechanical recycling is performed by shredding equipment and segregation of parts via magnetic separation, eddy current separation, and flotation. Efficient recovery of metals such as copper and aluminium is possible using these, whereas plastics and glass are recycled to secondary products (Thompson et al. 2021). Chemical recycling, also more commonly pyrometallurgical and hydrometallurgical, separates precious metals from intricate assemblies such as printed circuit boards (Li et al. 2018). Pyrometallurgy involves the use of high-temperature smelting to extract metals, and hydrometallurgy utilizes leaching of metals through aqueous solutions (Singh and Schwan 2011). Both processes depend on specialized equipment and technical expertise, stressing the necessity for investing in superior recycling facilities.

Technology innovations in recycling are allowing for higher recovery rates while lowering environmental footprints. Urban mining, for instance, recovering valuable elements from e-waste in urban areas, is becoming increasingly popular as a means of sustainable resource management (Ouro-Salim 2024). Sensor-based sorting, robotics, and artificial intelligence are improving the accuracy and effectiveness of recycling processes. Bioleaching, which employs microorganisms to remove metals from e-waste, is also being researched, with great hope for environmentally sound alternatives to chemical processing (Adetunji et al. 2023).

Apart from its significance, recycling is hampered by factors such as economic feasibility, policy impediments, and the fast pace of electronic devices' evolution. The declining precious metals content in newer products and the growing application of sophisticated composites make material recovery more difficult and costly. In addition, conflicting policy frameworks among countries result in illicit shipments of e-waste and slow the formation of global recycling standards. Global instruments such as the Basel Convention try to solve these problems by controlling transboundary movement of hazardous waste and ensuring environmentally sound management approaches (Alter 2000).

Combining repair, refurbishment, and recycling approaches in a circular economy structure maximizes resource productivity and reduces harm to the environment. Circular economy practices involve keeping products and materials in use for as long as possible, designing for disassembly and recyclability, and recovering materials at end-of-life. This can be achieved through collaboration along the value chain among designers, manufacturers, consumers, recyclers, and policymakers.

Repair and refurbishment design is a key enabler of sustainable e-waste management. Choosing standardized components, modular architecture, and long-lasting materials makes it easier for manufacturers to perform maintenance and extend product life. Transparency through documentation and open standards supports repair and refurbishment and minimizes the reliance on proprietary parts or equipment.

Producer responsibility policies, like extended producer responsibility (EPR), make producers responsible for ensuring that products' whole lifecycle, including take-back and end-of-life treatment, is handled. EPR schemes provide financial and organizational incentives for producers to create sustainable products and fund repair and recycling infrastructure. EPR has been mandated in some countries for electronics, leading to higher collection rates and investment in recycling technologies (Sin and Tueen 2023a, b).

Consumer behavior also plays a key role. Awareness campaigns, return incentives for old devices, and creating a culture of reuse can stimulate demand for repaired and refurbished goods. Online platforms that enable resale and exchange of electronics help extend device lifetimes and minimize waste (Sin and Tueen 2023a, b).

Finally, repair, refurbishment, and recycling are cornerstones of electronic waste management in a sustainable way. Repair extends device lifespan, suppressing premature replacements and minimizing environmental consequences linked with manufacturing. Refurbishment converts used devices into useful assets for new consumers, enhancing social fairness and resource preservation. Recycling recovers precious materials and disposes of hazardous materials safely, closing the e-products loop. When triangulated and buttressed by supportive policies, technological advancements, and customer participation, these strategies constitute an integrated strategy for minimizing the impending e-waste challenges and developing a more sustainable digital economy.

11.2.3 Policy Frameworks and Global Initiatives on E-Waste

The international problem of e-waste has stimulated the creation of multiple policy systems and global programs focusing on the sustainable management and minimizing the environmental and health hazards resulting from inappropriate disposal. E-waste with dangerous chemicals like lead, mercury, and cadmium, together with recoverable valuable substances, requires multi-jurisdictional action. At the top of global initiatives is the Basel Convention on the Control of Transboundary Movements of Hazardous Wastes and Their Disposal, signed in 1989 to control the transboundary movement of hazardous waste, including e-waste, and dump it in developing nations. Imposing its attention to the growing masses of e-waste, the Basel Convention made a groundbreaking amendment in 2022, being the Ban Amendment and the e-waste amendment, that encompasses all e-waste—both hazardous and non-hazardous—under stricter regulatory control by utilizing the Prior Informed Consent (PIC) process (Basel Convention Secretariat 2023). This amendment, which will become operative in 2025, mandates exporters to seek express approval from importing nations prior to exporting any e-waste and thereby seeks to check illegal exports and promote environmentally sound management worldwide.

Parallel to the Basel Convention's worldwide approach is the Bamako Convention, a regional agreement adopted in 1991 by African countries to regulate the transboundary movement and disposal of hazardous wastes in Africa. Unlike the Basel Convention, which allows controlled shipments under certain conditions, the Bamako Convention outright bans the import of hazardous waste into African countries, reflecting the continent's strong stance against becoming a dumping ground for e-waste generated elsewhere (Bamako Convention 1998). This regional setup highlights the environmental justice aspect of managing e-waste, safeguarding vulnerable groups from the harm posed by the squalid and iniquitous recycling that commonly infests developing areas.

In addition to these conventions, the Organisation for Economic Co-operation and Development (OECD) has revised its guidelines according to emerging international best practices. In 2024, OECD member nations updated their e-waste shipment regulations to conform their regulations to the Basel Convention's revised provisions to ensure notification and approval requirements for all e-waste exports (OECD 2024). This move is important in having transparency in e-waste trade between developed nations and ensuring commitments to proper recycling.

At the international level, nations have started to implement legislation institutionalizing extended producer responsibility (EPR), a policy framework where the responsibility is put on the manufacturers to deal with the entire life cycle of their products, including the collection and recycling of post-consumer wastes. India, being one of the rapidly growing e-waste generators in the world, adopted the E-Waste Management Rules in 2022, which require producers to meet certain collection targets and make sure that the e-waste gets treated in environmentally friendly facilities (Ministry of Environment, Forest and Climate Change 2022). The rules focus on the creation of producer responsibility organizations (PROs) that organize the collection and recycling, which holds them responsible and ensures effective recovery of resources.

Just like that of Japan, the overall legal system for the management of e-waste brings together the Law for the Promotion of Effective Utilization of Resources (LPUR) and the Law for the Recycling of Specified Kinds of Home Appliances (LRHA). Introduced in 2001 and amended several times, the LRHA requires manufacturers to take care of the recycling of some types of home appliances such as refrigerators, air conditioners, television sets, and washing machines (Ministry of environment 2023). This law lays down specific recycling goals, mandates retailers to collect discarded appliances, and charges consumers a recycling fee in order to have an integrated system that promotes circular resource flows and minimizes landfill disposal.

In the United States, there are very different e-waste management policies at the state level, with the Environmental Protection Agency (EPA) laying down general guidelines and voluntary programs. Of particular note is California's leadership through legislative means with the Electronic Waste Recycling Act of 2003, which limits the utilization of specific hazardous chemicals in electronics and requires a recycling fee imposed at the time of sale to support statewide e-waste collection programs (Dhanorkar and Muthulingam 2024). Subsequent states have established patchworks of legislation that further prioritize producer accountability, safe e-waste handling, and public education.

Industry-driven efforts also have a critical role to play in formulating sustainable e-waste management. Voluntary certification programs like the e-Stewards Standard, which was formulated by the Basel Action Network (BAN), offer a stringent set of guidelines for electronics recyclers to show compliance with environmental and ethical principles. The e-Stewards certification forbids the export of toxic e-waste to developing nations and requires safe recycling processes, protection of workers' health, and openness in operations (Basel Action Network 2025). Through the establishment of market differentiation for ethical recyclers, these certifications promote the enhancement of recycling facilities and foster consumer trust.

Notwithstanding these developments, e-waste management still has ongoing issues. Informal recycling industries, especially in nations like Ghana, Nigeria, and India, remain prevalent e-waste processing, driven by economic conditions and inadequate formal infrastructure. Rudimentary practices such as open burning and acid leaching are employed to harvest precious metals in these informal settings, thereby exposing workers and the population to harmful substances and severe environmental degradation (E-Waste Monitor. 2024). Closing the gap between formal and informal recycling involves deep-seated policies that incorporate informal workers, offering training and upgrading recycling equipment.

In addition, the accelerated rate of technology innovation brings shorter product lifespan and rising amounts of wasted electronics, making it challenging to enforce rules and recover resources. More recent devices include sophisticated materials and compact components that stretch current recycling technologies and lower the economic attractiveness of material extraction (International Energy Agency 2023). Solving these problems requires policy encouragement for product design innovations focused on durability, modularity, and recyclability in addition to motivating manufacturers to apply circular economy principles.

Future policy priorities focus on multi-stakeholder engagement and the convergence of international standards. Governments are urged to enhance EPR schemes through the establishment of challenging collection and recycling rates, enhancing e-waste flow traceability, and promoting R&D of new recycling technologies like urban mining and bioleaching (World Economic Forum 2024). Public education campaigns are important to encourage consumer involvement in take-back programs and sustainable consumption behaviors.

Global cooperation is essential to counter illegal exports and environmental injustice in e-waste trade. The mechanisms of information sharing and enforcement under the Basel Convention serve as a model for cooperative action, yet more resources and political will are required for covering compliance monitoring and capacity building in developing nations. The application of circular economy ideas in more general climate and resource policy also brings e-waste management into line with international sustainability objectives, such as the United Nations Sustainable Development Goals (SDGs) (UN Environment Programme 2023).

Overall, efficient e-waste management relies on a sophisticated interrelationship between international treaties, local laws, industry norms, and local movements. The Basel and Bamako Conventions, and OECD guidelines, establish the legal and moral basis for regulating hazardous waste transportation and ensuring environmentally friendly recycling. National policies involving EPR and product stewardship push responsibility to sustainable product life cycles, and industry certifications promote best practices for recycling. Addressing informal sector challenges, technological complexity, and enforcement shortfalls calls for continuous innovation and partnership. With increasing awareness of e-waste's social and environmental consequences, these policy guidelines and activities together offer a way forward to mitigate the global e-waste burden and create a more circular and sustainable digital economy.

11.2.4 Circular Design Principles for Electronics

Circular design principles are fundamental to transforming the electronics industry from a linear "take-make-dispose" model to a regenerative system that minimizes waste, maximizes resource efficiency, and extends product lifecycles. These principles are increasingly recognized as essential for reducing the environmental impact of electronic devices and promoting sustainability (Husain 2024). Modularity is a cornerstone of circular design, allowing individual components of electronic devices—such as batteries, screens, and processors—to be easily replaced or upgraded. This approach extends the device's lifespan and reduces the need for complete replacements. For instance, companies like Fairphone have pioneered modular smartphones that empower users to repair and upgrade their devices, thereby reducing electronic waste (Schischke et al. 2019).

Repairability goes hand-in-hand with modularity. Designing electronics to be easily disassembled and repaired encourages a shift from a disposable culture to one that values maintenance and longevity. This can involve using standardized screws, avoiding the use of glue, and providing clear documentation for repairs. Such practices not only facilitate repairs but also promote a repair culture among consumers (Sierra-Fontalvo et al. 2023).

Designing products for disassembly is crucial for enabling efficient recycling and reuse of materials. Devices that are difficult to take apart often end up in landfills, as they cannot be effectively recycled. By designing electronics with disassembly in mind—such as using snap-fit components or screws instead of glue—manufacturers can ensure that devices can be easily taken apart at the end of their life cycle, facilitating material recovery and reuse (Ostapska et al. 2024).

The choice of materials in electronic devices significantly impacts their environmental footprint. Opting for sustainable and recyclable materials, such as recycled plastics or biodegradable components, can reduce the environmental impact of electronics. Moreover, avoiding the use of hazardous substances ensures that devices can be safely recycled and do not pose health risks during disposal (Quinto et al. 2025).

11.3 Understanding and Minimizing Algorithmic Waste

11.3.1 Definition and Causes of Algorithmic Waste

Algorithmic waste describes the adverse effects and system inefficiencies that emerge unintentionally when algorithmic systems are deployed. Social, ethical, and environmental harms emerge because of biases in data and a lack of transparency alongside the excessive use of automated systems and substantial resource consumption. The term "algorithmic pollution" describes the harmful effects generated by algorithmic systems that spread through social environments in a way that resembles environmental pollutants. The framing stresses how these harms remain widespread and typically uncontrolled while underscoring the need for thorough surveillance and responsibility in the functioning of algorithmic systems (Marjanovic et al. 2018).

Data bias serves as the main source of inefficiency in algorithmic systems. Bias in training data leads algorithms to maintain societal prejudice which results in discriminatory algorithmic decisions. Studies reveal that facial recognition systems produce higher misidentification rates for marginalized groups compared to white individuals due to biases present in their training datasets. Technical errors in algorithms represent deeper systemic issues because they are rooted in historical and societal data which reinforce existing social inequalities while appearing objective (Ferrara 2023a, b).

Decision-making in algorithms lacks transparency which serves as another contributing factor. AI systems function as "black boxes" which complicates the task of understanding their decision-making processes (Wischmeyer 2019). The hidden nature of these systems prevents accountability while making the identification and correction of system errors or biases extremely difficult. Stakeholders lack the ability to challenge or improve systems without transparent understanding of algorithmic processes. Proprietary protections combined with complex algorithms worsen transparency issues because they protect companies from examination while obstructing ethical practice assurance efforts (Wischmeyer 2019).

Algorithmic waste occurs when organizations depend too heavily on automated systems. Organizations' growing reliance on automated systems for decision-making threatens to marginalize human judgment (Schultz et al. 2024). When organizations depend mainly on automated systems for making decisions this can result in flawed algorithms making essential choices without proper human supervision which may lead to harmful consequences. The criminal justice system faces unjust sentencing results when risk assessment algorithms operate without human oversight. Excessive reliance on automation results in individuals losing their human identity while institutional trust deteriorates (Schultz et al. 2024).

Algorithmic waste also includes the environmental consequences of AI systems operations. The development of extensive AI models demands high-level computational power which results in considerable energy use and carbon production. AI operations supported by data centers need massive water quantities for their cooling systems. The environmental impact of AI systems highlights the necessity for sustainable development practices that align technological progress with ecological stewardship (Alsabt et al. 2024).

Multiple interrelated factors such as data bias, transparency deficits, automation dependency, and environmental elements lead to the creation of algorithmic waste. To resolve these problems we need a comprehensive strategy that includes ethical data handling principles along with transparency in AI systems and balanced human oversight combined with environmentally sustainable development methods. We can develop fair and accountable algorithmic systems by identifying and reducing these underlying factors.

11.3.2 Impact of Inefficient Coding and Data Processing on Energy Use

In the digital era, the efficiency of software code and data processing methods significantly influences energy consumption. Inefficient coding practices and suboptimal data handling not only degrade system performance but also lead to increased energy usage, contributing to higher operational costs and environmental concerns.

One of the primary contributors to energy inefficiency in software systems is the presence of "code smells"—patterns in the code that indicate deeper problems. These can include redundant computations, unnecessary object creation, and poor loop constructs. A study by Poy et al. (2024) demonstrated that the removal of such code smells could lead to a measurable reduction in energy consumption, highlighting the importance of clean coding practices.

Code refactoring, the process of restructuring existing computer code without changing its external behavior, has been identified as a method to improve energy efficiency. Research by found that specific refactoring techniques could lead to significant energy savings, especially when applied to object-oriented programming languages like Java and C# (Sanlialp and Ozturk 2020). The choice of programming language itself can impact energy consumption. A study by Nicolas et al. (2024) analyzed various programming languages and found that some languages

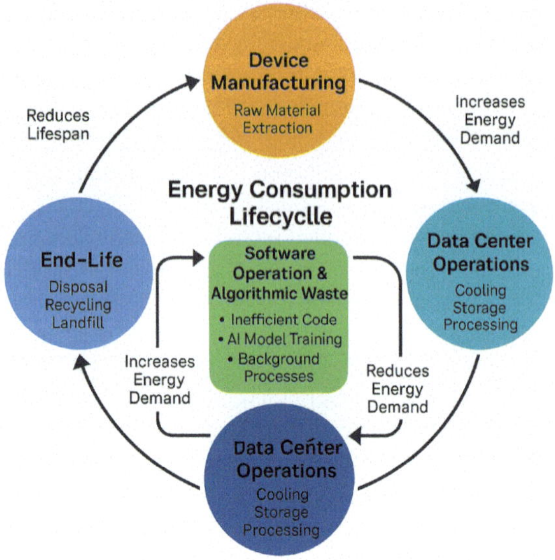

Fig. 11.2 Energy Consumption Lifecycle of the Digital Ecosystem. This diagram illustrates the interconnected stages of energy use in the digital economy, including device manufacturing, software operation and algorithmic waste, data center operations, data transmission, and end-of-life processing. It highlights how inefficient software and algorithmic processes increase energy demand across the system, ultimately contributing to shorter device lifespans and higher environmental impact

inherently consume more energy due to their design and runtime characteristics. This suggests that selecting an energy-efficient language is crucial for developing sustainable software (Fig. 11.2).

Beyond coding practices, data processing inefficiencies also play a significant role in energy consumption. Inefficient resource allocation in big data platforms can lead to idle resources that consume energy without performing useful work. Research by Song et al. (2018) emphasized the need for optimized resource scheduling to enhance energy efficiency in data processing tasks. Data centers, which are central to data processing, are significant energy consumers. According to the International Energy Agency (2023), data centers worldwide consumed about 460 TWh of electricity, accounting for approximately 2% of global electricity use. This underscores the importance of energy-efficient data processing and storage solutions (Khosravi et al. 2024).

Innovations in hardware design, such as in-memory processors, have been proposed to address data processing inefficiencies. A study by Overney (2023) introduced a processor that integrates data processing and storage, reducing the energy overhead associated with data movement between memory and processing units. In the realm of databases, the choice of database management systems (DBMS) and query optimization techniques can influence energy consumption. Furthermore, the adoption of green coding practices is essential for reducing the carbon footprint of software applications. The role of developer awareness and tooling in promoting eco-friendly coding is also highlighted, suggesting that integrating energy efficiency considerations into the development process can lead to more sustainable software (Salazar 2024).

11.3.3 Approaches to Optimize Software and Reduce Computational Overhead

Reducing computational overhead is a fundamental goal in designing efficient, scalable, and sustainable software systems. As digital systems grow in complexity and data processing demands increase, minimizing overhead becomes vital not only for performance but also for energy efficiency and environmental sustainability. One of the most direct ways to address this issue lies at the heart of software itself—through compiler-level optimizations that transform code to run more effectively.

Compiler optimizations play a pivotal role in reducing processing overhead by enhancing the efficiency of executable code. One of the most well-established techniques is *loop unrolling*, where the loop body is replicated multiple times to decrease the number of iterations and reduce the frequency of conditional checks. This transformation reduces the overhead associated with control flow, which can be particularly useful in time-critical applications. However, it can also lead to increased binary size, which may negatively impact instruction cache utilization, particularly on memory-constrained systems (Cooper and Torczon 2012). Thus, such optimizations must be carefully balanced to prevent performance regressions in other system areas.

Building upon basic loop-level transformations, *loop nest optimization* techniques such as *loop tiling* (or blocking) restructure nested loops to take advantage of spatial and temporal locality. By breaking down large loops into smaller, cache-friendly blocks, loop tiling significantly improves cache performance and reduces memory latency. This strategy is especially relevant in domains like high-performance computing (HPC), where multidimensional array computations are common, and cache misses can be extremely costly (Blaszynski et al. 2024). These compiler transformations not only reduce overhead but also boost execution speed without requiring hardware upgrades, making them highly cost-effective solutions.

Another technique that extends the power of compiler optimizations is *automatic vectorization*, which converts scalar operations into vector instructions that can process multiple data elements simultaneously. This is achieved using SIMD (Single Instruction, Multiple Data) instructions, which can drastically reduce execution time by parallelizing data-level operations. However, not all loops are amenable to vectorization—complex control structures or unpredictable branches can hinder the compiler's ability to generate efficient vector code. To overcome this, methods such as inserting *vector branches* or using conditional execution within vector lanes have been proposed, which help maintain performance while dealing with irregular data flows (Tayeb et al. 2023).

While code transformations improve software efficiency, optimizing how data is accessed and managed in memory is equally essential for reducing overhead. Efficient *memory management* ensures that data is available where and when it is needed, minimizing delays caused by memory access. One widely used technique is *prefetching*, where the system predicts future memory accesses and loads the corresponding data into the cache in advance. This proactive strategy reduces cache miss rates and enhances application responsiveness, especially in scenarios with predictable access patterns like multimedia processing or scientific simulations (Jeffers and Reinders 2013a, b).

In addition to improving cache usage, reducing memory duplication can also lower computational burden. In *cloud computing environments*, where multiple virtual machines often share similar workloads, *Static Memory Deduplication (SMD)* identifies identical memory pages and consolidates them. This not only decreases physical memory usage but also reduces the management overhead required to maintain separate memory copies. As a result, systems benefit from improved efficiency and lower energy consumption, making SMD an important technique in virtualized infrastructure (Jia et al. 2017).

A related concept, *in-memory processing*, further enhances performance by moving data storage and computation from disk to RAM. This eliminates the latency typically associated with disk I/O and is particularly effective in real-time analytics, financial services, and database systems. By keeping data close to the processor, in-memory architectures allow faster access and reduced overhead, which is crucial for workloads that require rapid decision-making (Jia et al. 2017).

Complementing compiler and memory optimizations, *parallel computing* offers another layer of efficiency by distributing tasks across multiple processing units. In *data parallelism*, subsets of a dataset are processed simultaneously by different processors executing the same operations. This model is especially suitable for matrix computations, simulations, and image processing, where uniform operations are applied across large datasets (Shrawankar and Joshi 2022). However, parallelism also introduces new challenges such as synchronization, thread contention, and inter-process communication delays. Poorly designed synchronization can negate the performance benefits of parallelism by creating bottlenecks and increasing overhead. Therefore, effective *scheduling algorithms*, minimized task dependencies, and efficient communication protocols are critical to realizing the full advantages of parallel computing (Shrawankar and Joshi 2022).

Beyond software and architectural optimizations, system-level improvements also play a crucial role in minimizing processing overhead. At the *operating system (OS) level*, managing context switches, streamlining system calls, and optimizing interrupt handling are key areas of focus. Each context switch incurs overhead as the OS saves and restores process states, which can become a performance bottleneck in systems with high concurrency. Reducing unnecessary switches and making system calls more efficient allows more CPU time to be allocated to application processes (Popiolek et al. 2021).

In *virtualized environments*, *performance monitoring tools* are indispensable for tracking resource usage and identifying bottlenecks. However, these tools can introduce their own overhead if not managed properly. A solution to this is *selective monitoring*, where only the most critical performance counters are tracked, thereby minimizing the impact of monitoring on system performance. By focusing on essential metrics, administrators can maintain observability without compromising efficiency (Popiolek et al. 2021).

Finally, hardware-level innovations such as *reconfigurable computing architectures* offer a promising approach to dynamically tailor computing resources to the needs of specific applications. Technologies like Field Programmable Gate Arrays (FPGAs) allow custom hardware configurations, which can be optimized for particular tasks such as signal processing or cryptographic computations. This flexibility reduces processing overhead by eliminating unnecessary general-purpose logic and executing only task-specific instructions, thereby maximizing performance per watt (Goldstein 1990).

Thus, reducing computational overhead is a multifaceted challenge that requires interventions at multiple layers of the computing stack—from compiler design and memory management to parallelism, OS optimization, and hardware customization. Each of these techniques contributes uniquely to minimizing latency, improving throughput, and reducing energy consumption. Together, they form a holistic strategy for building high-performance and environmentally sustainable computing systems. As the digital economy continues to grow, adopting such optimizations will be essential to meet the dual demands of computational intensity and ecological responsibility.

11.3.4 Case Studies on Green Software Engineering

Green software engineering is rapidly gaining traction as a key strategy to reduce the environmental impact of the digital sector. It focuses on designing, developing, and deploying software systems that are energy-efficient and environmentally sustainable. This discipline integrates practices such as optimizing code, selecting energy-efficient hardware, using renewable energy, and implementing sustainability throughout the software lifecycle. As digital infrastructures grow globally, these practices are not only beneficial but necessary. Numerous organizations and research groups have implemented green software engineering principles, and their case studies provide practical insights into its real-world application and benefits.

One illustrative example involves a prominent telecommunications company that collaborated with the Boston Consulting Group (BCG) and the Green Software Foundation to optimize its software systems. The initiative focused on streamlining the company's codebase and incorporating caching mechanisms, which resulted in a remarkable 30% reduction in energy consumption across its IT systems between 2019 and 2020. This effort effectively lowered the Software Carbon Intensity (SCI) score, a metric used to measure the environmental impact of software per unit of output. This case strongly underscores how seemingly small code-level optimizations can collectively lead to significant energy savings and operational efficiencies (Martin 2023).

Building on the importance of software and infrastructure alignment, the case of Hosterra and Kodama demonstrates how sustainable outcomes can be achieved through collaborative design choices. Hosterra, a French green web hosting provider, partnered with Kodama, an agency developing eco-designed websites. The companies chose to host on Scaleway's DC5 data center, which is renowned for its energy and water efficiency. Additionally, they implemented LiteSpeed web servers, known for using approximately 30% fewer CPU cycles than traditional servers. As a result, Kodama's websites achieved a 10–15% reduction in energy consumption. This example highlights the importance of aligning software performance optimization with eco-conscious infrastructure decisions to maximize energy savings (Martin 2023).

Further evidence of large-scale optimization can be seen in Google's continual enhancements to its data center energy efficiency. Through tactical physical reconfigurations such as relocating air vent tiles to the hottest zones, raising ambient temperatures to 27 °C, and using containment curtains to isolate uninterruptible power supply (UPS) areas, Google achieved a significant drop in energy usage per watt. These changes resulted in annual energy savings of 670 MWh and cost reductions of around $67,000 (Google 2023). This case demonstrates how physical data center management strategies can complement software-level efficiencies in achieving broader sustainability goals.

In a related effort to modernize IT infrastructure for energy optimization, energy provider E.ON partnered with Google Cloud to migrate its physical network infrastructure to the cloud. The migration enabled the company to enhance its scalability and operational resilience, achieving 99.9% system availability. More importantly, the shift improved the efficiency of E.ON's digital operations while allowing the company to offer customers more granular control over their energy use. Cloud computing, when executed with sustainability in mind, provides an opportunity for businesses to access shared computing resources that are more energy-efficient than isolated on-premise systems (Thomas 2024).

Artificial intelligence and intelligent thermal management have also been harnessed to drive energy efficiency in data centers, as seen in Huawei's case. The company introduced two AI-powered systems: iCooling, for thermal management, and iPower, for optimizing power distribution. These systems enabled Huawei to reach a Power Usage Effectiveness (PUE) ratio of 1.15 in its Ulanqab data center, which is among the most efficient globally. Additionally, the use of these AI-driven tools led to a 3.2% reduction in energy consumption at its Ningxia data center, amounting to a savings of 400,000 kWh annually (Thomas 2024). This example reveals how machine learning technologies can be employed to dynamically adjust operations, reducing environmental impact while maintaining system performance.

Beyond large tech firms, smaller organizations have also embraced green software engineering through thoughtful policy and tooling decisions. RJJ Software, for instance, implemented various sustainable development practices, such as using energy-efficient M2-series Mac hardware, refactoring code to reduce computational resource demand, and encouraging remote work to minimize emissions associated with commuting. Additionally, the company carefully selected cloud providers with strong sustainability credentials. This case exemplifies how even small businesses can contribute meaningfully to green IT efforts by making strategic decisions aligned with sustainability (RJJ Software 2024).

Cultural and organizational change is another crucial component of green software engineering. Shenyang Aircraft Corporation in China provides an example of how fostering a sustainability-oriented corporate culture can complement technical solutions. The company integrated environmental considerations into the design and production phases of its products and initiated training sessions to educate employees on green IT strategies. Furthermore, it established partnerships with academic institutions to advance sustainability research and practices. This holistic approach emphasizes the need for leadership commitment and employee involvement in driving sustainable outcomes (Zeng et al. 2018).

Academic and research communities have also contributed significantly to formalizing green software engineering. The Green Software Lab (GSL) is a prominent research project aiming to create a solid foundation for the discipline by developing methodologies and tools to evaluate and enhance the energy efficiency of software systems. Through empirical studies and tool development, GSL has promoted greater awareness and adoption of sustainable practices in software engineering education and industry (Abreu et al. 2021). Their work is paving the way for broader integration of environmental metrics in software performance assessments.

Supporting this academic momentum, the Green Algorithms framework offers researchers and developers a standardized method to estimate the carbon footprint of computational tasks. This tool calculates emissions based on variables like CPU/GPU usage time, hardware efficiency, and the carbon intensity of the data center's energy mix. Such frameworks are vital because they quantify software's environmental impact, which has traditionally been overlooked in performance evaluation. By providing visibility into carbon emissions, developers are empowered to make more environmentally responsible design choices (Lannelongue et al. 2020).

Together, these case studies reflect the multifaceted nature of green software engineering. From code-level changes and infrastructure adjustments to cultural shifts and research-driven tools, each contributes uniquely to making digital systems more sustainable. Importantly, they show that green software engineering is not limited to tech giants or research institutions—small businesses and public organizations can also make meaningful progress by adopting context-appropriate practices. As the global digital footprint continues to grow, these real-world applications offer a roadmap for building efficient, responsible, and sustainable software systems across all sectors.

11.4 Energy Consumption and Optimization in Data Centers

11.4.1 Energy Demands of Modern Data Centers

Modern data centers have become indispensable to the digital infrastructure, supporting services ranging from cloud computing to artificial intelligence (AI). However, this growing reliance has led to significant energy consumption, raising concerns about environmental sustainability. Understanding the energy demands of data centers and exploring optimization strategies is crucial for mitigating their environmental impact. Data centers consume substantial amounts of electricity to power servers, storage devices, and networking equipment (IEA 2024). According to the International Energy Agency (IEA), global data center electricity consumption in 2022 was estimated to be between 240 and 340 terawatt-hours (TWh), accounting for approximately 1–1.3% of global electricity demand. This figure excludes energy used for cryptocurrency mining, which adds an estimated 110 TWh. The proliferation of AI technologies has further escalated energy requirements. Research indicates that AI workloads could consume up to 20% of global data center electricity, potentially rising to nearly 50% by the end of 2025 (IEA 2024).The surge in AI applications necessitates high-performance computing resources, notably Graphics Processing Units (GPUs), which are more energy-intensive than traditional Central Processing Units (CPUs). State-of-the-art GPUs for AI tasks can consume up to 700 watts per chip, significantly increasing the power density within data centers. This escalation not only amplifies electricity consumption but also intensifies cooling requirements to maintain optimal operating temperature (Deloitte 2024a, b).

Cooling systems are critical components of data centers, ensuring that equipment operates within safe temperature ranges. Traditional air conditioning methods are energy-intensive, often accounting for 38–40% of a data center's total power consumption. Innovative cooling techniques, such as liquid cooling and immersion cooling, have emerged as more energy-efficient alternatives. Immersion cooling, for instance, involves submerging servers in thermally conductive dielectric fluids, effectively dissipating heat and reducing reliance on air-based cooling systems (Deloitte 2024a, b).

To assess and improve energy efficiency, data centers utilize the Power Usage Effectiveness (PUE) metric, which is the ratio of total facility energy consumption to the energy used by IT equipment. An ideal PUE is 1.0, indicating that all consumed energy is used for computing purposes. While the average PUE for U.S. data centers is around 2.0, state-of-the-art facilities have achieved PUEs as low as 1.01 through advanced cooling and energy management strategies (Shi et al. 2016).

Artificial Intelligence and machine learning are increasingly employed to optimize energy usage within data centers. By analyzing real-time data, AI systems can dynamically adjust cooling and power distribution, ensuring efficient operation. For example, AI-driven tools can predict server loads and modulate cooling systems accordingly, reducing unnecessary energy expenditure (Shi et al. 2016).

Renewable energy integration is another pivotal strategy for reducing the carbon footprint of data centers. Operators are investing in on-site renewable energy sources, such as solar and wind power, and entering Power Purchase Agreements (PPAs) to secure clean energy from external providers. These initiatives not only decrease reliance on fossil fuels but also stabilize energy costs. However, the intermittent nature of renewable sources necessitates the use of energy storage systems to ensure a consistent power supply (Shi et al. 2016),

Energy storage solutions, like Battery Energy Storage Systems (BESS), play a crucial role in managing energy consumption. They store excess energy generated during low-demand periods and supply it during peak loads or outages, enhancing the reliability and efficiency of data center operations. Moreover, leveraging energy storage can enable data centers to participate in energy markets, providing grid services and generating additional revenue (Shi et al. 2016).

Waste heat recovery is an innovative approach to improving energy efficiency. Data centers generate substantial heat, which can be captured and repurposed for heating nearby buildings or for industrial processes. For instance, some facilities in Europe have implemented systems that redirect waste heat to local heating networks, effectively reducing overall energy consumption and contributing to community heating needs (Zhou et al. 2025).

The concept of energy-proportional computing emphasizes designing systems where energy usage scales proportionally with workload. This approach ensures that servers consume minimal power during low-utilization periods, enhancing overall energy efficiency. Implementing energy-proportional principles requires advancements in hardware design and power management techniques (Katal et al. 2022). Therefore, the escalating energy demands of modern data centers, driven by the proliferation of AI and digital services, necessitate comprehensive optimization strategies. By adopting advanced cooling technologies, integrating renewable energy sources, utilizing AI for energy management, and implementing waste heat recovery, data centers can significantly reduce their environmental impact. Continued innovation and commitment to sustainability are essential to ensure that data centers meet the growing digital demands without compromising environmental integrity.

11.4.2 Technologies for Energy Efficiency: Cooling, Hardware, Renewable Integration

As data centers continue to expand rapidly to accommodate the accelerating demands of cloud computing, big data, and artificial intelligence (AI), their energy consumption has emerged as a critical concern. With the global volume of data doubling every few years and AI workloads becoming increasingly computationally intensive, energy use in these facilities is skyrocketing. This trend not only has financial implications for operators but also environmental consequences, contributing to greenhouse gas emissions and placing pressure on local power grids. In response, researchers, engineers, and data center operators are actively exploring and implementing a range of technologies aimed at improving energy efficiency. These efforts can be broadly categorized into three core areas: advanced cooling systems, energy-efficient hardware, and renewable energy integration (Yadav and Jain 2024; Jain 2025b).

One of the primary contributors to energy consumption in data centers is the cooling infrastructure required to maintain optimal operating temperatures for servers and networking equipment. Traditionally, air-based cooling systems have been used to dissipate heat generated by IT equipment. However, as computing densities increase, these methods are proving to be less effective and more energy-intensive. As a result, alternative approaches like immersion cooling are gaining traction. Immersion cooling involves submerging servers in non-conductive dielectric fluids that effectively absorb and dissipate heat. This technique has demonstrated the potential to reduce energy consumption associated with cooling by up to 50% when compared to conventional air cooling systems. Additionally, it allows for higher server density and reduced spatial requirements, making it a promising solution for hyperscale data centers facing space and energy constraints (Udinmwen 2025).

Complementing immersion cooling, another notable innovation is the use of rear-door heat exchangers, which provide an efficient and scalable method for heat removal. Dell's PowerCool Enclosed Rear Door Heat Exchanger (eRDHx), for instance, is designed to capture and dissipate 100% of the heat emitted by IT equipment directly at the rack level. This system utilizes water at elevated temperatures to remove heat, eliminating the need for power-hungry chillers and reducing cooling energy costs by as much as 60%. By enabling the use of higher-temperature cooling water, these systems align well with renewable thermal sources and further simplify infrastructure requirements (Udinmwen 2025). Such approaches demonstrate how innovative thermal design can yield substantial improvements in operational efficiency.

Beyond hardware-based solutions, intelligent cooling management powered by artificial intelligence is increasingly being adopted to optimize data center operations. AI-driven systems analyze real-time telemetry data, including temperature, humidity, and workload distribution, to dynamically adjust cooling parameters. Machine learning models, particularly deep reinforcement learning algorithms, have shown the ability to optimize these systems autonomously. For example, researchers have demonstrated that using AI to fine-tune cooling policies can lead to up to 15% energy savings without compromising performance or reliability (Li et al. 2017). These intelligent systems are not only capable of making decisions faster than human operators but can also adapt to changing environmental conditions and workload profiles, thereby continuously refining their strategies to maximize energy efficiency.

Parallel to cooling innovations, advancements in hardware design and power distribution are crucial to minimizing energy consumption in data centers. Modern servers now incorporate energy-efficient processors and chipsets that offer better performance-per-watt ratios. These components often come with built-in power management features that enable dynamic voltage and frequency scaling (DVFS), allowing systems to reduce energy usage during periods of low demand. In addition, virtualization technologies have revolutionized server utilization. By running multiple virtual machines on a single physical server, organizations can dramatically reduce the number of physical servers required, thereby decreasing the total energy footprint (Chidolue et al. 2024). This consolidation is especially valuable in legacy data centers where space and cooling capacity may be limited.

In terms of electrical architecture, the adoption of 400 V Direct Current (DC) power distribution systems is another promising development. Traditional Alternating Current (AC) systems often involve multiple stages of voltage conversion, each introducing inefficiencies and power losses. In contrast, 400 VDC systems minimize these conversion stages, resulting in more streamlined and efficient power delivery. This setup is particularly beneficial for data centers supporting AI and high-performance computing (HPC) workloads, which demand high power density and stable delivery. The move toward DC-based architectures also aligns with the use of renewable energy sources, many of which inherently produce DC power, thus avoiding the need for additional conversion equipment (Udinmwen 2025).

While improving internal efficiency is vital, integrating renewable energy sources into data center operations is equally important for reducing overall carbon emissions. Many data center operators are now investing in solar, wind, and hydroelectric power either through direct installation or long-term Power Purchase Agreements (PPAs). For instance, Digital Realty has committed to powering its facilities with 100% renewable energy and aims to reduce both direct and indirect emissions by 60% per square foot by 2030 (Mok 2025). This transition not only contributes to environmental sustainability but also protects operators from volatility in energy prices, adding a layer of economic resilience.

One particularly innovative renewable energy technology being explored is the photovoltaic thermal hybrid solar collector (PVT). These hybrid systems simultaneously generate electricity and heat, thereby maximizing the energy harvested from solar radiation. Compared to traditional photovoltaic systems, PVTs achieve higher overall energy conversion efficiency, making them well-suited for data centers that require both electrical and thermal energy. In regions with suitable solar insolation, PVT systems can significantly offset the energy demands of on-site cooling systems (Sreekumar et al. 2024).

Another sustainable practice gaining momentum is waste heat recovery, where excess heat generated by data centers is repurposed for other applications, such as district heating or industrial processes. A notable example is the Aquasar system at ETH Zurich, which uses warm-water cooling to collect waste heat from servers. This captured heat is then reused to supply heating for adjacent buildings, achieving up to 80% energy reuse. Such systems not only reduce the energy required for separate heating systems but also present a model for the circular economy within the tech industry (Miller 2024).

Overall, addressing the energy demands of data centers requires a multi-pronged approach that combines cooling efficiency, advanced hardware, and renewable energy adoption. From immersion cooling and AI-driven thermal management to efficient power distribution and innovative energy harvesting methods, each of these technologies contributes to a broader strategy of sustainable digital infrastructure. These solutions not only help reduce operational costs and environmental impact but also ensure that data centers can continue to scale without jeopardizing global climate goals. As data demand continues to grow, the widespread adoption of these energy-efficient technologies will be essential for building a greener, more resilient digital future (Saini et al. 2024).

11.4.3 Innovations in Data Center Design and Operation

As data centers evolve to meet the escalating demands of digital services and artificial intelligence (AI), innovations in design and operation have become pivotal for enhancing energy efficiency, scalability, and sustainability. These innovations are not isolated trends but rather reflect a broader shift in how organizations approach infrastructure, efficiency, and environmental responsibility. Technologies such as modular construction, AI-driven energy management, renewable energy integration, and strategic site selection are collectively transforming the data center landscape. With energy consumption rising due to increasing workloads and digital infrastructure requirements, these advancements offer a viable path toward a more sustainable and future-ready computing ecosystem (Khan et al. 2024).

Among the most transformative of these developments is modular construction. Unlike traditional data center builds that can span years and involve complex on-site logistics, modular construction uses prefabricated units that are manufactured off-site and assembled on location. This method dramatically shortens construction timelines, reduces labor costs, and ensures consistent quality across deployments. It also allows operators to scale operations incrementally, deploying modules only as needed in response to demand fluctuations. By matching capacity to real-time usage, this approach minimizes idle infrastructure and conserves both physical and energy resources. Digital Realty, one of the largest global data center operators, has adopted modular construction across several of its facilities to enhance energy performance and reduce construction-related emissions (Mok 2025). The flexibility inherent in modular construction also supports rapid geographic expansion, enabling data centers to be deployed in emerging markets or edge environments where traditional builds would be economically or logistically unfeasible.

While modular design addresses structural efficiency, the operational side of data center management has increasingly turned to artificial intelligence for further optimization. AI technologies offer the ability to monitor, predict, and control energy consumption in real-time, thereby optimizing operations beyond human capability. For instance, Digital Realty's Apollo AI platform harnesses real-time analytics to fine-tune cooling systems, balance server workloads, and reduce power consumption based on live environmental and system data. These adjustments are often imperceptible in individual actions but accumulate to generate substantial resource savings across large-scale facilities. Similarly, Meta has been pioneering the use of machine learning in its data center management strategies, particularly in advanced campuses like its Singapore facility, where AI models help regulate airflow and temperature with surgical precision (Jackson 2024). This AI-driven operational agility not only reduces energy use but also contributes to hardware longevity and operational resilience, reinforcing sustainability from multiple angles.

However, optimizing internal operations alone is not sufficient for long-term sustainability. As concerns over climate change intensify, data center operators are increasingly integrating renewable energy sources into their power supply strategies (Saini et al. 2025). Renewable integration serves two primary goals: reducing dependence on fossil fuels and stabilizing energy costs over time. Companies such as Digital Realty are leading this transition by committing to powering all of their global facilities with 100% renewable energy and reducing direct and indirect emissions per square foot by 60% by 2030. This goal aligns with broader corporate sustainability mandates and reflects growing investor and regulatory expectations for carbon accountability. Moreover, advanced technologies such as photovoltaic thermal hybrid solar collectors (PVT) are being adopted to maximize solar energy use. PVT systems combine solar photovoltaic electricity generation with solar thermal heating, resulting in higher total energy output and improved utilization of the solar spectrum (CMIC 2025). These systems not only supply electricity but can also power absorption chillers used for cooling, further lowering fossil energy consumption.

Where and how a data center is situated also plays a crucial role in both operational efficiency and environmental impact. Strategic site selection is a critical determinant of a data center's sustainability potential. Climate, energy infrastructure, local policies, and proximity to user bases all influence site desirability. For instance, building data centers in cooler climates can drastically reduce cooling requirements, while access to reliable renewable energy grids ensures cleaner operations. Mumbai has recently emerged as a global data center hotspot, ranking sixth in the world for under-construction capacity. Its robust infrastructure, energy availability, and connectivity make it a prime location for scaling digital services efficiently. Meanwhile, in the United States, Amazon's announcement of a $10 billion investment in North Carolina highlights the strategic importance of locating facilities in regions with favorable energy infrastructure and government incentives (Eanes 2025). By choosing optimal sites, companies can achieve both operational excellence and regulatory compliance more effectively.

As digital services become increasingly decentralized, the architecture of data centers is adapting accordingly. The rise of edge computing—where processing occurs closer to the source of data—has necessitated a redesign of data center infrastructure. Unlike traditional centralized data centers that serve vast geographies, edge data centers are smaller, modular, and often designed for rapid deployment. These facilities handle specific, localized workloads, reducing latency and bandwidth requirements. Edge data centers are also engineered to be energy-efficient, using modular technologies that align with renewable sources and intelligent management systems to optimize power use (CMIC 2025). They play a critical role in supporting emerging applications like autonomous vehicles, real-time analytics, and smart cities, which require near-instant data processing. As a result, edge computing is not merely a technical trend but a core driver of next-generation data center innovation.

Equally important to performance and efficiency is the issue of resilience. As dependency on digital infrastructure grows, the consequences of downtime become increasingly severe. Resilience and disaster recovery have, therefore, become paramount in data center design and operation. Modern data centers are engineered with multiple redundancies, including dual power sources, uninterruptible power supplies (UPS), advanced fire suppression systems, and even seismic isolation in earthquake-prone regions. These safeguards ensure operational continuity under adverse conditions. Vantage Data Centers, for example, builds its facilities to meet 99.999% uptime standards—a benchmark that translates to less than 5 min of downtime per year (Hutchinson 2024). Such resilience is achieved through layered infrastructure design, rigorous testing, and failover systems that activate automatically during disruptions. These capabilities are particularly critical in an age when even a few seconds of downtime can cost enterprises millions in lost revenue, reputational damage, or regulatory fines.

While these innovations are impressive on their own, their full value is realized when integrated into a comprehensive strategy for sustainability and performance. Rather than pursuing isolated upgrades, leading data center operators are adopting holistic design philosophies that embed sustainability into every layer of infrastructure. This includes optimizing airflow through computational fluid dynamics (CFD), using recycled water for cooling systems, adopting circular economy principles in equipment lifecycle management, and collaborating with local utilities to participate in demand response programs. These multidimensional approaches ensure that efficiency is not achieved at the expense of resilience, and vice versa (Ibn-Mohammed et al. 2024).

Furthermore, these strategies are increasingly being codified into sustainability reporting frameworks, such as those proposed by the Green Software Foundation and industry-specific guidelines under initiatives like the Climate Neutral Data Centre Pact. Compliance with these standards not only satisfies stakeholder expectations but also opens up access to green financing, tax incentives, and enhanced brand equity. The future of data center design will likely involve even deeper integration with smart grid technologies, AI-powered energy forecasting, and quantum computing innovations that could further reduce the power needs of large-scale computation (Gutterman 2024).

Overall, innovations in data center design and operation are not just responses to present challenges—they are proactive solutions shaping the future of digital infrastructure. From modular construction and AI-driven energy management to renewable integration, strategic siting, edge computing, and resilience engineering, these technologies collectively redefine what is possible in the realm of data center efficiency and sustainability. As demand for computing continues to rise exponentially, these innovations are essential not only for operational excellence but for building an infrastructure that aligns with global environmental goals. By continuing to invest in these forward-looking strategies, organizations can ensure that the backbone of the digital economy remains both resilient and responsible for generations to come.

11.4.4 Role of Cloud Computing and Edge Computing in Reducing Energy Load

As the digital landscape continues to evolve, the global reliance on data-driven services, artificial intelligence (AI), and internet-connected devices has grown exponentially. This transformation, while technologically empowering, has also contributed to a significant surge in energy demand, particularly within data centers—the foundational infrastructure of digital ecosystems. Data centers, which house servers and networking equipment, are estimated to consume between 1 and 2% of global electricity, a figure that is expected to rise as cloud services and AI applications scale further. In this context, the integration of energy-efficient technologies is not merely beneficial but essential for ensuring a sustainable digital future. Two of the most transformative technologies in this regard—cloud computing and edge computing—have emerged as powerful enablers of energy-efficient digital infrastructure. Each brings distinctive strengths to the table and, when strategically integrated, offers a synergistic model for reducing the energy load across the computing continuum (Himeur et al. 2023).

Cloud computing is perhaps the most visible component of this energy transformation. It centralizes storage, computing, and networking resources within large-scale, professionally managed data centers. These facilities are designed to maximize efficiency and reduce waste by consolidating workloads from numerous clients onto shared infrastructure. This aggregation allows cloud service providers to significantly improve server utilization rates and reduce the overall number of servers needed, thereby decreasing total energy consumption (Katal et al. 2021). Centralized facilities are also better equipped to deploy advanced thermal management systems, such as AI-controlled cooling, immersion cooling, and modular containment systems, which can lower energy used for temperature control—a major component of data center power consumption (Katal et al. 2021).

One of the most compelling advantages of cloud computing is its scalability and adaptability. Virtualization technologies, container orchestration, and elastic resource provisioning enable cloud platforms to dynamically adjust computing resources based on demand. This elasticity ensures that servers are not operating at full capacity unnecessarily, thereby reducing idle energy consumption. Furthermore, cloud providers are making significant investments in renewable energy procurement and power purchase agreements (PPAs) to run their data centers on low-carbon electricity. For example, both Google and Microsoft have committed to powering their operations with 100% renewable energy, a move that not only improves environmental performance but also sets new industry standards for sustainable IT practices (Reddy et al. 2024). These strategies enable cloud computing to offer economies of scale that are not only financially advantageous but also environmentally efficient.

Despite the benefits of centralized cloud infrastructure, not all workloads are best suited for remote data centers—particularly those requiring ultra-low latency or local data processing. This is where edge computing plays a crucial role. Unlike cloud computing, which relies on centralized data processing, edge computing brings computation and data storage closer to the point of data generation. In practical terms, this could mean deploying micro data centers or intelligent gateways near factories, retail locations, vehicles, or urban intersections equipped with IoT sensors. By doing so, edge computing reduces the need for data to travel long distances to centralized servers, significantly cutting down the energy required for data transmission across networks (George et al. 2023).

The energy-saving potential of edge computing is particularly pronounced in applications requiring real-time decision-making, such as autonomous vehicles, telemedicine, smart manufacturing, and predictive maintenance. These systems cannot tolerate the delays associated with round-trip communication to distant data centers. By enabling localized data processing, edge computing not only meets these latency requirements but also minimizes the energy costs linked to high-volume data transfer. Additionally, localized systems can be engineered for efficiency using low-power processors and purpose-built AI accelerators, which are optimized for specific tasks and consume less energy than general-purpose hardware (Kolapo et al. 2024). This level of task-specific optimization is particularly important as the number of IoT devices and edge applications continues to grow.

However, neither cloud nor edge computing offers a complete solution in isolation. The most energy-efficient and performant infrastructures increasingly rely on hybrid architectures that blend the strengths of both paradigms. In a well-orchestrated hybrid model, edge computing handles latency-sensitive and context-aware processing tasks locally, while less time-critical data is offloaded to the cloud for deeper analysis, storage, or backup. This division of labor ensures that computational tasks are distributed to the most appropriate location based on urgency, data volume, and processing requirements. It not only optimizes energy usage but also enhances system resilience and reduces congestion on core networks (Narayana 2024).

This balance between cloud and edge is increasingly being enhanced by advances in artificial intelligence and machine learning. AI plays a key role in monitoring workloads, predicting performance bottlenecks, and dynamically allocating computational resources in both cloud and edge environments. Machine learning models can analyze real-time data on server loads, energy prices, and environmental conditions to determine the most energy-efficient strategy for processing and distributing workloads. For example, AI can shift computation from cloud to edge during peak energy pricing hours or when renewable energy availability is high locally. Such intelligent workload orchestration ensures continuous performance optimization while aligning with energy efficiency objectives (Harris 2024).

Moreover, cloud-edge synergy enables innovations in demand-side energy management. In smart grids, for instance, edge nodes can locally manage energy-intensive tasks based on real-time energy availability, while cloud systems analyze regional consumption patterns to make macro-level adjustments. Together, these technologies form a distributed energy-aware computing fabric that adapts to both user needs and environmental constraints (Sathupadi et al. 2024).

While technological capability is essential, policy frameworks and industry standards also influence the energy impact of cloud and edge computing. Governments and regulatory bodies are increasingly encouraging data center efficiency through carbon reporting requirements, energy benchmarking, and green certification programs. These measures, coupled with market-driven pressure for sustainability, are prompting data center operators and cloud providers to adopt more aggressive efficiency goals and invest in greener infrastructure (Silva et al. 2023).

In summation, cloud computing and edge computing are not merely technological conveniences—they are fundamental to the quest for sustainable digital infrastructure. Cloud computing brings unparalleled economies of scale, operational flexibility, and renewable integration capabilities (Amajuoyi et al. 2024). Edge computing, on the other hand, offers localized intelligence, reduced network energy use, and enhanced performance for latency-sensitive applications. Together, they form a complementary system that balances centralized efficiency with distributed agility (Nair 2023).

As the scale and complexity of global computing systems continue to grow, the integration of cloud and edge architectures will become not only more prevalent but more vital. Their combined strengths—when orchestrated with AI-driven management, powered by renewables, and supported by regulatory frameworks—create a blueprint for an energy-aware digital future. This integrated approach promises not only to meet the computational demands of tomorrow's digital society but to do so in a way that is aligned with our urgent need for environmental stewardship. In this sense, cloud and edge computing are not just technological shifts—they are essential pillars of digital sustainability (Table 11.2).

11.5 Climate Change Implications and Mitigation Strategies

11.5.1 Contribution of Digital Infrastructure to Greenhouse Gas Emissions

Digital infrastructure—including data centers, cloud platforms, and communication networks—has become integral to modern society, enabling everything from online communication to advanced data analytics. However, this digital expansion comes with significant environmental implications, particularly concerning greenhouse gas (GHG) emissions (Bhardwaj, et al. 2025; Agarwal and Jain 2024; Jain 2025a, b).

Data centers are central to digital infrastructure and are substantial energy consumers. They require continuous power to operate servers and maintain optimal temperatures through cooling systems. According to the International Energy Agency (IEA), data centers and data transmission networks accounted for approximately 330 million tonnes of CO_2 equivalent in 2020, representing about 0.9% of energy-related GHG emissions. This figure underscores the significant carbon footprint of digital operations (IEA, 2024).

The rapid growth of digital services, particularly cloud computing and artificial intelligence (AI), exacerbates this issue. AI applications are notably energy-intensive,

Table 11.2 Summary of key strategies for sustainable digital ecosystems

Strategy/Intervention	Domain	Key benefits	References
Right-to-repair laws and repair cafés	E-waste management	Extends device life, reduces premature disposal, empowers consumers	George et al. (2024), Dangal et al. (2025)
Modular and circular product design	Product manufacturing	Improves repairability, facilitates recycling and reduces material waste	Van Der Velden (2014), Sierra-Fontalvo et al. (2023)
Refurbishment programs (e.g., Apple Renew)	Consumer electronics	Reduces demand for new products, lowers emissions, enhances digital access	Wallner et al. (2024), Belbağ and Belbağ (2023)
Bioleaching and urban mining	E-waste recycling	Enhances recovery of rare metals with reduced environmental impact	Adetunji et al. (2023), Ouro-Salim (2024)
Compiler-level and memory optimization techniques	Algorithmic efficiency	Lowers software energy demand, improves processing speed	Cooper and Torczon (2012), Jeffers and Reinders (2013a; b)
Green software engineering (e.g., SCI score use)	Software development	Reduces carbon intensity of code, encourages sustainable development practices	Martin (2023), Lannelongue et al. (2020)
AI-driven data center cooling (e.g., iCooling)	Data center operations	Enhances energy efficiency, reduces thermal load	Thomas (2024)
Immersion cooling and liquid cooling systems	Data center cooling	Cuts cooling energy use up to 50%, supports higher server density	Udinmwen (2025)
Renewable energy and battery storage integration	Infrastructure/ Operations	Decreases carbon footprint, increases energy resilience	Mok (2025), Shi et al. (2016)
Waste heat recovery and circular heat use	Sustainable infrastructure	Reuses energy for heating, promotes energy circularity	Miller (2024)
Cloud + Edge computing	Distributed computing	Reduces data transit energy, supports real-time processing	George et al. (2023), Himeur et al. (2023)
Extended Producer Responsibility (EPR) laws	Governance and policy	Holds manufacturers accountable, encourages eco-friendly product lifecycle	Ministry of Environment, Forest and Climate Change (2022), Basel Convention Secretariat (2023)

requiring substantial computational power and, consequently, more energy. Reports indicate that the electricity usage of data centers is expected to double by 2026, with AI contributing to a 160% increase in demand for data center power. This surge in energy consumption poses challenges for sustainability efforts (World Economic Forum 2025).

Furthermore, the environmental impact of digital infrastructure extends beyond energy consumption. The manufacturing, maintenance, and disposal of hardware components contribute to the sector's overall carbon footprint. Embodied emissions from the production of servers, cooling systems, and networking equipment add to the environmental burden. A study highlighted that data centers currently consume about 1% of global electricity demand and contribute around 2% of global GHG emissions, with projections indicating emissions could reach 1.9 gigatonnes of CO_2 per year by 2025 (Helaimia 2024).

In response to these challenges, several mitigation strategies are being implemented. One approach involves enhancing energy efficiency within data centers. This includes adopting advanced cooling technologies, optimizing server utilization, and implementing AI-driven energy management systems. For instance, Digital Realty, a global data center operator, is focusing on sustainability by using renewable energy, improving cooling systems, and incorporating liquid cooling into future data center designs (Mok 2025).

Transitioning to renewable energy sources is another critical strategy. By powering data centers with solar, wind, or hydroelectric energy, operators can significantly reduce their carbon emissions. However, it's essential to ensure that these renewable energy sources are genuinely additive and not merely offsetting emissions through certificates that may not lead to actual reductions. Critics argue that reliance on renewable energy certificates (RECs) can obscure the true environmental impact if they lack "additionality"—meaning they don't necessarily lead to new renewable energy projects (Mundy 2025).

Moreover, policy interventions play a vital role in driving sustainable practices. Governments and regulatory bodies are beginning to enforce energy-use transparency, set efficiency standards, and create specific rate structures for data centers. For example, Californian lawmakers are proposing legislation to mandate energy-use transparency and set efficiency standards for data centers, aiming to prevent the public from bearing the costs associated with the infrastructure needed to support these facilities (Johnson 2025).

While digital infrastructure is indispensable for modern society, its environmental implications, particularly concerning GHG emissions, cannot be overlooked. Through a combination of technological innovation, renewable energy adoption, and policy interventions, it's possible to mitigate the environmental impact of digital infrastructure and move towards a more sustainable digital future.

11.5.2 Carbon Footprint Assessment and Reporting

Carbon footprint assessment and reporting have become integral components of global efforts to mitigate climate change. As governments, corporations, and individual stakeholders intensify efforts to limit global warming to below 1.5 °C—as outlined in the Paris Agreement—the ability to quantify and monitor greenhouse gas (GHG) emissions has become not only a strategic necessity but also a regulatory expectation. With increasing transparency demands and pressure from investors, regulators, and consumers, robust methodologies and transparent reporting practices are essential to assess environmental performance accurately and guide informed decision-making (Geneidy et al. 2021).

The concept of a carbon footprint refers to the total amount of GHG emissions caused directly and indirectly by an individual, organization, event, or product throughout its lifecycle. These emissions are typically measured in terms of carbon dioxide equivalents (CO_2e), encompassing other potent GHGs such as methane (CH_4), nitrous oxide (N_2O), and fluorinated gases, each normalized to the global warming potential of CO_2. This measurement provides a common metric for comparing emissions from different activities, sectors, or products. Importantly, carbon footprinting goes beyond just counting emissions—it forms the foundation for identifying emission hotspots, prioritizing mitigation strategies, and setting emission reduction targets (Wiedmann and Minx 2008).

One widely adopted methodology for carbon footprint assessment is the Life Cycle Assessment (LCA) approach. LCA evaluates the environmental impacts associated with all stages of a product's life—from raw material extraction and manufacturing to usage and end-of-life disposal. Its strength lies in its holistic perspective: emissions from upstream and downstream processes are considered, helping to avoid the common pitfall of problem-shifting, where emission reductions in one part of the lifecycle cause increases elsewhere. For example, a study on the carbon footprint of office paper found significant variations in results depending on whether ISO 14040/14044, PAS 2050, or the CEPI framework was applied (Scrucca et al. 2020). These discrepancies underline the importance of standardizing assessment methodologies to ensure consistency, comparability, and credibility across sectors and regions.

Beyond LCA, carbon accounting frameworks are extensively used by organizations to quantify, categorize, and report emissions in a more manageable format. The Greenhouse Gas Protocol, developed by the World Resources Institute (WRI) and the World Business Council for Sustainable Development (WBCSD), is one of the most recognized standards for organizational carbon accounting. It categorizes emissions into three scopes: Scope 1 refers to direct emissions from owned or controlled sources; Scope 2 includes indirect emissions from the consumption of purchased electricity, steam, heating, and cooling; and Scope 3 encompasses all other indirect emissions that occur in the value chain, such as those from purchased goods, transportation, waste, and business travel (Jha et al. 2021). This classification system allows for a structured and detailed understanding of where emissions are generated, enabling targeted reduction strategies and more robust sustainability disclosures.

However, significant challenges persist in carbon footprint assessment and reporting, particularly concerning methodological inconsistencies. The lack of harmonized standards and varying assumptions can lead to divergent results even when evaluating the same process or product. A comparative study of carbon footprint standards emphasized that differences in system boundaries, functional units, and emission factor databases can produce substantially different results, impeding cross-organizational and cross-sectoral comparability (Gao et al. 2013). This variability not only undermines confidence in reported figures but also complicates benchmarking, regulatory compliance, and third-party verification.

In addition to methodological disparities, the reliability of carbon footprint assessments is heavily dependent on the quality, granularity, and timeliness of input data. Incomplete, outdated, or approximated data can result in significant overestimations or underestimations of emissions, thereby distorting sustainability performance and misleading stakeholders. For instance, a study analyzing municipal emissions inventories in the United States discovered that many American cities tend to underreport their carbon footprints due to data limitations and inconsistent accounting practices, leading to discrepancies between self-reported inventories and top-down scientific estimates (Gurvich and Creamer 2021). This highlights the urgent need for standardized data collection protocols, integration with real-time monitoring systems, and improved data governance.

To address these limitations and improve accuracy, emerging technologies are being integrated into carbon footprinting practices. One of the most promising developments is the use of artificial intelligence (AI) and machine learning algorithms to process large and complex datasets, identify patterns, and enhance predictive capabilities. These technologies enable real-time tracking of emissions and can fill data gaps with high-confidence estimates. A notable example includes the development of AI-powered models for evaluating the carbon footprint of alternative cement materials. Researchers demonstrated that such models could significantly expedite the analysis of various cement compositions, enabling the rapid assessment of lower-emission alternatives (Al-Fakih et al. 2024). These tools are particularly relevant in construction, manufacturing, and other high-emission sectors where material substitution can offer significant climate benefits.

In parallel with the application of AI, cloud-based digital platforms are emerging as practical tools to help organizations manage their carbon footprint in an integrated and scalable way. These platforms offer dashboards for emission tracking, scenario analysis, target setting, and report generation. Plan A, a Berlin-based sustainability software company, exemplifies this approach. Their platform allows organizations to input emissions data across operations and supply chains, assess it against science-based targets, and identify interventions that align with global climate goals (Plan A 2024). By providing actionable insights in an accessible format, such digital tools lower the barrier to entry for companies seeking to enhance their environmental accountability.

Moreover, the integration of carbon footprint assessment with corporate sustainability reporting is gaining momentum. Investors and regulators increasingly demand disclosures aligned with frameworks such as the Task Force on Climate-related

Financial Disclosures (TCFD), the Sustainability Accounting Standards Board (SASB), and the upcoming International Sustainability Standards Board (ISSB). These frameworks often require companies to report Scope 1, 2, and increasingly Scope 3 emissions, and to disclose climate-related risks and opportunities. Carbon footprint assessments thus serve not only as an environmental metric but also as a financial risk indicator, informing investment decisions and corporate valuations.

From a policy perspective, several governments have introduced or are piloting mandatory carbon reporting for large organizations. The European Union's Corporate Sustainability Reporting Directive (CSRD), for instance, will require around 50,000 companies to provide detailed climate-related disclosures starting in 2024. Similarly, California's Climate Corporate Data Accountability Act will mandate companies operating in the state and earning over $1 billion in revenue to report their emissions, including Scope 3. These regulatory developments are pushing carbon footprinting from a voluntary initiative into a compliance obligation, further increasing the importance of reliable methodologies and robust data infrastructure.

Looking forward, advances in blockchain technology may also offer promising pathways for improving the transparency and traceability of carbon footprint data. Blockchain's decentralized ledger can be used to securely log emissions data, create immutable audit trails, and verify the authenticity of carbon credits and offsets. While still in its nascent stages, the intersection of carbon accounting and blockchain holds potential to enhance trust in carbon markets and prevent issues such as double-counting or greenwashing (Thanasi-Boçe and Hoxha 2025).

Therefore, carbon footprint assessment and reporting are essential instruments for climate action. They provide the foundation for measuring progress, setting credible targets, and informing stakeholders about environmental performance. Although methodological challenges and data quality issues remain, technological innovations and policy developments are rapidly addressing these barriers. The increasing alignment of carbon footprinting with regulatory standards, financial disclosures, and digital tools suggests that it will continue to play a central role in the global transition to a low-carbon economy. By embracing accurate, transparent, and technology-driven approaches to carbon accounting, businesses and institutions can not only reduce their environmental impact but also position themselves competitively in an increasingly climate-conscious world.

11.5.3 Policy Measures, Corporate Commitments, and Carbon Offsetting

The escalating urgency of the climate crisis has galvanized a multifaceted response encompassing governmental policies, corporate strategies, and market-based mechanisms like carbon offsetting. While each component plays a distinct role, their interplay determines the efficacy of global efforts to achieve net-zero greenhouse gas

(GHG) emissions. This essay delves into the intricacies of policy measures, corporate commitments, and carbon offsetting, examining their synergies, challenges, and the path forward (Etukudoh et al. 2024).

Governments worldwide are enacting policies to drive emissions reductions and promote sustainable practices. Incorporating carbon offsetting into national climate strategies can enhance the effectiveness of emission reduction efforts. For instance, the European Scientific Advisory Board on Climate Change has advised the EU to exclude international carbon credits from its climate targets for 2040, warning that their inclusion could undermine efforts to significantly reduce domestic emissions. The board recommends a 90–95% cut in greenhouse gas emissions from 1990 levels by 2040. Board chair Ottmar Edenhofer emphasized the urgency of domestic reductions and criticized carbon credits as risky and potentially misleading due to concerns over greenwashing. This stance conflicts with Germany's position, which supports counting international carbon offsets for up to 3% of its emissions goals (ESABCC 2025). France's support hinges on business compensation measures, while political polarization within EU member states further complicates consensus. Despite these challenges, the EU remains on track for its 2030 target of a 55% emissions reduction and views the 2040 goal as a crucial milestone toward achieving net-zero emissions by 2050. The 2040 target will influence the EU's 2035 commitment to the UN ahead of COP30 and is seen as a key indicator of the EU's ongoing climate leadership and global commitments, particularly in light of previous U.S. policy reversals under President Trump (Hancock 2025).

Corporations are increasingly setting ambitious climate targets, but the quality and transparency of these commitments vary. Many companies have announced net-zero targets; however, the exclusion of significant emission sources, such as Scope 3 emissions, undermines the comprehensiveness of these goals. A study by that some corporate net-zero pledges are not as robust as they appear, often omitting critical emission sources. While carbon offsets can play a role in corporate climate strategies, overreliance on them without substantial internal emission reductions raises concerns about the authenticity of commitments (Kim et al. 2024). Research indicates that some companies use offsets strategically for certification and ranking benefits rather than genuine decarbonization. The lack of standardized reporting and verification mechanisms hampers the ability to assess the true impact of corporate climate actions. The Oxford Principles for Net Zero Aligned Carbon Offsetting advocate for prioritizing emission reductions, ensuring transparency, and promoting long-term carbon removal solutions. For corporate commitments to be effective, they must be grounded in science-based targets, encompass all emission scopes, and be accompanied by transparent reporting and third-party verification (Kim et al. 2024).

Carbon offsetting allows entities to compensate for their emissions by investing in projects that reduce or remove greenhouse gases elsewhere. While offsetting can be a valuable tool, it presents several challenges. The effectiveness of carbon offsets depends on the credibility of the projects. Issues such as additionality, permanence, and leakage must be addressed to ensure genuine emission reductions (Mustapha 2022). A study by the Grantham Research Institute highlighted concerns over the misallocation of carbon offsets in the Clean Development Mechanism, questioning their actual impact on emission reductions. The voluntary carbon market has faced criticism for a lack of transparency and the proliferation of low-quality credits. Strengthening market standards and oversight is essential. The Berkeley Carbon Trading Project emphasizes the need for robust standards to address quality issues in carbon offset programs. Experts emphasize that carbon offsetting should complement, not replace, direct emission reduction efforts (LSE 2024). The Oxford Principles underscore that offsetting should be used only after all feasible internal emission reductions have been achieved. To enhance the credibility of carbon offsetting, initiatives like the Oxford Principles for Net Zero Aligned Carbon Offsetting advocate for prioritizing emission reductions, ensuring transparency, and promoting long-term carbon removal solutions (Oxford University 2024).

The path to net zero requires a synergistic approach that combines effective policy measures, authentic corporate commitments, and credible carbon offsetting practices. Policymakers must establish robust frameworks that incentivize genuine emission reductions and ensure the integrity of offset markets. Corporations should commit to comprehensive, science-based targets and prioritize internal emission reductions before resorting to offsets. By aligning policies, corporate actions, and offsetting strategies, stakeholders can collectively advance towards a sustainable and climate-resilient future.

11.5.4 Future Directions for Sustainable Digital Ecosystems

As the digital revolution accelerates, the imperative to align technological advancement with environmental sustainability becomes increasingly urgent. Digital ecosystems—comprising interconnected technologies, platforms, and infrastructures—are integral to modern life, yet they also contribute significantly to environmental challenges. Addressing these issues requires a multifaceted approach that encompasses energy efficiency, responsible innovation, ethical governance, and inclusive practices.

One of the primary concerns in digital sustainability is the substantial energy consumption of data centers. These facilities, which power cloud computing and data storage, currently account for approximately 1.5% of global electricity use, with projections indicating a potential doubling by 2030. The integration of artificial intelligence (AI) further exacerbates this demand, particularly due to the intensive cooling requirements of AI hardware. For instance, some AI-related cooling systems consume up to 5 million gallons of water per day per site. To mitigate these impacts, companies like Digital Realty are implementing strategies such as utilizing renewable energy sources, enhancing cooling systems, and incorporating liquid cooling technologies into future data center designs. Their goals include a 60% reduction in direct and indirect emissions per square foot and a 24% reduction in supply chain emissions by 2030. These efforts underscore the importance of cross-department collaboration and early-stage planning for clean energy integration (Mok 2025).

Beyond infrastructure, the concept of digital sustainability encompasses the responsible use and development of digital technologies. This involves designing digital artifacts and systems that not only minimize environmental impact but also promote social and economic well-being. The integration of sustainable practices into digital transformation initiatives is crucial for building a more sustainable and responsible digital future. This includes adopting principles of corporate digital responsibility (CDR) and transitioning from Industry 4.0 to Industry 5.0, which emphasizes human-centric approaches and human-AI partnerships (Pappas et al. 2023).

The role of digital platforms in fostering sustainable innovation is also gaining attention. Digital platform ecosystems, such as those operated by major technology companies, enable new forms of collaboration and value creation. These platforms can be leveraged to drive sustainable innovation by facilitating the development and dissemination of environmentally friendly technologies and practices. For example, platforms can support the sharing economy, promote circular business models, and enable more efficient resource utilization (Calabrese et al. 2021).

In the realm of policy and governance, the development of sustainable digital ecosystems necessitates a collaborative approach involving various stakeholders, including governments, industry, academia, and civil society. Initiatives like the German Plattform Industrie 4.0's vision for 2030 emphasize the importance of shaping digital ecosystems globally with sustainability as a key pillar. This vision outlines development paths towards a digital, networked, and sustainable manufacturing industry, highlighting the need for multi-actor, multi-sector, and multi-level collaboration to achieve broad policy impacts (Sautter 2021).

Emerging technologies such as AI, big data analytics, cloud computing, and the Internet of Things (IoT) hold significant potential for advancing sustainable development goals. These technologies can enhance efficiency, reduce waste, and enable more informed decision-making across various sectors. For instance, AI can optimize energy consumption in buildings, big data analytics can improve supply chain sustainability, and IoT devices can monitor environmental conditions in real-time. However, the deployment of these technologies must be guided by ethical considerations and a commitment to minimizing their environmental footprint (Bocean 2025).

The concept of "digital sustainability" also involves rethinking economic models to prioritize environmental and social outcomes alongside financial performance. This includes embracing the notion of "Sustainomy," which integrates prosperity, people, and the planet into a cohesive system promoting equitable wealth generation, environmental stewardship, and inclusive development. Achieving this transformation requires continuous innovation, collaboration, and a shift away from traditional profit-centric business practices (Deloitte 2024a, b).

Therefore, the future of sustainable digital ecosystems hinges on our ability to harmonize technological advancement with environmental and social responsibility. This entails not only improving the energy efficiency of digital infrastructures but also fostering responsible innovation, ethical governance, and inclusive practices. By adopting a holistic approach that integrates sustainability into every facet of the digital ecosystem, we can ensure that technological progress contributes positively to the well-being of both people and the planet.

11.6 Conclusion

The acceleration of digital transformation in the twenty-first century has generated profound benefits across communication, productivity, and service delivery but has simultaneously unleashed substantial environmental costs. This chapter has explored the multifaceted dimensions of the digital economy's ecological footprint, focusing on the interlinked challenges of e-waste generation, algorithmic inefficiency, and massive energy consumption in data centers. Through an integrated analysis of current practices, technological innovations, policy interventions, and future directions, it becomes clear that only a holistic and coordinated strategy involving hardware redesign, software optimization, regulatory frameworks, and stakeholder engagement can usher in a sustainable digital future.

A key challenge examined is the explosive growth of electronic waste, driven by consumer habits, short device lifecycles, and planned obsolescence. With global e-waste expected to reach 82 million tonnes by 2030 (UNITAR 2024a, b), and only 17.4% being formally recycled (Forti 2024), the environmental and health repercussions are staggering. The toxic composition of e-waste—laden with lead, mercury, and flame retardants—poses severe risks, especially in low-income nations where informal recycling practices dominate (Bakhiyi et al. 2017). Concurrently, the fragmentation of global electronics supply chains makes lifecycle tracking and accountability increasingly difficult (Buchholz 2021). These challenges are compounded by economic structures that incentivize replacement over repair, a lack of infrastructure for recycling, and inadequate regulatory enforcement in many regions (Kavil et al. 2025).

Addressing these issues requires a systemic shift toward circular economy principles. The chapter illustrates that strategies such as repair, refurbishment, and recycling serve as the cornerstone of sustainable e-waste management. Encouraging the "right to repair" and mandating manufacturers to design products with modularity and recyclability significantly extends product lifespans (George et al. 2024). Community repair initiatives, corporate take-back schemes like Apple Renew, and transparent grading systems for refurbished goods all contribute to reducing the volume of discarded electronics (Wallner et al. 2024; Belbağ and Belbağ 2023). Recycling, although resource-intensive, remains indispensable for recovering valuable materials like gold and palladium while preventing environmental pollution (Ankit et al. 2021a, b). Innovations such as bioleaching and urban mining further improve recovery efficiency and environmental performance (Adetunji et al. 2023; Ouro-Salim 2024). Nonetheless, these efforts must be supported by robust infrastructure, regulatory oversight, and public awareness.

Complementing the hardware dimension is the often-overlooked issue of algorithmic waste. Inefficient coding practices, bloated software, and resource-intensive algorithms significantly elevate energy consumption and hardware strain (Bharany et al. 2022). The training of advanced AI models alone can rival the lifetime emissions of multiple vehicles (Ueda et al. 2024), while user-facing applications plagued by software bloat reduce device lifespan and drive premature replacement (Islam and Huda 2018). The chapter highlights compiler-level optimizations, intelligent data processing, and efficient memory management as effective solutions to mitigate computational overhead. Techniques such as loop tiling, automatic vectorization, and in-memory computing optimize system performance without necessitating new hardware (Cooper and Torczon 2012; Jeffers and Reinders 2013a, b). Software sustainability is further enhanced through green engineering practices, including low-power language selection, energy-aware development tools, and awareness initiatives among programmers (Salazar 2024).

Case studies exemplify these principles in action. Telecommunications firms, cloud platforms, and eco-conscious developers have demonstrated that code-level improvements, coupled with sustainable hosting infrastructure, can deliver notable energy reductions—up to 30% in some systems (Martin 2023). Tools like the Green Algorithms framework quantify software emissions, empowering developers to make informed choices (Lannelongue et al. 2020). Such evidence reiterates that software optimization is not merely a technical concern but a pivotal environmental issue.

Equally pressing is the energy consumption of modern data centers, which underpin the digital economy. These facilities are projected to consume up to 8% of global electricity by 2030 if no action is taken (Thangam et al. 2024), with AI workloads contributing increasingly to power demand (IEA 2024). Conventional cooling systems alone account for nearly 40% of total energy use, intensifying carbon emissions when powered by fossil fuels (Deloitte 2024a, b). Additionally, their reliance on rare earth elements and heavy metals further complicates the sustainability equation by linking operational emissions with extractive environmental degradation (Goel et al. 2024).

This chapter provides an in-depth examination of technological innovations that promise to reduce data center energy footprints. Efficient cooling technologies, such as immersion cooling and AI-driven thermal management, have demonstrated potential savings of up to 50% compared to traditional methods (Udinmwen 2025). Energy-efficient processors, dynamic voltage scaling, and server virtualization enhance performance-per-watt ratios (Chidolue et al. 2024). Renewable energy integration through Power Purchase Agreements (PPAs) and on-site solar or wind installations represents a strategic move toward decarbonizing digital infrastructure (Mok 2025). Companies like Google and Microsoft aim for 100% carbon-free operations by 2030, aligning business growth with climate responsibility (Joppa 2021).

Furthermore, operational innovations such as modular data center construction, edge computing, and waste heat recovery are redefining sustainability in digital infrastructure. Modular designs enable scalable, resource-efficient deployments with lower construction emissions (Mok 2025), while edge computing reduces data transmission energy by processing information closer to the source (George et al. 2023). Heat recovery systems, such as Aquasar in Zurich, illustrate circular principles in infrastructure by using server-generated heat for building heating (Miller 2024). These approaches ensure not just energy efficiency but also resilience and geographic flexibility.

Critically, the chapter emphasizes that technological solutions alone are insufficient without aligned policy frameworks and multi-stakeholder cooperation. International treaties such as the Basel and Bamako Conventions establish a legal foundation for regulating e-waste trade and promoting environmentally sound management (Basel Convention Secretariat 2023). National policies—such as Japan's LRHA law and India's E-Waste Management Rules—mandate producer accountability through Extended Producer Responsibility (EPR), reinforcing the design-for-recyclability ethos (Ministry of Environment 2023a, b; Ministry of Environment, Forest and Climate Change 2022). Industry standards and certifications, like the e-Stewards program, support ethical recycling practices by banning toxic exports and ensuring worker safety (Basel Action Network 2025).

Nonetheless, challenges persist. Informal recycling sectors in countries like Ghana, India, and Nigeria continue to dominate e-waste processing due to economic disparities and weak enforcement (E-Waste Monitor 2024). The complexity of new devices, with compact components and composite materials, complicates recycling economics and limits recovery yields (International Energy Agency 2023). Bridging the gap between formal and informal systems requires investments in infrastructure, training, and policy that incorporates marginalized workers into safe, regulated recycling systems. Moreover, international cooperation is essential to curb illegal exports, standardize recycling protocols, and build capacity in the Global South.

An integrated approach—combining hardware design, software efficiency, policy governance, and public engagement—is thus indispensable for meaningful progress. Hardware innovations must prioritize modularity, repairability, and non-toxic materials (Van Der Velden 2014). Software development must internalize energy impacts and prioritize lean, efficient coding practices from the ground up (Sanlialp and Ozturk 2020). Data centers must be reimagined not just as technical infrastructure but as dynamic ecosystems that balance performance, sustainability, and equity. Simultaneously, policies must harmonize international and national regulations to promote traceability, ethical production, and equitable waste distribution. Education and awareness campaigns are essential for cultivating digital citizenship and empowering consumers to make sustainable choices—from prolonging device use to participating in recycling programs (Dwivedi et al. 2021).

Stakeholders across the digital value chain must assume specific responsibilities to enable this transformation. For industry actors, particularly manufacturers and data center operators, the imperative is to invest in sustainable R&D, adopt transparent reporting mechanisms, and internalize environmental externalities into product pricing and lifecycle management. Policymakers must enforce right-to-repair laws, incentivize eco-innovation through subsidies and tax breaks, and ensure equitable global waste governance. Consumers play a pivotal role by shifting consumption patterns, advocating for sustainable products, and participating in circular practices. Software developers and digital architects must embed environmental consciousness into design frameworks, treating code not just as logic but as infrastructure with environmental implications.

Looking forward, the future of sustainable digital ecosystems depends on embracing innovation while honoring planetary boundaries. The convergence of artificial intelligence, renewable energy, advanced recycling, and circular design offers unprecedented potential to align digital growth with ecological stewardship. Yet realizing this vision requires moving beyond piecemeal efforts and embracing cross-sectoral collaboration, systems thinking, and long-term accountability. The integration of sustainability into every layer of digital operations—from transistor to cloud server, from algorithm to supply chain—will define the digital era's legacy on Earth.

In sum, this chapter revealed a deep interdependencies among e-waste, algorithmic waste, and data center energy consumption, advocating for a cohesive response grounded in integrated approaches. By combining technological ingenuity with policy coherence and cultural transformation, we can mitigate the environmental toll of the digital age. Achieving sustainable digital ecosystems is not a matter of technical feasibility alone—it is a question of collective will, ethical vision, and systemic change. As we navigate the future of our hyperconnected world, the decisions made today will determine whether technology becomes a tool for planetary regeneration or a driver of irreversible degradation.

References

Abalansa S, Mahrad BE, Icely J, Newton A (2021) Electronic waste, an environmental problem exported to developing countries: the GOOD, the BAD and the UGLY. Sustainability 13(9):5302. https://doi.org/10.3390/su13095302

Abera Y (2023) Optimizing construction waste recycling: strategies, technologies, and environmental impacts. Research Square (Research Square). https://doi.org/10.21203/rs.3.rs-3497942/v1

Abreu R, Couto M, Cruz L, Cunha J, Fernandes JP, Pereira R, Perez A, Saraiva J (2021) Green software lab: towards an engineering discipline for green software. https://arxiv.org/abs/2108.03028

Adetunji AI, Oberholster PJ, Erasmus M (2023) Bioleaching of metals from E-waste using microorganisms: a review. Minerals 13(6):828. https://doi.org/10.3390/min13060828

Agarwal S, Jain H (2024) Investigating the potential of floral waste as a vermicompost and dual-functional biosorbent for sustainable environmental management. Water Air Soil Pollut 235:322. https://doi.org/10.1007/s11270-024-07144-y

Agarwal S, Jain H, Dhupper R, Mathur A (2025) Optimizing methyl orange dye removal from aqueous solutions using white chrysanthemum floral waste-derived bioadsorbent: a study of kinetics, thermodynamics, isotherms, and RSM optimization. Int J Environ Res 19(3):104

Aldoseri A, Al-Khalifa KN, Hamouda AM (2024) AI-powered innovation in digital transformation: key pillars and industry impact. Sustainability 16(5):1790. https://doi.org/10.3390/su16051790

Al-Fakih A, Al-Wajih E, Saleh RA, Muhit IB (2024) Ensemble machine learning models for predicting the CO_2 footprint of GGBFS-based geopolymer concrete. J Clean Prod 472:143463. https://doi.org/10.1016/j.jclepro.2024.143463

Alsabt R, Alkhaldi W, Adenle YA, Alshuwaikhat HM (2024) Optimizing waste management strategies through artificial intelligence and machine learning—an economic and environmental impact study. Clean Waste Syst 8:100158. https://doi.org/10.1016/j.clwas.2024.100158

Alter H (2000) Environmentally sound management of the recycling of hazardous wastes in the context of the basel convention. Resour Conserv Recycl 29(1–2):111–129. https://doi.org/10.1016/s0921-3449(99)00061-0

Amajuoyi NCP, Nwobodo NLK, Adegbola NMD (2024) Transforming business scalability and operational flexibility with advanced cloud computing technologies. Comput Sci IT Res J 5(6):1469–1487. https://doi.org/10.51594/csitrj.v5i6.1248

Amankwah-Amoah J (2016) Global business and emerging economies: towards a new perspective on the effects of e-waste. Technol Forecast Soc Chang 105:20–26. https://doi.org/10.1016/j.techfore.2016.01.026

Ankit N, Saha L, Kumar V, Tiwari J, Sweta N, Rawat S, Singh J, Bauddh K (2021a) Electronic waste and their leachates impact on human health and environment: global ecological threat and management. Environ Technol Innov 24:102049. https://doi.org/10.1016/j.eti.2021.102049

Ankit N, Saha L, Kumar V, Tiwari J, Sweta N, Rawat S, Singh J, Bauddh K (2021b) Electronic waste and their leachates impact on human health and environment: global ecological threat and management. Environ Technol Innov 24:102049. https://doi.org/10.1016/j.eti.2021.102049

Annamalai M, Gurumurthy K (2020) Characterization of end-of-life mobile phone printed circuit boards for its elemental composition and beneficiation analysis. J Air Waste Manag Assoc 71(3):315–327. https://doi.org/10.1080/10962247.2020.1813836

Apple (2025) Apple expands the use of recycled materials across its products. Apple Newsroom (India). https://www.apple.com/in/newsroom/2022/04/apple-expands-the-use-of-recycled-materials-across-its-products/

Bakhiyi B, Gravel S, Ceballos D, Flynn MA, Zayed J (2017) Has the question of e-waste opened a Pandora's box? an overview of unpredictable issues and challenges. Environ Int 110:173–192. https://doi.org/10.1016/j.envint.2017.10.021

Bamako Convention (1998) Bamako convention on the ban of the import into Africa and the control of transboundary movement and management of hazardous wastes within Africa | African Union. https://au.int/en/treaties/bamako-convention-ban-import-africa-and-control-transboundary-movement-and-management

Basal Action Network (2025) Electronics stewardship—basel action network. Basel Action Network. https://www.ban.org/e-stewardship

Basel Convention (2023) Overview. Copyright 2011 by Basel Convention. https://www.basel.int/Implementation/Ewaste/EwasteAmendments/Overview/tabid/9266/Default.aspx

Belbağ AG, Belbağ S (2023) Remanufactured products: a systematic review from the consumer perspective. J Remanuf 13(2):207–241. https://doi.org/10.1007/s13243-023-00125-0

Bharany S, Sharma S, Khalaf OI, Abdulsahib GM, Humaimeedy ASA, Aldhyani THH, Maashi M, Alkahtani H (2022) A Systematic survey on energy-efficient techniques in sustainable cloud computing. Sustainability 14(10):6256. https://doi.org/10.3390/su14106256

Bhardwaj LK et al (2025) Resilience in the digital age: emerging technologies for climate adaptation. In: Choudhury M, Dixit G, Majumdar S (eds) Sustainable synergy: harnessing ecosystems for climate resilience. Climate change management. Springer, Cham. https://doi.org/10.1007/978-3-031-77957-2_11

Blaszynski P, Palkowski M, Bielecki W, Poliwoda M (2024) Efficiency of various tiling strategies for the zuker algorithm optimization. Mathematics 12(5):728. https://doi.org/10.3390/math12050728

Bocean CG (2025) Sustainable development in the digital age: harnessing emerging digital technologies to catalyze global SDG achievement. Appl Sci 15(2):816. https://doi.org/10.3390/app15020816

Bovea MD, Pérez-Belis V (2018) Identifying design guidelines to meet the circular economy principles: a case study on electric and electronic equipment. J Environ Manag 228:483–494. https://doi.org/10.1016/j.jenvman.2018.08.014

Buchholz K (2021) The countries producing the most E-waste. Statista Daily Data. https://www.statista.com/chart/24291/e-waste-by-country/

Calabrese M, La Sala A, Fuller RP, Laudando A (2021) Digital platform ecosystems for sustainable innovation: toward a new meta-organizational model? Admin Sci 11(4):119. https://doi.org/10.3390/admsci11040119

Chidolue NO, Ohenhen NPE, Umoh NAA, Ngozichukwu NB, Fafure NAV, Ibekwe NKI (2024) Green data centers: sustainable practices for energy-efficient it infrastructure. Eng Sci Technol J 5(1):99–114. https://doi.org/10.51594/estj.v5i1.730

Ciampoli P (2024) Report Details Challenges to Virginia Power Grid Tied to Growth of Data Centers | American Public Power Association. https://www.publicpower.org/periodical/article/report-details-challenges-virginia-power-grid-tied-growth-data-centers

CMIC (2025) Key Data Center Construction Trends in 2025. https://cmicglobal.com/resources/article/data-center-construction-trends-in-2025?

Cooper KD, Torczon L (2012) Scalar optimizations. In: Elsevier eBooks, pp 539–596. https://doi.org/10.1016/b978-0-12-088478-0.00010-4

Dangal S, Sandez S, Arriola JB, Faludi J, Balkenende R (2025) Empirical evaluation of reparability scoring systems for validity and reliability. Resour Conserv Recycl 218:108211. https://doi.org/10.1016/j.resconrec.2025.108211

Deloitte (2024a) Sustainability and abundance? a futurist explains how it can happen. WSJ. https://deloitte.wsj.com/sustainable-business/sustainability-and-abundance-a-futurist-explains-how-it-can-happen-131b498e

Deloitte (2024b) As generative AI asks for more power, data centers seek more reliable, cleaner energy solutions. Deloitte Insights. https://www2.deloitte.com/us/en/insights/industry/technology/technology-media-and-telecom-predictions/2025/genai-power-consumption-creates-need-for-more-sustainable-data-centers.html

Dhanorkar S, Muthulingam S (2024) Recycling standards, green inventions, and spillover: evidence from California's electronic waste recycling act (EWRA). Manuf Serv Oper Manag. https://doi.org/10.1287/msom.2023.0444

Dutta D, Rautela R, Gujjala LKS, Kundu D, Sharma P, Tembhare M, Kumar S (2022) A review on recovery processes of metals from E-waste: a green perspective. Sci Total Environ 859:160391. https://doi.org/10.1016/j.scitotenv.2022.160391

Dwivedi YK, Hughes L, Kar AK, Baabdullah AM, Grover P, Abbas R, Andreini D, Abumoghli I, Barlette Y, Bunker D, Kruse LC, Constantiou I, Davison RM, De R, Dubey R, Fenby-Taylor H, Gupta B, He W, Kodama M et al (2021) Climate change and COP26: are digital technologies and information management part of the problem or the solution? an editorial reflection and call to action. Int J Inf Manag 63:102456. https://doi.org/10.1016/j.ijinfomgt.2021.102456

Eanes Z (2025). Amazon to invest $10 billion into AI data centers in North Carolina. Axios. https://www.axios.com/local/raleigh/2025/06/04/amazon-ai-data-centers-in-north-carolina-richmond-county

ESABCC (2025) Staying the course on climate action essential to EU security and competitiveness. https://climate-advisory-board.europa.eu/news/staying-the-course-on-climate-action-essential-to-eu-security-and-competitiveness

Etukudoh EA, Umoh AA, Hamdan A, Ibekwe KI, Sonko S (2024) Carbon emission reduction strategies: a global policy review and analysis of effectiveness. Deleted J 3(2):69–75. https://doi.org/10.26480/egnes.01.2024.69.75

E-Waste Monitor (2024) The global E-waste monitor 2024—E-waste monitor. https://ewastemonitor.info/the-global-e-waste-monitor-2024/

Faiz NF, Ninduwezuor-Ehiobu NN, Adanma NUM, Solomon NNO (2024) Blockchain for sustainable waste management: Enhancing transparency and accountability in waste disposal. Compr Res Rev Sci Technol 2(1):045–069. https://doi.org/10.57219/crrst.2024.2.1.0032

Ferrara E (2023a) Fairness and bias in artificial intelligence: a brief survey of sources, impacts, and mitigation strategies. Sci 6(1):3. https://doi.org/10.3390/sci6010003

Ferrara E (2023b) Fairness and bias in artificial intelligence: a brief survey of sources, impacts, and mitigation strategies. Sci 6(1):3. https://doi.org/10.3390/sci6010003

Forti V (2024) Global electronic waste up 21% in five years, and recycling isn't keeping up. World Economic Forum. https://www.weforum.org/stories/2020/07/global-electronic-waste-recycling-management/

Gao T, Liu Q, Wang J (2013) A comparative study of carbon footprint and assessment standards. Int J Low-Carbon Technol 9(3):237–243. https://doi.org/10.1093/ijlct/ctt041

Gaur TS, Yadav V, Mittal S, Sharma MK (2023) A systematic review on sustainable E-waste management: challenges, circular economy practices, and a conceptual framework. Manag Environ Qual Int J 35(4):858–884. https://doi.org/10.1108/meq-05-2023-0139

Geneidy SE, Baumeister S, Govigli VM, Orfanidou T, Wallius V (2021) The carbon footprint of a knowledge organization and emission scenarios for a post-COVID-19 world. Environ Impact Assess Rev 91:106645. https://doi.org/10.1016/j.eiar.2021.106645

George AS, Baskar T, Baskar T (2024) Repairing the future: the global fight for accessible fixes and sustainable tech. ResearchGate. https://doi.org/10.5281/zenodo.10953185

George A, George A, Baskar T (2023) Edge computing and the future of cloud computing: a survey of industry perspectives and predictions. Zenodo (CERN European Organization for Nuclear Research). https://doi.org/10.5281/zenodo.8020101

Goel A, Masurkar S, Pathade GR (2024) An overview of digital transformation and environmental sustainability: threats, opportunities, and solutions. Sustainability 16(24):11079. https://doi.org/10.3390/su162411079

Goldstein SC (1990) Computing without processors. https://www.cs.cmu.edu/~seth/papers/goldstein04-ersa04.html

Google (2023) Google's green data centers: network POP case study. In: Google's green data centers: network POP case study, pp 2–10. https://static.googleusercontent.com/media/www.google.com/en//corporate/datacenter/dc-best-practices-google.pdf

Grandhi SP, Dagwar PP, Dutta D (2024) Policy pathways to sustainable E-waste management: a global review. J Hazard Mater Adv 16:100473. https://doi.org/10.1016/j.hazadv.2024.100473

Gurvich A, Creamer GG (2021) Overallocation and correction of carbon emissions in the evaluation of carbon footprint. Sustainability 13(24):13613. https://doi.org/10.3390/su132413613

Gutterman AS (2024) Sustainability reporting frameworks, standards, instruments, and regulations. ResearchGate. https://www.researchgate.net/publication/350787634_Sustainability_Reporting_Frameworks_Standards_Instruments_and_Regulations

Hancock A (2025) EU science advisers put clamp on carbon credits to meet 2040 climate target. Financial Times. https://www.ft.com/content/f3999de3-0be3-4767-af81-72e1bdb8d990

Harris L (2024) Artificial intelligence and edge computing for energy-efficient cloud services. ResearchGate. https://doi.org/10.13140/RG.2.2.20804.23683

Hazelwood DA, Pecht MG (2021) Life extension of electronic products: a case study of smartphones. IEEE Access 9:144726–144739. https://doi.org/10.1109/access.2021.3121733

Helaimia R (2024) Environmental footprint of data centers. ResearchGate. https://www.researchgate.net/publication/379477916_Environmental_Footprint_of_Data_Centers

Himeur Y, Sayed AN, Alsalemi A, Bensaali F, Amira A (2023) Edge AI for internet of energy: challenges and perspectives. Internet of Things 25:101035. https://doi.org/10.1016/j.iot.2023.101035

Huang X, Liu Y, Hwang S, Kang S, Patnaik D, Cortes JF, Rogers JA (2014) Biodegradable materials for multilayer transient printed circuit boards. Adv Mater 26(43):7371–7377. https://doi.org/10.1002/adma.201403164

Husain NN (2024) Designing for the circular economy: how product designers can create circular products. Institute of Sustainability Studies. https://instituteofsustainabilitystudies.com/insights/guides/how-product-designers-can-create-circular-products/

Hutchinson E (2024) How can data centres balance operational resilience with the demand for sustainability, and what key innovations will drive this?–Intelligent Data Centres. https://www.intelligentdatacentres.com/2024/11/11/how-can-data-centres-balance-operational-resilience-with-the-demand-for-sustainability-and-what-key-innovations-will-drive-this/

Ibn-Mohammed T, Yamoah F, Acquaye A, Omoteso K, Koh S (2024) Enhancing life cycle product design decision-making processes: insights from normal accident theory and the satisficing framework. Resour Conserv Recycl 205:107523. https://doi.org/10.1016/j.resconrec.2024.107523

IEA. (2024). Data centres and networks—IEA. https://www.iea.org/energy-system/buildings/data-centres-and-data-transmission-networks

Ilankoon I, Ghorbani Y, Chong MN, Herath G, Moyo T, Petersen J (2018) E-waste in the international context–a review of trade flows, regulations, hazards, waste management strategies and technologies for value recovery. Waste Manag 82:258–275. https://doi.org/10.1016/j.wasman.2018.10.018

International Energy Agency (2023) World Energy Outlook 2023—Analysis—IEA. IEA. https://www.iea.org/reports/world-energy-outlook-2023

Islam MT, Huda N (2018) Reverse logistics and closed-loop supply chain of Waste Electrical and Electronic Equipment (WEEE)/E-waste: a comprehensive literature review. Resour Conserv Recycl 137:48–75. https://doi.org/10.1016/j.resconrec.2018.05.026

Jackson A (2024) Top 10: data centre innovations. Data Centre Magazine. https://datacentremagazine.com/top10/top-10-data-centre-innovations?

Jain H (2025a) Synergizing carbon reduction technologies and nature-based solutions for a climate-neutral future. Mitig Adapt Strateg Glob Change 30:42. https://doi.org/10.1007/s11027-025-10237-6

Jain H (2025) Data analytics enabled by the Internet of Things and artificial intelligence for the management of Earth's resources. In: Data analytics and artificial intelligence for earth resource management, pp 19–36. Elsevier

Jain M, Kumar D, Chaudhary J, Kumar S, Sharma S, Verma AS (2023) Review on E-waste management and its impact on the environment and society. Waste Manag Bull 1(3):34–44. https://doi.org/10.1016/j.wmb.2023.06.004

Jeffers J, Reinders J (2013a) Lots of data (Vectors). In: Elsevier eBooks, pp 107–164. https://doi.org/10.1016/b978-0-12-410414-3.00005-0

Jeffers J, Reinders J (2013b) Lots of data (Vectors). In: Elsevier eBooks, pp 107–164. https://doi.org/10.1016/b978-0-12-410414-3.00005-0

Jha G, Soren S, Mehta KD (2021) Carbon footprint assessment with LCA methodology. In: Environmental footprints and eco-design of products and processes, pp 1–34. https://doi.org/10.1007/978-981-33-4373-3_1

Jia G, Han G, Wang H, Yang X (2017) Static memory deduplication for performance optimization in cloud computing. Sensors 17(5):968. https://doi.org/10.3390/s17050968

Johnson K (2025) Crackdown on power-guzzling data centers may soon come in California. San Francisco Chronicle. https://www.sfchronicle.com/bayarea/article/crackdown-power-guzzling-data-centers-soon-come-20173899.php

Joppa L (2021) Made to measure: sustainability commitment progress and updates—the official Microsoft Blog. The Official Microsoft Blog. https://blogs.microsoft.com/blog/2021/07/14/made-to-measure-sustainability-commitment-progress-and-updates/

Justus D, Brennan J, Bonner S, McGough AS (2018) Predicting the computational cost of deep learning models. IEEE Int Conf Big Data (Big Data) 2021:3873–3882. https://doi.org/10.1109/bigdata.2018.8622396

Katal A, Dahiya S, Choudhury T (2021) Energy efficiency in cloud computing data center: a survey on hardware technologies. Clust Comput 25(1):675–705. https://doi.org/10.1007/s10586-021-03431-z

Katal A, Dahiya S, Choudhury T (2022) Energy efficiency in cloud computing data centers: a survey on software technologies. Clust Comput 26(3):1845–1875. https://doi.org/10.1007/s10586-022-03713-0

Kavil YN, Alshemmari H, Alkasbi MM, Alelyani SS, Orif MI, Al-Farawati RK (2025) Electronic waste and environmental diplomacy: how GCC E-waste management interfaces with the stockholm convention. J Hazard Mater Adv 100610. https://doi.org/10.1016/j.hazadv.2025.100610

Kez DA, Foley AM, Laverty D, Del Rio DF, Sovacool B (2022) Exploring the sustainability challenges facing digitalization and internet data centers. J Clean Prod 371:133633. https://doi.org/10.1016/j.jclepro.2022.133633

Khan MI, Yasmeen T, Khan M, Hadi NU, Asif M, Farooq M, Al-Ghamdi SG (2024) Integrating industry 4.0 for enhanced sustainability: pathways and prospects. Sustain Prod Consumption. https://doi.org/10.1016/j.spc.2024.12.012

Khosravi A, Sandoval OR, Taslimi MS, Sahrakorpi T, Amorim G, Pabon JJG (2024) Review of energy efficiency and technological advancements in data center power systems. Energy Build 114834. https://doi.org/10.1016/j.enbuild.2024.114834

Kim S, Li T, Wu Y (2024) Corporate carbon offsets: decarbonization or transition-washing? (M. Flannery & J. Ritter, Eds.) [Journal-article]. https://sustainablefinancealliance.org/wp-content/uploads/2024/06/GRASFI_2024_paper_99.pdf

Kolapo NR, Kawu NFM, Abdulmalik NAD, Edem NUA, Young NMA, Mordi NEC (2024) Edge computing: revolutionizing data processing for IoT applications. Int J Sci Res Arch 13(2):023–029. https://doi.org/10.30574/ijsra.2024.13.2.2082

Lannelongue L, Grealey J, Inouye M (2020) Green algorithms: quantifying the carbon footprint of computation. https://arxiv.org/abs/2007.07610

Li H, Eksteen J, Oraby E (2018) Hydrometallurgical recovery of metals from waste printed circuit boards (WPCBs): current status and perspectives–a review. Resour Conserv Recycl 139:122–139. https://doi.org/10.1016/j.resconrec.2018.08.007

Li Y, Wen Y, Guan K, Tao D (2017) Transforming cooling optimization for green data center via deep reinforcement learning. https://arxiv.org/abs/1709.05077

LSE (2024) What are carbon offsets?—Grantham Research Institute on climate change and the environment. Grantham Research Institute on Climate Change and the Environment. https://www.lse.ac.uk/granthaminstitute/explainers/what-are-carbon-offsets/

Machín A, Morant C, Márquez F (2024) Advancements and challenges in solid-state battery technology: an in-depth review of solid electrolytes and anode innovations. Batteries 10(1):29. https://doi.org/10.3390/batteries10010029

Marjanovic O, Cecez-Kecmanovic D, Vidgen R (2018) Algorithmic pollution: understanding and responding to negative consequences of algorithmic decision-making. In: IFIP advances in information and communication technology, pp 31–47. https://doi.org/10.1007/978-3-030-04091-8_4

Martin J (2023) How can engineers make IT more sustainable? Part 3: green IT case studies. Scaleway Blog. https://www.scaleway.com/en/blog/how-can-engineers-make-it-more-sustainable-part-3/

Miljövetenskap GU/IFBO (2024) Flows of waste and secondhand items from developed to developing countries: a case study of e-wastes in Nigeria and China. https://gupea.ub.gu.se/handle/2077/80832

Miller R (2024) IBM's hot-water supercomputer goes live. https://www.datacenterknowledge.com/supercomputers/ibm-s-hot-water-supercomputer-goes-live

Ministry of Environment, Forest and Climate Change (2022) E-waste (Management) rules, 2022. In: *भारत का राजपत्र*. . https://www.mppcb.mp.gov.in/proc/E-Waste-Management-Rules-2022-English.pdf

Ministry of Environment (2023a) Laws: waste and recycling. Ministry of the Environment, Government of Japan. https://www.env.go.jp/en/laws/recycle/index.html

Ministry of Environment (2023b) Laws: waste and recycling. Ministry of the Environment, Government of Japan. https://www.env.go.jp/en/laws/recycle/index.html

Mok A (2025) How a data center operator is upgrading its services for AI—and trying to stay green. Business Insider. https://www.businessinsider.com/digital-realty-ai-infrastructure-data-centers-sustainability-strategy-2025-6?

Moss S (2024) Irish data centers used more than 21% of nation's total metered electricity. DCD. https://www.datacenterdynamics.com/en/news/irish-data-centers-used-more-than-21-percent-of-nations-total-metered-electricity/

Mundy S (2025) The trouble with Microsoft's carbon numbers. Financial Times. https://www.ft.com/content/9fd2fbc1-ae30-4fe7-9e2b-e6f268992335

Mustapha C (2022) The burden of a ton CO_2! Emission trading systems and the air transport business. In: Elsevier eBooks, pp 27–53. https://doi.org/10.1016/b978-0-323-91522-9.00002-6
Nair SS (2023) Beyond the cloud - unraveling the benefits of edge computing in IOT. ResearchGate. https://www.researchgate.net/publication/375417262_BEYOND_THE_CLOUD_-_UNRAVELING_THE_BENEFITS_OF_EDGE_COMPUTING_IN_IOT
Narayana SS (2024) Synergizing edge computing and cloud integration for enhanced efficiency in the oil and gas industry. ResearchGate. https://www.researchgate.net/publication/379734115_SYNERGIZING_EDGE_COMPUTING_AND_CLOUD_INTEGRATION_FOR_ENHANCED_EFFICIENCY_IN_THE_OIL_AND_GAS_INDUSTRY
National Academies Press (US) (2011) Management of waste. Prudent Practices in the Laboratory - NCBI Bookshelf. https://www.ncbi.nlm.nih.gov/books/NBK55885/
Nicolas VK, Kwon H, Nguyen DT, Berger ED (2024) It's not easy being green: on the energy efficiency of programming languages. https://arxiv.org/abs/2410.05460
Noorakma AW, Arep H, Sultan AAM (2024) Household e-waste management: a comprehensive review. Multi Rev 7(11):2024256. https://doi.org/10.31893/multirev.2024256
OECD (2024) OECD countries update rules on international shipments of electronic waste. https://www.oecd.org/en/about/news/announcements/2024/06/oecd-countries-update-rules-on-international-shipments-of-electronic-waste.html
Ostapska K, Rüther P, Loli A, Gradeci K (2024) Design for disassembly: a systematic scoping review and analysis of built structures designed for disassembly. Sustain Prod Consumption 48:377–395. https://doi.org/10.1016/j.spc.2024.05.014
Ouro-Salim O (2024) Urban mining of e-waste management globally: literature review. Clean Waste Syst 9:100162. https://doi.org/10.1016/j.clwas.2024.100162
Overney J (2023) Redefining energy efficiency in data processing. EPFL. https://actu.epfl.ch/news/redefining-energy-efficiency-in-data-processing/
Oxford university (2024) The Oxford offsetting principles. Smith School of Enterprise and the Environment. https://www.smithschool.ox.ac.uk/research/oxford-offsetting-principles
Pappas IO, Mikalef P, Dwivedi YK, Jaccheri L, Krogstie J (2023) Responsible digital transformation for a sustainable society. Inf Syst Front 25(3):945–953. https://doi.org/10.1007/s10796-023-10406-5
Plan A (2024) ESG reporting software by Plan A. Plan A. https://plana.earth/report-sustainability
Popiolek PF, Machado KDS, Mendizabal OM (2021) Low overhead performance monitoring for shared infrastructures. Expert Syst Appl 171:114558. https://doi.org/10.1016/j.eswa.2020.114558
Poy O, Moraga MÁ, García F, Calero C (2024) Impact on energy consumption of design patterns, code smells and refactoring techniques: a systematic mapping study. J Syst Softw 112303. https://doi.org/10.1016/j.jss.2024.112303
Priyashantha AKH, Pratheesh N, Pretheeba P (2022) E-waste scenario in South-Asia: an emerging risk to environment and public health. Environ Anal Health Toxicol 37(3):e2022022. https://doi.org/10.5620/eaht.2022022
Quinto S, Law N, Fletcher C, Le J, Jose SA, Menezes PL (2025) Exploring the E-waste crisis: strategies for sustainable recycling and circular economy integration. Recycling 10(2):72. https://doi.org/10.3390/recycling10020072
Rajesh R, Kanakadhurga D, Prabaharan N (2022) Electronic waste: a critical assessment on the unimaginable growing pollutant, legislations and environmental impacts. Environ Challenges 7:100507. https://doi.org/10.1016/j.envc.2022.100507
Reddy VD, Rao GSVRK, Aiello M (2024) Energy efficient resource management in data centers using imitation-based optimization. Energy Inform 7(1). https://doi.org/10.1186/s42162-024-00370-y
RJJ Software (2024) Embracing sustainable software development: a case study in green computing at RJJ software. RJJ Software Ltd. https://rjj-software.co.uk/case-studies/embracing-sustainable-software-development-a-case-study-in-green-computing-at-rjj-software/

Saini J, Gupta AK, Shrivastava A, Dhupper R (2024) Study of urban warming in Chandigarh union territory, India: geospatial approach: J. Saini et al. Theor Appl Climatol 155(8):8245–8261

Saini J, Gupta AK, Dhupper R, Shrivastava A (2025) Spatio-temporal study of urban dynamics with implications on land surface temperature of Gurugram City India. Environ Monit Assess 197(8):1–25

Sakao T, Bocken N, Nasr N, Umeda Y (2024) Implementing circular economy activities in manufacturing for environmental sustainability. CIRP Ann 73(2):457–481. https://doi.org/10.1016/j.cirp.2024.06.002

Salazar J (2024) What is green coding and its impact on ethical web development? IEEE Comput Soc. https://www.computer.org/publications/tech-news/trends/green-coding

Sanlialp I, Ozturk MM (2020) Investigating the impact of code refactoring techniques on energy consumption in different object-oriented programming languages. In: Lecture notes on data engineering and communications technologies, pp 142–152. https://doi.org/10.1007/978-3-030-36178-5_12

SAP (2024) What is EPR? The Answer to Sustainability | SAP. https://www.sap.com/hk/products/scm/responsible-design-and-production/what-is-extended-producer-responsibility.html

Sathupadi K, Achar S, Bhaskaran SV, Faruqui N, Abdullah-Al-Wadud M, Uddin J (2024) Edge-cloud synergy for AI-enhanced sensor network data: a real-time predictive maintenance framework. Sensors 24(24):7918. https://doi.org/10.3390/s24247918

Sautter B (2021) Shaping digital ecosystems for sustainable production: assessing the policy impact of the 2030 vision for Industrie 4.0. Sustainability 13(22), 12596. https://doi.org/10.3390/su132212596

Schischke K, Proske M, Nissen NF, Schneider-Ramelow M (2019) Impact of modularity as a circular design strategy on materials use for smart mobile devices. MRS Energy Sustain 6(1). https://doi.org/10.1557/mre.2019.17

Schultz MD, Clegg M, Hofstetter R, Seele P (2024) Algorithms and dehumanization: a definition and avoidance model. AI and Soc. https://doi.org/10.1007/s00146-024-02123-7

Scrucca F, Barberio G, Fantin V, Porta PL, Barbanera M (2020) Carbon footprint: concept, methodology and calculation. In: Environmental footprints and eco-design of products and processes, pp 1–31. https://doi.org/10.1007/978-981-15-9577-6_1

Shi Y, Xu B, Zhang B, Wang D (2016) Leveraging energy storage to optimize data center electricity cost in emerging power markets. https://arxiv.org/abs/1606.01536

Shrawankar U, Joshi M (2022) Overhead management in multi-core environment. https://arxiv.org/abs/2202.06744

Shrivastava A, Abhishek K, Gupta AK, Jain H, Kumari M, Patel M, Sharma P (2024) Removal of micro-and nano-plastics from aqueous matrices using modified biochar–a review of synthesis, applications, interaction, and regeneration. J Hazard Mater Adv 16:100518

Sierra-Fontalvo L, Gonzalez-Quiroga A, Mesa JA (2023) A deep dive into addressing obsolescence in product design: a review. Heliyon 9(11):e21856. https://doi.org/10.1016/j.heliyon.2023.e21856

Silva BV, Holm-Nielsen JB, Sadrizadeh S, Teles MP, Kiani-Moghaddam M, Arabkoohsar A (2023) Sustainable, green, or smart? pathways for energy-efficient healthcare buildings. Sustain Cities Soc 100:105013. https://doi.org/10.1016/j.scs.2023.105013

Sin LT, Tueen BS (2023a) International policies of plastic use and consumption. In: Elsevier eBooks, pp 255–296. https://doi.org/10.1016/b978-0-12-824489-0.00009-x

Sin LT, Tueen BS (2023b) International policies of plastic use and consumption. In: Elsevier eBooks, pp 255–296. https://doi.org/10.1016/b978-0-12-824489-0.00009-x

Singh S, Schwan A (2011) Sulfur metabolism in plants and related biotechnologies. In: Elsevier eBooks, pp 257–271. https://doi.org/10.1016/b978-0-08-088504-9.00268-3

Song J, Ma Z, Thomas R, Yu G (2018) Energy efficiency optimization in big data processing platform by improving resources utilization. Sustain Comput Inform Syst 21:80–89. https://doi.org/10.1016/j.suscom.2018.11.011

Sreekumar S, Pugsley A, Chakrabarti S, Hewitt N, Mondol JD, Shah N (2024) Experimental investigation on the performance of MXene/C-dot hybrid nanofluid-based photovoltaic/thermal system: an energy, exergy, and enviro-economic analysis. Sol Energy Mater Sol Cells 272:112904. https://doi.org/10.1016/j.solmat.2024.112904

Swinhoe D (2021) Microsoft, Amazon, and Google operate half the world's 600 hyperscale data centers. DCD. https://www.datacenterdynamics.com/en/news/microsoft-amazon-and-google-operate-half-the-worlds-600-hyperscale-data-centers/

Sydor M, Stańczyk K (2025) Analyzing joinery for furniture designed for disassembly. J Manuf Mater Process 9(5):162. https://doi.org/10.3390/jmmp9050162

Tayeb H, Paillat L, Bramas B (2023) Autovesk: automatic vectorized code generation from unstructured static kernels using graph transformations. ACM Trans Archit Code Optim 21(1):1–25. https://doi.org/10.1145/3631709

Thanasi-Boçe M, Hoxha J (2025) Blockchain for sustainable development: a systematic review. Sustainability 17(11):4848. https://doi.org/10.3390/su17114848

Thangam D, Muniraju H, Ramesh R, Narasimhaiah R, Khan NMA, Booshan S, Booshan B, Manickam T, Ganesh RS (2024) Impact of data centers on power consumption, climate change, and sustainability. In: Advances in computational intelligence and robotics book series, pp 60–83. https://doi.org/10.4018/979-8-3693-1552-1.ch004

Thomas (2024) Green computing: 5 strategies and case studies for data center sustainability. Technology Curated. https://technologycurated.com/editorial/green-computing-5-strategies-and-case-studies-for-data-center-sustainability/

Thompson D, Hyde C, Hartley JM, Abbott AP, Anderson PA, Harper GD (2021) To shred or not to shred: a comparative techno-economic assessment of lithium ion battery hydrometallurgical recycling retaining value and improving circularity in LIB supply chains. Resour Conserv Recycl 175:105741. https://doi.org/10.1016/j.resconrec.2021.105741

Udinmwen E (2025) Microsoft, Google, and Meta have borrowed EV tech for the next big thing in data centers: 1MW watercooled racks. TechRadar. https://www.techradar.com/pro/microsoft-google-and-meta-have-borrowed-ev-tech-for-the-next-big-thing-in-data-center-1mw-watercooled-racks

Ueda D, Walston SL, Fujita S, Fushimi Y, Tsuboyama T, Kamagata K, Yamada A, Yanagawa M, Ito R, Fujima N, Kawamura M, Nakaura T, Matsui Y, Tatsugami F, Fujioka T, Nozaki T, Hirata K, Naganawa S (2024) Climate change and artificial intelligence in healthcare: review and recommendations towards a sustainable future. Diagn Interv Imaging 105(11):453–459. https://doi.org/10.1016/j.diii.2024.06.002

UNEP (2023) Sustainable future of E-waste. International Environmental Technology Centre. https://www.unep.org/ietc/news/story/sustainable-future-e-waste

UNITAR (2024a) Global e-waste monitor 2024: electronic waste rising five times faster than documented E-waste recycling. https://unitar.org/about/news-stories/press/global-e-waste-monitor-2024-electronic-waste-rising-five-times-faster-documented-e-waste-recycling

UNITAR (2024b) GEM 2020 - E-waste monitor. E-waste monitor. https://ewastemonitor.info/gem-2020/

Van Der Velden M (2014) Re-politicising participatory design: what can we learn from fairphone. In: Ninth international conference on culture and technology and communication (CaTaC). https://d1wqtxts1xzle7.cloudfront.net/38809927/maja_van_der_velden_fairphone2-libre.pdf?1442606991=&response-content-disposition=inline%3B+filename%3DRe_politicising_Participatory_Design_Wha.pdf&Expires=1748456713&Signature=CJ6D837IikFGRfdWbBZeshtX7Xni~MRgIt0RohS4kblri7J-uR5uo-714HxwL8HFoNnWHV55zFJcuZisEBx5F4k8VCSFTaCLL4pFP9s~~WWNg-dOwpxKaw7j0SVSUtoGEaeH0d7d1D7hN6qwhEY~LhEjf3JotkFxACt7sYRgnM29SpKqzoPKRQyyrEtPf82x07sZGOYzydGmI74a~yvpIb9fe-JsX8gkrjBuSi8Utn2ar9stHPnE37cw6YZUzLKxYIIoI0g4zikUZAix5uKfeG4LntWiC1PsWJQgF4r0lWxB~keH2qkfPnaPXHyCtPuiiJ2Ss7QYBPilR~nMDY5YRA__&Key-Pair-Id=APKAJLOHF5GGSLRBV4ZA

Velenturf AP, Purnell P (2021) Principles for a sustainable circular economy. Sustain Prod Consumption 27:1437–1457. https://doi.org/10.1016/j.spc.2021.02.018

Wallner TS, Haslbeck JM, Magnier L, Mugge R (2024) A network analysis of factors influencing the purchase intentions for refurbished electronics. Sustain Prod Consumption 46:617–628. https://doi.org/10.1016/j.spc.2024.03.009

Wiedmann T, Minx JC (2008) A definition of carbon footprint. *ResearchGate*. https://www.researchgate.net/publication/247152314_A_Definition_of_Carbon_Footprint

Windisch-Kern S, Gerold E, Nigl T, Jandric A, Altendorfer M, Rutrecht B, Scherhaufer S, Raupenstrauch H, Pomberger R, Antrekowitsch H, Part F (2021) Recycling chains for lithium-ion batteries: a critical examination of current challenges, opportunities and process dependencies. Waste Manag 138:125–139. https://doi.org/10.1016/j.wasman.2021.11.038

Wischmeyer T (2019) Artificial intelligence and transparency: opening the black box. In: Springer eBooks, pp 75–101. https://doi.org/10.1007/978-3-030-32361-5_4

World Economic Forum (2024) E-waste management key to sustainable digital transformation. World Economic Forum. https://www.weforum.org/stories/2024/11/least-developed-countries-sustainable-ewaste-management/

World Economic Forum (2025) 6 ways data centres can cut their emissions: a case study. https://www.weforum.org/stories/2025/01/6-ways-data-centres-can-cut-emissions/

Yadav K, Jain H (2024) Integrating urban master planning with environmental impact assessment for sustainable and resilient cities. In: Cities of tomorrow: urban resilience and climate change preparedness. Springer Nature Singapore, Singapore, pp 3–23

Zeng D, Fu X, Ouyang T (2018) Implementing green IT transformation for sustainability: a case study in China. Sustainability 10(7):2160. https://doi.org/10.3390/su10072160

Zhang Q, Meng Z, Hong X, Zhan Y, Liu J, Dong J, Bai T, Niu J, Deen MJ (2021) A survey on data center cooling systems: technology, power consumption modeling and control strategy optimization. J Syst Architect 119:102253. https://doi.org/10.1016/j.sysarc.2021.102253

Zhou C, Hu Y, Liu R, Liu Y, Wang M, Luo H, Tian Z (2025) Energy performance study of a data center combined cooling system integrated with heat storage and waste heat recovery system. Buildings 15(3):326. https://doi.org/10.3390/buildings15030326

Chapter 12
Entrenching Inequality: Climate Neutrality in Developed, Developing, and Underdeveloped Countries

Karina Bhatia Kakkar and Yanshik Khatri

12.1 Introduction

Achieving climate neutrality, where human activities produce net-zero greenhouse gas emissions, is one of the most pressing challenges of our time. It is a goal that unites nations under the common objective of mitigating climate change, yet the path to achieving this goal varies greatly across countries. In developed countries, the journey toward climate neutrality is both advanced and complex. These nations typically have the financial resources, technological innovations, and infrastructure needed to make significant strides in reducing emissions and transitioning to renewable energy sources, aligning with SDG 7. However, the high levels of industrial activity and energy consumption inherent in these economies pose substantial challenges. For example, while countries like Germany and Sweden are leading in renewable energy adoption, they also face the challenge of retrofitting existing infrastructure and industries to align with climate-neutral goals. Furthermore, the focus on SDG 8, promoting sustained economic growth, often results in increased consumption and waste generation, complicating efforts to achieve SDG 13 (Climate Action). Developed nations also have a significant impact on global ecosystems, as their consumption patterns and industrial activities often contribute to biodiversity loss (SDG 15) and the degradation of marine environments (SDG 14).

In contrast, developing countries are in a transitional phase, where economic growth is paramount, yet environmental sustainability cannot be overlooked. These nations are striving to improve their energy infrastructure, moving towards cleaner energy sources as outlined in SDG 7, but often lack the necessary financial and technological resources. Countries like India and Brazil exemplify this struggle, where rapid urbanization and industrialization drive economic growth (SDG 8) but also increase emissions and environmental degradation. The challenge for these nations

K. B. Kakkar (✉) · Y. Khatri
School of Management, GD Goenka University, Sohna, Haryana, India
e-mail: kareenabhatia@gmail.com

H. Jain et al. (eds.), *The Circular Fallacy: Closing the Loop in a Consumerist World*, Circular Economy and Sustainability, https://doi.org/10.1007/978-3-032-22687-7_12

is to balance their development needs with climate action (SDG 13), often requiring external support in the form of technology transfer, financial aid, and capacity building. Moreover, the protection of ecosystems (SDGs 14 and 15) is critical in these regions, as their rich biodiversity is often threatened by unchecked industrialization and deforestation, exacerbating the impacts of climate change. Underdeveloped countries, meanwhile, face the most significant hurdles in the pursuit of climate neutrality. These nations often struggle with severe socio-economic challenges, including poverty, a lack of infrastructure, and limited access to clean energy. For instance, many sub-Saharan African countries are heavily reliant on traditional biomass for energy, which not only hampers progress towards SDG 7 but also contributes to deforestation and air pollution, undermining SDG 15. The need to prioritize basic human needs, such as food, water, and shelter, often takes precedence over environmental concerns, making it difficult for these countries to invest in climate action (SDG 13). Moreover, underdeveloped nations are disproportionately affected by the impacts of climate change, such as rising sea levels and extreme weather events, which further strain their limited resources and hinder progress. towards SDG 14 (Life Below Water) and SDG 15.

The differences in how developed, developing, and underdeveloped countries approach climate neutrality are stark, driven by their unique economic, technological, and environmental contexts. While developed nations are focused on mitigating the environmental impacts of their industrialized economies, developing countries are striving to achieve a balance between growth and sustainability. Underdeveloped countries, however, are often left grappling with immediate survival needs, making climate neutrality a distant goal. The global pursuit of climate neutrality, therefore, requires tailored strategies that account for these diverse challenges, ensuring that all nations can contribute to and benefit from a sustainable future. The objective of the research is to compare and contrast the developed, developing, and underdeveloped countries around the globe on the basis of the seven critical factors. In order to understand the impact of each on the achievement of SDGs, which will lead to carbon neutrality. The findings of the study reveal the role of development on carbon emissions and other critical factors. In the study, fifteen nations were compared and contrasted; many other countries can also be selected to understand the role of development on carbon emissions. The future scope of the study,

Objective of the Study

This study draws on the relationship between GDP and the alignment of a country with the SDGs. The understanding herein is that economic growth, which is measured by the parameter of GDP, has a correlational relationship with the dimensions of sustainable development: social, environmental, and economic. The diagram comparing GDP and SDGs was created to visually highlight the similar veins between economic progress and the efforts of sustainable development. This diagram maps their intersections to further illustrate how improvements in clean energy—a compact phrasing of SDG 7, decent work—a compact phrasing of SDG 8, climate action—a compact phrasing of SDG 13, and ecosystem preservation—a compact phrasing of SDGs 14 and 15—can both drive and result from a country's economic performance.

The comparison is to be done so as to highlight the fact that sustainable development is not apart from but an integral part of economic growth and therefore is presented as part of the holistic view of the progress of a nation.

Research Methodology

The study employs a literature review methodology by analysing articles, reports, websites, and periodicals from reputable academic sources, such as Scopus and Web of Science journals. Te emphasis is placed on exploring the alignment of nations with Sustainable Development Goals (SDGs) to understand their progress toward climate neutrality. The analysis focuses on five developed, five developing, and five underdeveloped countries as classified by the Global Competitiveness Index (GCI). It evaluates the relationship between economic growth, greenhouse gas emissions, and environmental protection through the lens of specific SDGs, SDG 7 that is affordable and clean Energy, SDG 8 that is decent work and economic growth, SDG 13 that is climate action, SDG 14 that is life below water and SDG 15 that is life on land.

The countries are chosen based on GDP rankings and GCI scores to ensure diversity and represent the economic, social, and environmental challenges across different development levels. The key metrics include GDP, carbon emissions, renewable energy adoption, greenhouse gas emissions, and indicators related to the selected SDGs. The comparative analysis of energy consumption, environmental degradation, and economic growth are carried out to highlight disparities and challenges among countries. The critical success factors (CSF) approach is used to assess the contribution of the selected SDGs toward climate neutrality. The study evaluates the interplay between economic activities and their environmental impacts, considering varying capacities and challenges among developed, developing, and underdeveloped nations. This methodology underpins the paper's comparative analysis, aiming to emphasize the global disparities and inequities in achieving climate neutrality while addressing SDG-driven sustainability goals. The steps of the research in the study is shown in Fig. 12.1.

12.2 Literature Review

12.2.1 Climate Neutrality

Climate neutrality simply means that as much GHG is taken up as is emitted, so net zero is the change in carbon emissions. Mitigation involves reducing as many emissions as possible to reach climate neutrality through the modification in energy production, industry, and transport, among other sectors, while decreasing the rest of the emissions through measures such as reforestation, CCS technology, or carbon credits. This, according to Wolf et al. (2021), is the hub of all global climate actions because this part of the Paris Agreement aims to stop the rise in global temperature well below 2 °C above the pre-industrial levels, while the ambition is limiting this to

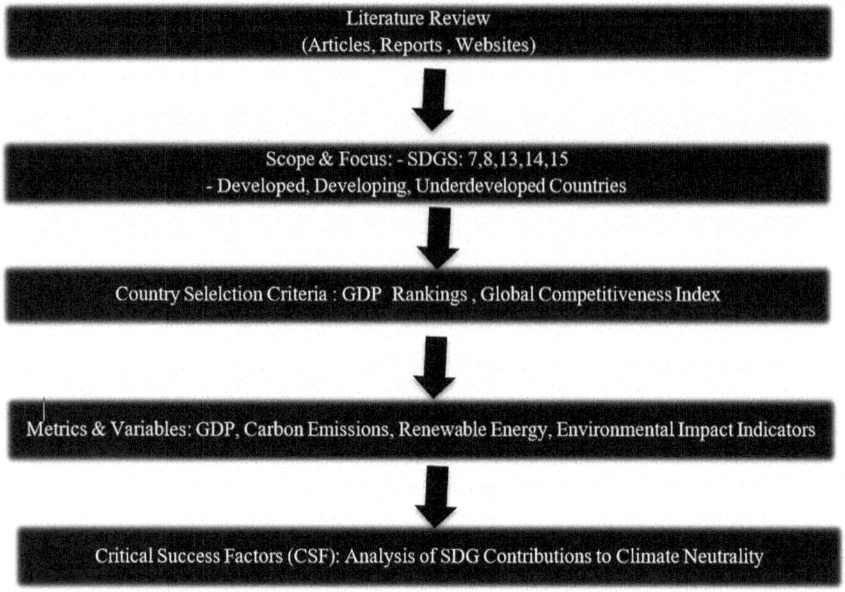

Fig. 12.1 Research methodology of the study

1.5 °C. It is now high time countries took the initiative to achieve climate neutrality, complemented by companies and other organizations (Saini et al. 2024). They can realize such changes by shifting from dependence on fossil fuels to renewable sources of energy, using energy efficiently, and maintaining sustainable practices in all aspects of life. Generally, climate neutrality is defined in the context of wider sustainability goals, which include those of the United Nations SDGs, particularly SDG 13, Climate Action. Climate neutrality is viewed as one of the most important steps toward fighting climate change and preserving the future of our planet (Ciambra et al. 2023).

12.2.2 Climate Neutrality in Developed Countries

For many years, the developed world—particularly North America and Europe-has been responsible for the vast bulk of emissions. Indeed, industrialization based on heavy use of coal and other fossil fuels as a source of fuel energy laid the foundations both for modern economies and the current climate changes and rise of carbon emissions (Saini et al. 2025). While these countries are making their way towards climate neutrality, they represent themselves as leading in the race of climate action. At the same time, however, their underlying narratives reflect the big role they have played hitherto in creating this problem in the first place. As Salvia et al. (2021) argue, the two decades of emissions produced by them play an important role in showing climate equalities. The goal of developed countries has been to achieve complete climate neutrality, with most of them aspiring to net zero by 2050, based on huge financial endowments, advanced technologies, and infrastructure. For example, the European Union Green Deal targeted making Europe the first climate-neutral continent by 2050, using an uptick in renewable energy resources, carbon pricing, and innovation in green technologies as underpinnings. While these efforts are commendable, they often rely heavily on carbon-heavy mechanisms that allow countries to compensate for their emissions by investing in carbon reduction projects. This has been criticized because this approach allows the developed countries to further their high levels of consumption and thereby create carbon emissions while shifting the burden of emission reduction to other regions, mostly to developing and underdeveloped countries (Zhao et al. 2022).

Economic and technological development of a country allows it to achieve climate neutrality with relatively minor disruptions to its economy. Advancements in renewable energy technologies, for example, are being pursued aggressively to enable these countries to move away from dependence on fossil fuels without significant sacrifices in economic performance. Another advantage brought by the large availability of financial capital is the investment in sustainable green technologies and infrastructure, further easing the transition. This advantage also points to one important inequality between developed and underdeveloped or developing countries. Whereas the developed countries can invest in clean energy because they can afford to do so, in the case of the developing or underdeveloped world, great difficulties arise in being able to meet basic energy needs—let alone invest in expensive renewable technologies (Wang et al. 2021). Notwithstanding these advances, developed countries are taken to task over how they pursue climate neutrality. Carbon offset and the unequal distribution of climate responsibility have cast debates on the equity issue in global climate policy. Critics say that it only serves the interests of wealthy nations at the expense of developing countries. But even beyond that, financial commitments from developed countries not met—as evidenced in the $100 billion annually of climate finance promised to developing countries—continue to drive global disparity further upward.

12.2.3 Climate Neutrality in Developing Countries

For the developing world, pursuit of climate neutrality is a complex challenge: economic development with a reduction of emissions. Unlike in developed countries, which have already achieved economic benefits from industrialization, in developing countries their economies are still building and trying to lift their populations out of poverty. A combination of these two aspects builds a challenging task in achieving climate neutrality while pursuing these goals. Most developing countries need much more time using fossil fuels to reach appreciable levels of economic growth before arriving at climate neutrality. This view emanates from the very principle of "common but differentiated responsibilities" (Alola and Adebayo 2023), which means that though shared by all countries, the responsibility to deal with the climate problem is different in capabilities and historical contributions. One solution that has been proposed for developing countries is "leapfrogging," whereby countries bypass traditional fossil fuel-based development and move directly to convert to renewable energy. Yet, it also has its own problems. Most developing countries lack the financial resources, technology, and infrastructure to undertake large-scale renewable energy projects. Additionally, the requirement for rapid economic growth will have an impact on the pace at which transactions in renewable energy and climate neutrality take place within a developing country. Take India, for example: its targets for renewable energy are ambitious; by 2030, 40% of its electricity generation is to come from non-fossil sources. But coal remains an indispensable source of energy for the country-a healthy reminder that each day's task of shifting into renewable energy in a fast-growing population and economy will not be easy. According to Khan et al. (2024), developing countries require international partners if they are ever to achieve climate neutrality. Financial aid and technology transfer go hand in hand to enable such countries to invest in clean energy and reduce their emissions. Yet, the support presently coming from developed countries is below what is required, and countries cannot even meet their financial commitments. The Green Climate Fund, set up to assist developing countries in mitigation and adaptation efforts, has been underfunded and slow to disburse (Chen et al. 2024).

12.2.4 Climate Neutrality in Underdeveloped Countries

In poorer countries, the emphasis is less on climate neutrality than on adequate supplies of basic energy services. Most of these countries, in fact, have poor supplies of steady electricity; in Sub-Saharan Africa, for example. To such countries, longer-term use of fossil fuels may be justified by pressing development needs: powering hospitals, schools, and industries. The global discourse on climate neutrality, at times, can consider the condition of these underdeveloped countries. They are the countries that contribute least to global emissions while having a high effect on climate change (Wang et al. 2022). The international community needs to support them in providing the basic needs and balance economic growth with climate neutrality. The underdeveloped countries account for a small fraction of the global carbon emissions. For that reason, they are still subjected to the effects of climate change, which are very severe: droughts, floods, and food insecurity. This is the tragic irony: that those who have contributed least to the problem suffer most of its effects. The global drive for climate neutrality should take into account the very minimal contribution by underdeveloped countries to global emissions. Climate policies are designed in a way that these countries can attain their economic challenges and address climate goals without inflicting any harm on the economy (Zhao et al. 2024).

The underdeveloped nations rely on international support to bridge mitigation and adaptation needs. So far, financial mechanisms like the $100 billion climate fund by developed nations have failed to materialise. Most underdeveloped countries are being denied the resources needed to create resilience against the impacts of climate change or even invest in renewable energy. The reliance on international aid really underlines the vulnerabilities of underdeveloped countries in the global climate framework. Unless these are changed with significant shifts in these distributions of climate finance, countries like those will always be playing catch-up toward achieving neutrality in their climate.

12.2.5 Global Inequality and Climate Neutrality

It follows from the principle of "common but differentiated responsibilities" central to the international framework on climate: all countries must take action in view of the challenge, yet their capacities and historical contributions to the problem are very different. However, this principle has been applied in an inconsistent fashion, as attested to by the gross inequalities in the way climate responsibilities are shared between developed and developing nations. It is expected that the developed countries, having better resources and historical emissions, take the lead and initiative to reduce carbon emissions and support other nations in efforts to reduce carbon emissions. Realities are always different from expectations. Developing and underdeveloped countries pass burdens very frequently which are above their capacities and show global inequalities Emmerling et al. (2024). Developed countries failed to meet their financial commitments. It also severely underestimates how much carbon is emitted throughout the entire life cycle of cement. The $100 billion a year of climate finance that was supposed to help developing nations mitigate and adapt to climate change has not been delivered. The shortfall provides huge prongs to global inequality in that the developing and underdeveloped nations that needed to achieve neutrality-for that matter, resilience—against climate impacts are without resources. It also contradicts the global stand against climate change by increasing existing inequalities due to a lack of adequate financial support. Without substantial and secured funding, developing and underdeveloped countries cannot mobilize the activities to be undertaken to fulfill their pledges on the climate issue (Giannetti et al. 2020).

Among the more controversial issues linked to carbon neutrality is carbon offsetting. These offsets are often what net-zero means for a developed country, which involves investing in carbon-reduction projects in other countries rather than reducing emissions at home. While one-sided, offsets can be a useful tool; they also present significant equity concerns. The developed countries thereby continue using more carbon, putting the burden of reduction on the developing and underdeveloped nations. The problem is, this practice does not only undermine the integrity of the net-zero commitment; it also deepens global inequalities by imposing unjustified pressure on poorer countries to attain emissions reductions (Emmerling et al. 2024).

12.2.6 SDGs Designed by the United Nations

The SDGs consist of 17 goals that are globally nature-designed by the United Nations in 2015 to guide development up to 2030. These goals serve as a common vision for creating a more fair, prosperous, and sustainable world. The SDGs deal with different, but interrelated, problems-poverty, inequality, climate change, and environmental degradation, peace, justice. Certain targets and indicators accompany each goal to accomplish; that forms a critical framework for action and monitoring progress. The SDGs are designed to be universal in application for all countries regardless of their level of development. They capture participation from various stakeholders at the global level (Soergel et al. 2021a, b). On this basis, the focus falls on selected SDGs, including: SDG 7: Affordable and Clean Energy; SDG 8: Decent Work and Economic Growth; SDG 13: Climate Action; SDG 14: Life Below Water; and SDG 15: Life on Land. The selected goals are central to addressing big environmental and economic challenges that also have a large bearing on sustainable development. By focusing on these SDGs, we ensure that global priorities balance economic growth with climate neutrality (Giannetti et al. 2020).

12.2.7 SDGs Related with Climate Neutrality

The Fig. 12.2 shows SDG 7 (Affordable and Clean Energy), SDG 8 (Decent Work and Economic Growth), SDG 13 (Climate Action), SDG 14 (Life Below Water), and SDG 15 (Life on Land) respectively represented with icons-a lightbulb for SDG 7, a briefcase or gear for SDG 8, a globe or climate icon for SDG 13, an ocean wave for SDG 14, and a tree for SDG 15. Such icons reinforce their thematic areas of concentration: sustained energy, economic growth, taking care of the climate, life in the ocean, and life on land. The diagram underlines the fact that these objectives are intertwined-these goals are advanced by each other-so this will mark a holistic approach toward sustainable development: equilibria of economic growth with environmental care and climate resilience.

The following illustration represents the SDGs: SDG 7: Affordable and Clean Energy; SDG 8: Decent Work and Economic Growth; SDG 13: Climate Action; SDG 14: Life Below Water; and SDG 15: Life on Land. It will visually explain in detail each goal with an iconic figure: a lightbulb for clean energy, a briefcase for economic growth, Earth for climate action, an ocean wave for marine conservation, and a tree for terrestrial ecosystems. This diagram illustrates exactly what each SDG means by drawing a picture of their focus areas: sustainable energy access, inclusive economic growth, climate change mitigation, marine resource protection, and ecosystem preservation. It also points to how connected these goals are, and together they facilitate a balanced approach toward sustainable development; thus, efforts should be taken forward integrally to realise a healthy, more equable planet.

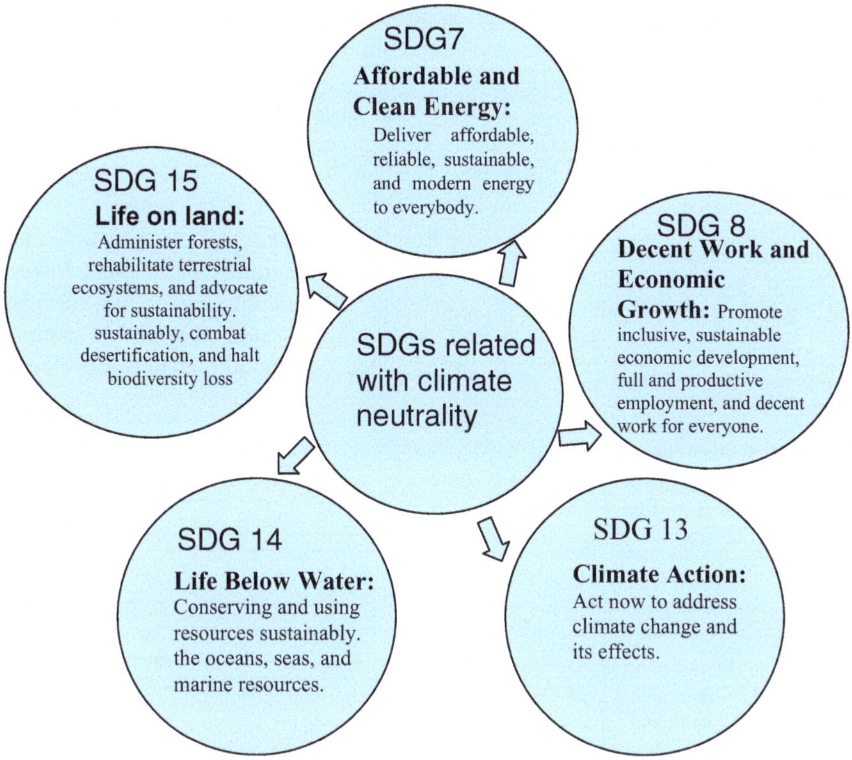

Fig. 12.2 Sustainable development goals (SDGs) used in the study (United Nation (2015))

Table 12.1 considers five SDGs. The table summarises key SDGs taken into consideration in this report and elaborates the factors evaluated and the definition taken into account in considering each goal. In particular, SDG 7 guarantees access to affordable, reliable, sustainable, and modern energy, which is proven to be absolutely relevant in accelerating economic growth and contributing to the improvement of quality of life in every country. SDG 8 seeks to achieve the end of economic growth that is inclusive, productive, employment, and decent work for everyone and sustainable. All this is what is meant by poverty reduction and prosperity. It is the most important goal to be seen in the developing and underdeveloped nations (Saxena et al. 2021). SDG 13 calls for immediate action to combat climate change and its impacts because climate disruption presents humanity with a global threat. SDG 14 defines the importance of conservation and sustainable use of oceanic and marine resources, which are vital in maintaining ecological balance and supporting livelihoods, creating a balance in the ecosystem. Finally, SDG 15 indicates the protection and restoration of terrestrial ecosystems, including forests managed sustainably, halting land degradation and biodiversity loss for preserving natural resources underpinning environmental health. Each of these SDGs is important in addressing certain challenges and promoting a balanced approach to sustainable development, integrating economic, social, and environmental dimensions that are vital at the global level (Kreinin and Aigner 2022).

The factors affecting climate neutrality are deeply connected with the socio-economic status of a nation (Developed, Developing or underdeveloped), as highlighted by the United Nations Sustainable Development Goals (SDGs). In particular, SDG 7 (Affordable and Clean Energy), SDG 8 (Decent Work and Economic Growth), SDG 13 (Climate Action), SDG 14 (Life Below Water), and SDG 15 (Life on Land) serve as key indicators of progress towards climate neutrality. These comparisons and analysis are drawn upon insights from reputable academic sources, like Springer Nature, Elsevier, Taylor & Francis, and various Scopus-indexed books, to explore how these factors manifest differently in developed, developing, and underdeveloped

Table 12.1 Critical success factors selected from SDGs

S. No	SDGs	Factors evaluated	Definition	Citations
1	SDG 7	Affordable and clean energy	Deliver affordable, reliable, sustainable, and modern energy to everybody	Wang et al. (2020) Carpejani et al. (2020); Chovancová and Vavrek (2022)
2	SDG 8	Decent work and economic growth	Advocate for inclusive, sustainable economic development, comprehensive and productive employment, and dignified labour for all individuals	Kreinin, and Aigner (2022) Leal Filho et al. (2021) Bieszk-Stolorz, and Dmytrów (2023)
3	SDG 13	Climate action	Take immediate action to mitigate climate change and its consequences	Soergel et al (2021a, b) Gjorgievski et al. (2022) Arora and Mishra (2023)
5	SDG 14	Life below water	Utilise oceans, seas, and marine resources in a sustainable manner	Küfeoğlu (2022) Gulseven (2020) Leal Filho et al. (2022)
6	SDG 15	Life on land	Preserve terrestrial ecosystems, manage forests judiciously, combat desertification, and halt biodiversity decline figure	Huan and zhu (2023) Foláyan et al. (2024) Fitawek and Hendriks (2024)

Source United Nation (2015), Tan et al. (2022), Versey (2021), Huyer et al. (2020), Kılkış et al. (2020), Lakshani et al (2023)

countries. On the basis of GCI, five countries of each type have been selected for the analysis.

12.3 Classification of Countries Based on Scales

Countries are classified as developed, developing, or underdeveloped based on various economic, social, and structural indicators. The scales and classifications that are used in this analysis are Gross Domestic Product (GDP) and the Global Competitiveness Index (GCI).

12.3.1 Gross Domestic Product (GDP)

Gross Domestic Product refers to the overall value of all the products produced and services provided within the nation's boundaries within a certain period or year. Gross Domestic Product is generally adopted when examining the country's economic performance and ranking various countries based on economic strength. Gross Domestic Product encompasses broad measures and consists of four significant constituents explaining the economy of any nation. Firstly, Consumption refers to spending by household consumers on goods and services; examples include durable goods, nondurable goods, and services ranging from health and education to leisure activities (World Health Organization 2021). Secondly, investment involves spending on capital businesses, buildings, and inventories that add to economic growth by increasing productive capacity and earning income in the future. Thirdly, government Spending involves money spent by the government in creating goods and services in infrastructures, public services, and also in defence, which is needed to assist in developing the country's economy and improving the welfare of people (Gulseven 2020). These three components together tend to provide the economic health and performance of a country that basically help in influencing policies and economic strategies. Gross Domestic Product, GDP, is one of the important indicators of economic health as it provides insights into the general or overall performance of an economy (He et al. 2023). A high level of GDP normally tends to point to a healthy and growing economy as it is reflections of increasing economic activities and prosperity. On the other hand, a low or negative level of GDP may indicate such economic ills as recession or stagnation, pinpoints the areas that must be taken care of. In addition, there is a close link between GDP and the living standard for the people of that country; the higher the rate of GDP, so better living standards, as it generally would imply that there is more wealth and resources to go around within the populace. (Sobkowiak et al. 2020). This may not necessarily be the case because GDP growth does not automatically translate into a fair distribution of wealth in an economy. It is the basis upon which fiscal and monetary policies are contemplated on issues to do with taxation, government expenditure, and interest rates. Through the analysis of the trends of GDP, policymakers can strategize on how to create an enabling environment for growth to take place, manage inflationary pressures, and mitigate economic challenges; hence, GDP becomes fundamental in economic planning and policy formulation (Tan and Chew 2020).

12.3.2 Global Competitiveness Index (GCI)

The GCI is one in-depth tool created by the World Economic Forum to examine competitiveness within a country. Competitiveness, in general, refers to the set of institutions, policies, and factors that determine the level of productivity of a country and, in that way, propels a nation toward economic growth and development. GCI components: Institutions Infrastructure, Macroeconomic Stability, Health and Primary Education, Higher Education and Training, Goods Market Efficiency, Labour Market Efficiency, Financial Market Development, Technological Readiness, Market Size, Business Sophistication, Innovation Capability Components defining the competitiveness of a country Institutions represent the legal and administrative frameworks governing interactions between individuals, businesses, and governments; they make for stability and support economic activities (Gajdosova 2023). Infrastructure refers to basic facilities that involve such things as roads, ports, and telecommunications, which is fundamental to business efficiency and economic development. Macroeconomic Stability reveals the balance in the economy, where aspects such as inflation rates and fiscal balance cut down uncertainty and thus help in sustainable development. Health and Primary Education: It assesses the general health of the workforce and the level of basic education; both are of major importance to have a productive population (Ortiz and Roser 2023). Higher Education and Training: This ensures the quality of work and higher economic progress. Goods Market Efficiency: This shows the quality of the production and distribution of goods, hence competitive markets ensure better quality products. Labour Market Efficiency: addressing flexibility, efficiency and fairness of labour practices, which enables needed matching of skills to jobs, and acceptance to economic changes. Financial Market Development discusses availability and access to financial services that support business investment and economic stability (Cracknell et al. 2020). Technological Readiness refers to how far a country can adopt and use new technology; it is very important for innovation and productivity. Market Size encompasses domestic and international market potentials, affording opportunities for businesses to expand and realize economies of scale. Business Sophistication signals the quality of business networks, strategies, and operations that a country has, which make it more competitive in global markets. Finally, Innovation Capability captures the ability of a country to develop and introduce new products and services-a critical factor for long-term competitiveness and improved lifestyle globally. The following are some reasons why the Global Competitiveness Index is important. It acts as an important long-term growth indicator whereby it identifies key factors that will spur continued economic growth and prosperity. Understanding these factors better equips countries to be ready for the challenges ahead (Meng et al. 2020). The GCI is an excellent benchmarking tool whereby countries are able to compete with each other regarding competitiveness. Such comparison may bring strengths and weaknesses to light on which improvements can be made. Furthermore, GCI provides further policy guidance for policymakers in the design and implementation of economic policies that enhance the level of productivity and competitiveness. It thus enables countries to make more informed decisions in an effort to promote a more dynamic and resilient economy (Ardaillon and Roebel 2020).

12.4 Analysis of Selected Countries based on SDGs

The country selection would include the United States, China, Japan, Germany, the United Kingdom, India, Brazil, Indonesia, Mexico, Turkey, Afghanistan, Haiti, the Central African Republic, Madagascar, and Niger, based on their ranking in GDP and GCI. The GCI provides useful insight into factors like infrastructure, macroeconomic stability, health, education, and innovation that underpin productivity and thus economic growth. Therefore, these countries have been selected considering both GDP and GCI rankings:

12.4.1 Developed Countries

The United States, China, Japan, Germany, and the United Kingdom lead both on economic output per person and competitiveness, serving as benchmarks for economic performance. The United States maintains its top-ranking position, being the largest global economy with highly developed, diversified economic structures. Innovation capability, business dynamism, and market size show high GCI scores underpinned by leading in technology and higher education. Second to none in the list of the largest economies, China is an export powerhouse in manufacturing, having scored well in market size and macroeconomic stability, while rapid technological adoption and infrastructure development further enhance its competitiveness (Wannisinghe et al. 2023). Japan was the third-largest economy in the world, boasting highly advanced technology in manufacturing, while it received high scores in the sub-indices concerning infrastructure, health, and innovation capability by means of technological readiness and sophistication when conducting business. The largest economy in Europe, Germany is supported by a large, diversified industrial base; in most of the factors, especially infrastructure, innovation capability, and macroeconomic stability, it scores high, with its engineering and manufacturing sectors being truly globally competitive. Finally, the United Kingdom is a leading global financial centre with a diversified economy, scoring well in the financial system, business dynamism, and market size, underpinned by its strong legal and institutional frameworks (Sinha et al. 2020).

12.4.2 Developing Countries

India, Brazil, Indonesia, Mexico, and Turkey are emerging economies in respective industrialization and economic growth. Their GCI scores indicate improvements on infrastructure, education, and market efficiencies (Destek et al. 2021). The flourishing economy and a huge population of youthful citizens propel India to the top of this list in market size and innovation capability, which is also being enabled by technological advancement and infrastructure development. Brazil is considered a leading economy in the South American continent, offering several key advantages that include an abundance of natural resources, significant domestic market size, and good macroeconomic stability, although infrastructure and institutional quality remain poor. Indonesia is considered the leading economy in SouthEast Asia and boasts a gain in its consumer base with enhanced macro-economic stability, though remains to be challenged to reinforce infrastructure and business conditions. Apart from the U.S. (Khalid et al. 2021). Mexico is an integral part of the North American economy, with leading market size and macroeconomic stability, strong infrastructure, and financial systems promising further improvement. Turkey is on the way to becoming a dynamically developing economy at the crossroads between Europe and Asia, demonstrating serious strength in market size and infrastructure and striving to improve its innovative capability and business sophistication with the purpose of sustaining its competitive position (Khalid et al. 2021).

12.4.3 Underdeveloped Countries

Afghanistan, Haiti, the Central African Republic, Madagascar, and Niger bear many challenges in both economic output and competitiveness. Low scores of GCI identify those countries with alarming areas of development (Paprotny 2021). Afghanistan is regarded as one of the world's poorest countries, and instability due to conflict hinders improvement, evidenced by very low scores in GCI: in institution, infrastructure, and health, which impede its competitive economy a lot (Armin Razmjoo et al. 2020). Haiti is the poorest country in the Western Hemisphere, with very limited economic development, low levels of GCI scores in infrastructure, health, and education-all further heightened by poor institutional environments and rampant natural disasters. The fragile economy of Central Africa has been characterized by low GCI pillar scores, especially its institutions, infrastructure, and the macroeconomic environment, as it remains plagued with problems of governance and overall development.

On the other side, (Penglong et al. 2022). Madagascar is an example of a low-income country with an agrarian economy, really with poor scores of infrastructure, health, and education that could help explain why such countries are not very competitive due to their weak institutions and poor infrastructures. The same thing can be said regarding Niger: one of the poorest countries in the world and highly dependent on agriculture, with extremely low scores regarding infrastructure, health, and education; besides, it has less resource availability and a very hostile climate. As such, most of those countries have to catch up with development in those important areas in order to improve their economic and competitive positioning (Miao et al. 2023).

12.4.4 GDP of Selected Countries

The United States, for instance, had a GDP of $23.2 trillion and government expenditures amounting to $6.2 trillion or 29% of its GDP on defense, healthcare, and social security; its investment expenditures reached $3.8 trillion or 18% of its GDP on business infrastructure, real estate, and technology. High consumption, and investment nonetheless, the country registers a trade deficit amounting to $1.1 trillion. China with a GDP amounting to 19.4 trillion, investment in China stands heavy at $8.8 trillion or 43% of GDP in the manufacturing, real estate, and technology sectors. Government spending is pegged at 3.6 trillion or 19% of GDP for the funding of infrastructures and social programs. The trade surplus, therefore, is at $0.5 trillion. That is to say, out of Japan's GDP of $4.9 trillion, government spending on social welfare and public services was at $2.4 trillion, or 30 percent of GDP; investment was at $1.1 trillion, or 13 percent of GDP, with a bias for technology and infrastructure; while a trade surplus keeps it at $0.2 trillion. With a GDP of $4.4 trillion, it spends $1.3 trillion for social services, education and infrastructure, or 26 percent of its GDP, while it invests $0.9 trillion, or 18 percent of its GDP, in high-tech industries and green energy. It thereby can generate a trade surplus of $0.3 trillion.

Table 12.2 depicts wide disparities in the level of government spending and investment across the world. The UK, with a GDP of US$3.3 trillion, spends US$1.1 trillion-22% of its GDP-on healthcare, education, and public services, and invests an additional US$0.9 trillion, or 17% of its GDP, in financial services and technology; all this while maintaining a trade deficit of US$0.2 trillion. The National Income: correspondingly, India's GDP is $3.8 trillion; government spending is $0.8 trillion, 18% of GDP; investment in infrastructure and industries is $0.9 trillion, 20% of GDP, while the trade deficit is $0.1 trillion. Similarly, Brazil's GDP is $2 trillion, with government spending of $0.5 trillion, 25% of GDP, and investment of $0.4 trillion, 20% of GDP, with a trade deficit of $0.05 trillion. GDP of Indonesia is $1.4 trillion, government spending of $ 0.2 trillion or 15% of GDP, and investment of $ 0.3 trillion or 25% of GDP leave a small trade surplus of $ 0.02 trillion. GDP of Mexico is also $1.4 trillion, the country spends 15% of its GDP on public services with $0.2 trillion and invests 20% of GDP at $0.3 trillion and has a trade deficit of $ 0.1 trillion. At the same time, however, an economy like Turkey's has 1 trillion-dollar

GDP, which allocated 0.15 trillion dollars (15% of its GDP) to pay for government programs and invested 0.3 trillion dollars (25% of its GDP), leaving a trade deficit of 0.05 trillion dollars.

The share of government expenditure to the GDP is very high for the less developed countries, lying in the range of more than 40 to 60 percent. The Afghan government spends around \$0.015 trillion, constituting about 65 percent of its GDP; Haiti constitutes around 40 percent with government spending of \$0.005 trillion; the government spending of the Central African Republic accounts for about 60 percent of its GDP, with a value amounting to \$0.008 trillion. This indicates their dependence on extraneous aids and government support. The investment is minimum, as Afghanistan is investing \$0.005 trillion, 7% of GDP, while Haiti is doing \$0.002 trillion, 10% of GDP, and the Central African Republic \$0.001 trillion, 8% of GDP. All the three countries run immense trade deficit, further showing their inability to put a proper and balanced trade and economic development in place.

12.4.5 Critical factors of SDG

The diagram below Fig. 12.3 is an application of the SDGs by mapping key drivers in relation to the following key metrics: GDP, carbon emission annually, affordable and clean energy, economic growth, greenhouse gas emission, life below water, and life on land. It serves as a plot for how these elements intersect each other in their driving force. In this case, GDP and economic growth are the economic drivers for progress, while simultaneously affecting environmental elements. The carbon emission and greenhouse gas emission are showing the environmental issues from economic activities. Then, affordable and clean energy are portrayed as very important factors that could provide low emissions to support sustainable development. Life below Water and Life on Land stand for environmental outcomes influenced by the economic and energy undertakings. It suggests that an integrated approach to managing economic growth, reducing emissions, and managing ecosystems will be necessary in order to attain the agreed-upon targets for sustainability in respect of marine and terrestrial ecosystems.

The diagram in Fig. 12.3 unfolds the meaning of Gross Domestic Product and Economic Growth as indicators of the country's economic performance and growth, while Carbon Emission annually and Greenhouse Gas Emission show the environmental impact which results from this economic activity. Affordable and Clean Energy stresses that the world needs to go for sustainable energy alternatives that reduce emissions and foster development. Moreover, Life below Water and Life on Land set their prism to the conservation of marine and terrestrial ecosystems, respectively, thus underlining protection from the adverse effects that economic growth and pollution might have on such environments. The diagram integrates these to show interconnectivity between economic health, environmental sustainability, and ecosystem preservation.

Table 12.2 Comparative analysis of GDP components of selected countries

Country	GDP	Consumption	Investment	Government spendings	Net export
United States	$ 23.2.trillion	$14.3 trillion (about 61% of GDP)	$3.8 trillion (about 16% of GDP)	$6.2 trillion (about 26% of GDP)	(−$1.1) trillion (negative balance)
China	$19.4 trillion	$6.6 trillion (about 34% of GDP)	$8.8 trillion (about 45% of GDP)	$3.6 trillion (about 19% of GDP)	$0.4 trillion (positive balance)
Japan	$4.9 trillion	$3 trillion (about 61% of GDP)	$0.7 trillion (about 14% of GDP)	$1 trillion (about 20% of GDP)	$0.2 trillion (positive balance)
Germany	$4.4 trillion	$2.4 trillion (about 55% of GDP)	$1 trillion (about 22% of GDP)	$0.7 trillion (about 16% of GDP)	$0.3 trillion (positive balance)
United Kingdom	$3.3 trillion	$2.1 trillion (about 64% of GDP)	$0.6 trillion (about 17% of GDP)	$0.7 trillion (about 22% of GDP)	−$0.06 trillion (negative balance)
India	$3.8 trillion	$2.4 trillion (about 63% of GDP)	$1.2 trillion (about 31% of GDP)	$0.4 trillion (about 10% of GDP)	−$0.2 trillion (negative balance)
Brazil	$2.0 trillion	$1.2 trillion (about 65% of GDP)	$0.4 trillion (about 20% of GDP)	$0.5 trillion (about 25% of GDP)	−$0.05 trillion (negative balance)
Indonesia	$1.4 trillion	$0.8 trillion (about 60% of GDP)	$0.3 trillion (about 25% of GDP)	$0.2 trillion (about 15% of GDP)	$0.02 trillion (positive balance)
Mexico	$1.4 trillion	$1.0 trillion (about 70% of GDP)	$0.3 trillion (about 20% of GDP)	$0.2 trillion (about 15% of GDP)	−$0.1 trillion (negative balance)
Turkey	$1 trillion	$0.6 trillion (about 65% of GDP)	$0.3 trillion (about 25% of GDP)	$0.1 trillion (about 15% of GDP)	−$0.05 trillion (negative balance)
Afghanistan	$16 billion	$0.013 trillion (about 80% of GDP)	$0.005 trillion (about 30% of GDP)	$0.003 trillion (about 18% of GDP)	−$0.005 trillion (negative balance)

(continued)

Table 12.2 (continued)

Country	GDP	Consumption	Investment	Government spendings	Net export
Haiti	$17 billion	$0.01 trillion (about 60% of GDP)	$0.002 trillion (about 10% of GDP)	: $0.006 trillion (about 35% of GDP)	−$0.001 trillion (negative balance)
Central African Republic	$3.8 billion	$0.0024 trillion (about 65% of GDP)	$0.001trillion (about 25% of GDP)	$0.0006 trillion (about 15% of GDP)	−$0.0002 trillion (negative balance)
Madagascar	$15 billion	$0.01 trillion (about 65% of GDP)	$0.002 trillion (about 15% of GDP)	$0.004 trillion (about 25% of GDP)	−$0.001 trillion (negative balance)
Niger	$14 billion	$0.01 trillion (about 70% of GDP)	$0.002 trillion (about 15% of GDP)	$0.004 trillion (about 30% of GDP)	−$0.002 trillion (negative balance)

Source Gajdosova (2023), Jakovljevic et al. (2020), Tan and Chew (2020), Ortiz-Ospina and Roser (2023), Lakshani et al. (2023)

12.4.6 Critical Success Factors of SDGs

The critical factors influencing the attainment of SDGs are very different for developed, developing, and underdeveloped countries as a reflection of their standing in economic status and environmental impact. Since they are more developed with high incomes and advanced technologies, they are in a better position to invest in clean energy and implement effective climate strategies (Chen et al. 2022). Historically, they are very carbon-emission high, but their efforts in sustainability and green technologies are leading gradually to a decrease in these emissions. They also face quite significant challenges with ensuring marine and terrestrial ecosystems, SDGs 14 and 15, due to their industrial activities, though they normally have good conservation programs in place. Developing countries have medium GDPs, faster economic growth-they are beginning to be industrialised-and, therefore, their carbon emissions are growing (Jakovljevic et al. 2020). While access to clean energy is improving, many still depend on fossil fuels. The Critical factors are shown in Fig. 12.3.

The economic growth is robust, but rather often in conflict with the environmental sustainability on issues of life below water and life on land. Underdeveloped countries, generally with low GDPs and scarce resources, face rather substantial challenges across the board. Their carbon emissions are relatively low per capita, but they are highly vulnerable to the impacts of climate change and lack access to affordable clean energy. For example, (Gajdosova 2023) note that whereas economic growth is slow, environmental challenges such as deforestation and degradation of marine and

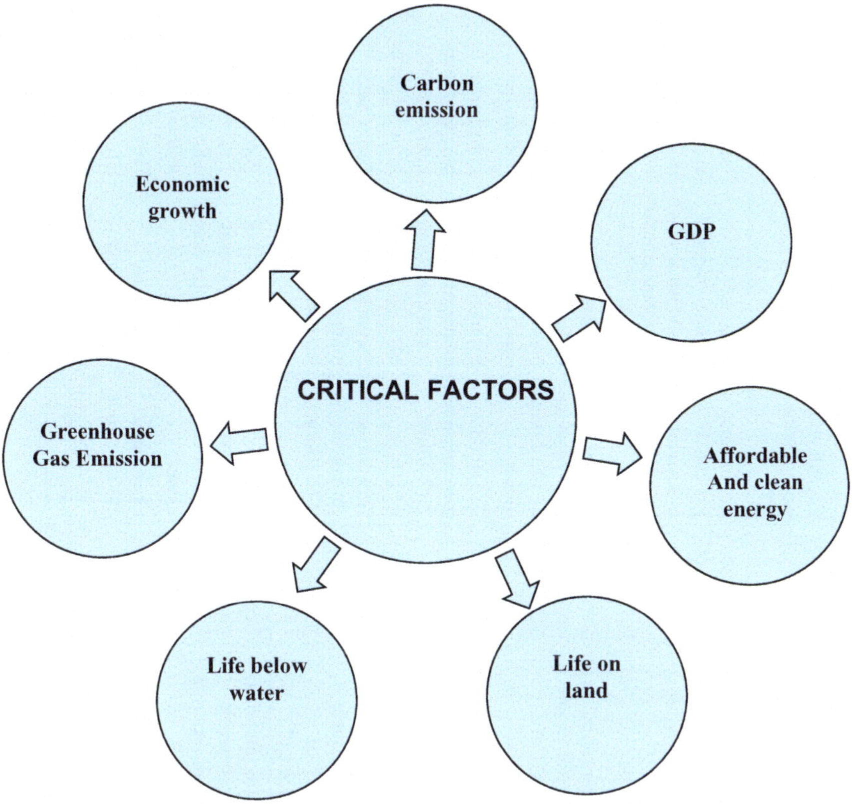

Fig. 12.3 Critical factors influencing the attainment of SDGs. *Source* (Gajdosova 2023; Jakovljevic et al. 2020; Tan and chew 2020; Ortiz-Ospina and roser 2023; Lakshani et al. 2023)

terrestrial ecosystems are serious due to a lack of proper infrastructural and conservationist provisions. This then means that each country has its very own economic and environmental pathways to the attainment of SDGs, which are therefore different and need appropriate strategies and support in addressing such inequities (Tan and Chew 2020).

In Table 12.3, GDP and carbon emissions across developed, developing, and underdeveloped countries provide a critical lens through which global economic indicators are analysed, environmental impacts, and progress toward the Sustainable Development Goals (SDGs).

12.4.6.1 Economic Power and Carbon Emissions

The correlation of the country's GDP with its carbon emission is rather discernible. The developed countries with high GDP generally have remarkably high carbon

Table 12.3 Comparative analysis of critical success factors of SDG of selected countries

Country	Type	GDP	Carbon emission (annually)	Affordable and clean energy	Economic growth	Greenhouse gas emission	Life below water	Life on land
United States	Developed	$26.9 trillion	5.0 billion metric tons	Electricity consumption of 4,000 TWh. Renewable energy demand is 23%	Economic growth of 2.1%	Greenhouse gas emissions is decreased by 1.8%	Approx. 2.5 million tonnes of plastic waste enter the oceans annually	Forest area is approx 310 million hectares which is 33% of total land area
China	Developed	$19.4 trillion	11.7 billion metric tons	Electricity consumption of 8,640 TWh. Renewable energy is 29.4% of electricity demand	Economic growth of 5.5%	Greenhouse gas emissions of 10.1 billion tonnes	Estimated 2.4 million tonnes of plastic waste enter oceans annually	Forest area: 22.3% of total land area; 208 million hectares
Japan	Developed	$4.9 trillion	1.1 billion metric tons	Electricity consumption: 950 TWh renewable energy: 20% of electricity demand	Economic growth: 1.5%	Greenhouse gas emissions: 1.12 billion tonnes	Estimated 200,000 tonnes of plastic waste enter oceans annually	Forest area: 68.5% of total land area; 25.1 million hectares
Germany	Developed	$4.2 trillion	0.7 billion metric tons	Electricity consumption: 550 TWh. Renewable Energy: 47.2% of electricity demand	Economic growth: 0.2%	Greenhouse gas emissions: 0.72 billion tonnes	Estimated 150,000 tonnes of plastic waste enter oceans annually	Forest area: 32% of total land area; 11.4 million hectares
United Kingdom	Developed	$3.4 trillion	0.4 billion metric tons	Electricity consumption: 310 TWh. Renewable energy: 40% of electricity demand	Economic growth: 0.4%	Greenhouse gas emissions: 0.41 billion tonnes	Approx. 170,000 tonnes of plastic waste enter oceans annually	Forest area: 13.2% of total land area; 3.2 million hectares
India	Developing	$3.5 trillion	2.8 billion metric tons	Electricity consumption: 1,570 TWh. Renewable energy: 20% of electricity demand	Economic growth: 6.4%	Greenhouse gas emissions: 2.6 billion tonnes	Approx. 1.0 million tonnes of plastic waste enter oceans annually	Forest area: 21.7% of total land area; 72.1 million hectares

(continued)

Table 12.3 (continued)

Country	Type	GDP	Carbon emission (annually)	Affordable and clean energy	Economic growth	Greenhouse gas emission	Life below water	Life on land
Brazil	Developing	$2.0 trillion	0.4 billion metric tons	Electricity consumption: 630 TWh. Renewable energy: 84% of electricity demand (primarily hydropower)	Economic growth: 2.9%	Greenhouse gas emissions: 0.45 billion tonnes	Approx. 130,000 tonnes of plastic waste enter oceans annually	Forest area: 59% of total land area; 497 million hectares
Indonesia	Developing	$1.4 trillion	0.6 billion metric tons	Electricity consumption: 300 TWh. Renewable Energy: 12% of electricity demand	Economic growth: 5.2%	Greenhouse gas emissions: 0.65 billion tonnes	Approx. 620,000 tonnes of plastic waste enter oceans annually	Forest area: 49.7% of total land area; 91.2 million hectares
Mexico	Developing	$1.3 trillion	0.4 billion metric tons	Electricity consumption: 290 TWh Renewable Energy: 24% of electricity demand	Economic growth: 2.5%	Greenhouse gas emissions: 0.47 billion tonnes	Approx. 150,000 tonnes of plastic waste enter oceans annually	Forest area: 33% of total land area; 64.8 million hectares
Turkey	Developing	$0.9 trillion	0.5 billion metric tons	Electricity consumption: 305 TWh Renewable energy: 44% of electricity demand	Economic growth: 3.5%	Greenhouse gas emissions: 0.51 billion tonnes	Approx. 90,000 tonnes of plastic waste enter oceans annually	Forest area: 29.2% of total land area; 22.3 million hectares
Afghanistan	Underdeveloped	$15 billion	2–3 million metric tons	Electricity consumption: 6 TWh Renewable energy: 75% of electricity demand (mostly hydro)	Economic growth: -3.6%	Greenhouse gas emissions: 0.02 billion tonnes	Approx. 5,000 tonnes of waste in water bodies	Forest area: 2% of total land area; 1.3 million hectares

(continued)

Table 12.3 (continued)

Country	Type	GDP	Carbon emission (annually)	Affordable and clean energy	Economic growth	Greenhouse gas emission	Life below water	Life on land
Haiti	Underdeveloped	$18 billion	3–4 million metric tons	Electricity consumption: 0.5 TWh Renewable energy: 70% of electricity demand	Economic growth: -1.7%	Greenhouse gas emissions: 0.01 billion tonnes	Approx. 10,000 tonnes of waste in water bodies	Forest area: 3% of total land area; 312,000 hectares
Central African Republic	Underdeveloped	$2 billion	Less than 1 million metric tons	Electricity Consumption: 0.1 TWh Renewable energy: 90% of electricity demand (mostly hydro)	Economic growth: 3.0%	Greenhouse gas emissions: 0.01 billion tonnes	Approx. 7,000 tonnes of waste in water bodies	Forest area: 36% of total land area; 22.5 million hectares
Madagascar	Underdeveloped	$15 billion	4–5 million metric tons	Electricity consumption: 0.5 TWh Renewable energy: 65% of electricity demand	Economic growth: 4.2%	Greenhouse gas emissions: 0.02 billion tonnes	Approx. 30,000 tonnes of waste in water bodies	Forest area: 21.6% of total land area; 12.9 million hectares
Niger	Underdeveloped	$14 billion	1–2 million metric tons	Electricity consumption: 0.3 TWh Renewable energy: 85% of electricity demand	Economic growth: 7.3%	Greenhouse gas emissions: 0.01 billion tonnes	Approx. 5,000 tonnes of waste in water bodies	Forest area: 2.6% of total land area; 3.1 million hectares

Source Wiedenhofer et al. (2020), Dong et al. (2022a, b), Tiwari (2021), Da Rosa and Ordonez (2021), Mantero et al. (2020)

emissions. For instance, the United States and China, being the world's two largest economies, are also the two largest emitters of CO_2 (Hu et al. 2021). This is largely because of the areas of industrial activities, high use of energy, and reliance on fossil fuel. All these countries are highly contributing to global emissions, though these are also in the frontline in technological innovations to reduce the carbon footprint-including renewable energy and energy efficiency advancement. But, it is quite challenging to balance economic growth with sustainability (Hailemariam et al. 2020).

12.4.6.2 Developing Countries: A Double-Edged Sword

For developing countries like India, Brazil, and Indonesia, it is a mixed bag. Their GDPs are growing, but they are also arising as contributors of carbon emissions. India has the fastest-growing economies, yet there are some very serious challenges in terms of energy access and sustainability. These countries industrialise, but with every passing day, their growing emissions add much to global climate change. However, many developing countries are also leading in some areas of the SDGs. Indeed, Chen et al. (2022). remark that Brazil, while facing challenges related to deforestation, has achieved major strides with renewable energy-sourced mainly from hydropower. Indonesia struggles with its peatland fires and deforestation, but it is also striving to make its environmental policies better (Dong et al. 2022a, b).

12.4.6.3 Underdeveloped Countries: Low Emissions but High Vulnerability

Underdeveloped countries like Afghanistan, Haiti, and the Central African Republic, each having very small GDPs and emitting only minimal amount of carbons independently. These countries add very little to global climate change; at the same time, they are generally quite vulnerable to its effects. For example, Haiti and Madagascar are experiencing huge environmental degradation, further worsened by their limited resources to mitigate or adapt to climate-related impacts (Caetano et al. 2020). These countries also face challenges in basic economic development, and their improvement in the SDGs is particularly difficult. In addition, with limited energy access, high rates of poverty, and weak institutional capacity, both the needs for economic growth and those for environmental sustainability are far from being successfully met (Soergel et al. 2021a, b).

The Fig. 12.4 illustrates the differences and similarities between Sustainable Development Goals (SDG) and Gross Domestic Production (GDP). It realises that there are different focuses and points of conjunction. GDP is represented as a narrow economic indicator that only concerns the monetary value of the goods and services produced, while the environmental and social concerns are not taken into consideration. In contrast, the SDGs are far-reaching and cover goals of social equity and environmental sustainability in addition to economic growth for such SDGs as

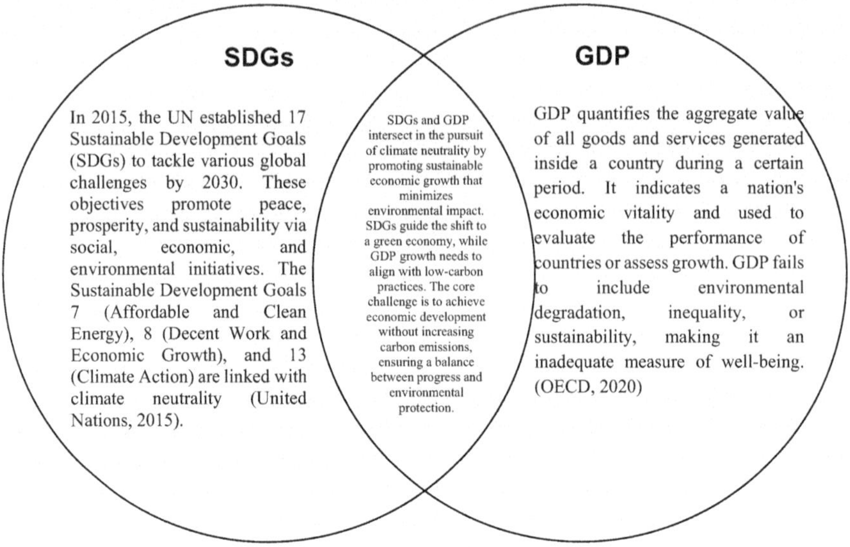

Fig. 12.4 Differences and similarities between sustainable development goals (SDG) and gross domestic production (GDP). 2020 *Source* Organisation for Economic Co-operation (OECD) (); (United Nations, 2015)

Affordable and Clean Energy (SDG 7) and Climate Action (SDG 13). However, the diagram also shows their intersecting points-sustainable economic growth, or SDG 8, also includes GDP to the extent that they both attempt to trace the progress of nations, except that the SDGs do so more comprehensively by integrating sustainability into economic development.

12.5 Discussion

The Sustainable Development Goals (SDGs), particularly SDGs 7, 8, 13, 14, and 15, connect economic status with environmental impact, and provide a foundation for global climate efforts. Developed countries hold the greatest responsibility for reducing carbon emissions without compromising their economic prosperity. They must lead by transitioning to clean energy, setting high sustainability standards, and inspiring others to follow suit. Economic growth and environmental sustainability must be balanced, particularly by countries in development. These nations require international support in adopting clean technologies and sustainable practices, which can ensure that their development does not contribute to global environmental issues. Financial, technological, and capacity-building assistance will enable developing countries to pursue their development goals while mitigating environmental impacts. The least developed countries, which heavily rely on foreign aid and capacity-building programs, require special assistance to build resilience against climate change and pursue sustainable growth paths. This support is vital for helping them achieve their SDGs, even with limited resources. While developed nations such as the United States and China have significant economic capacity, their environmental impacts—especially in energy consumption, renewable energy adoption, greenhouse gas emissions, and waste management—remain high. Developing nations like India and Brazil face challenges in greenhouse gas emissions and plastic waste but are increasingly adopting renewable energy, with Brazil leading in renewable energy adoption for electricity. Meanwhile, underdeveloped nations like Afghanistan and Haiti, despite having low economic outputs and energy consumption, rely heavily on renewable energy sources like hydropower to meet their energy needs.

12.6 Findings

The comparison between developed, developing, and underdeveloped countries reveals significant disparities in economic capacity, energy consumption, renewable energy adoption, greenhouse gas emissions, plastic waste management, and forest cover. Developed nations like the United States and China, with GDPs of $26.9 trillion and $19.4 trillion respectively, have high energy consumption (8,640 and 4,000 TWh). However, their renewable energy adoption is relatively low, with the US at 23% and China trailing behind. These nations are also among the highest emitters of greenhouse gases, with China emitting 11.7 billion metric tons and the US emitting 5.0 billion metric tons annually. Furthermore, they are major contributors to ocean plastic pollution, with China and the US releasing 2.4 million and 2.5 million tonnes of plastic waste each year, respectively. In contrast, developing nations like India and Brazil, with GDPs of $3.5 trillion and $2.0 trillion, have lower energy consumption (1,570 TWh for India, significantly lower for Brazil), but India's renewable energy adoption is at 20%, and Brazil leads with 84% of its electricity sourced from renewables. Despite their renewable energy adoption, both countries face challenges, with India emitting 2.8 billion metric tons of greenhouse gases and contributing 1 million tonnes of plastic to the oceans, while Brazil emits 0.45 billion metric tons and contributes 130,000 tonnes of plastic waste. Underdeveloped countries like Afghanistan and Haiti, with much lower GDPs ($15 billion and $18 billion respectively), consume far less energy (6 and 0.5 TWh), but rely heavily on renewable energy sources (75% in Afghanistan and 70% in Haiti). These nations have minimal greenhouse gas emissions but still face environmental challenges, with plastic waste being an issue (5,000 tonnes from Afghanistan and 10,000 tonnes from Haiti). Additionally, forest cover varies significantly, with the US at 33%, China at 22.3%, and Brazil at 59%, while Afghanistan and Haiti have minimal forest cover at 2 and 3%, respectively. Despite these differences, all countries share common environmental challenges, such as greenhouse gas emissions, ocean plastic pollution, and the need for forest preservation. The findings highlight the necessity for differentiated climate action approaches tailored to the specific capacities, challenges, and environmental impacts of each country, ensuring that global strategies are both equitable and effective (Bhardwaj et al. 2025; Agarwal and Jain 2024; Jain 2025).

12.7 Practical Implementation

Achieving climate neutrality requires the adoption of differentiated climate policies that move away from one-size-fits-all frameworks and instead consider each nation's economic capacity, technological development, and historical responsibility for greenhouse gas emissions. Climate action must be guided by principles of justice and equity, ensuring that developing and underdeveloped countries are not subjected to the same regulatory expectations as wealthier nations, given their limited contributions to climate change and urgent developmental needs. Developed nations must lead by example in reducing emissions and play a proactive role in providing financial and technological assistance to countries with fewer resources. This necessitates a comprehensive restructuring of global climate agreements to establish flexible, country-specific targets that support sustainable progress across all regions. Furthermore, strong international coordination and cooperation are essential, with developed nations actively partnering with vulnerable countries to support both climate adaptation and mitigation efforts.

12.8 Future Scope of the Study

Future research on climate neutrality should focus on the development of quantifiable equity indicators that can guide policymaking and ensure fairness in international climate commitments. Advanced policy simulation models are needed to assess the outcomes of differentiated responsibilities across regions, helping to refine global strategies for more effective action. Additionally, impact studies evaluating the current climate finance mechanisms will be crucial in identifying their effectiveness and exploring innovative funding solutions to support sustainable transitions in the Global South. Cross-sectoral research is essential to understand how climate neutrality goals can be integrated into broader development sectors, such as education, healthcare, infrastructure, and energy access, to promote holistic progress. Longitudinal case studies of countries at various levels of development will provide critical insights into the real-world implications of differentiated climate approaches, offering valuable guidance for creating more just and impactful global climate strategies.

1. **Intersection of Climate Policy and Inequality**

The whole world striving for climate neutrality underlines sharp contrasts between developed, developing, and underdeveloped countries. Already developed nations have a leading position in historical GHG emissions and are also much better endowed with finance and technology on the path to climate neutrality. However, such leadership gives relevance to rather nagging questions about equity within global climate policy, which all too often puts uniform demands on countries with striking differences in capacities and historical responsibilities. Climate neutrality is a dilemma for all developing and underdeveloped countries because, on one hand, these countries feel the pressure from the international community to reduce emissions, whereas, on the other hand, still in the throes of economic development, fossil fuel comes out to be a major driver for improvement in the living standard. It does not consider that they are supposed to be bound by the same climate targets as richer countries, but with poor access to renewable energy technologies and with financial constraints.

2. **Ethical Dimensions of Climate Neutrality**

The questions of ethics will also be profound regarding climate neutrality policies: should countries which have contributed the least to global emissions throughout history be raised to the same standards as those which have benefited from several centuries of industrialization? In accordance with the principle of "common but differentiated responsibilities," the obligations of climate policy must be graduation proportional to each country's historical emissions and current capabilities. In practice, the international regime seldom implements this principle; as a result, policies are designed that could reinforce old disparities. The developing and underdeveloped nations demand the protection of a rightful per capita emission so that they can strive for growth in economic terms and poverty alleviation even at the cost of having to retain a certain dependence on fossil fuels during the coming few decades. This perspective is anchored on the understanding that climate justice invites different stages of development and flexibility in the manner countries achieve their climate objectives.

3. **Power Dynamics in Global Climate Governance**

The power relationship is also a significant context in which global climate governance takes place. In general, the interests of developed countries, which are better represented through international negotiations, set the agenda for climate policies at the expense of the interests of developing nations. This is evident sometimes when policies reflect the priorities of wealthier nations, such as reliance on carbon offsetting, which allows them to meet climate targets without necessarily overhauling consumption patterns. In such discussions, the voices of developing and underdeveloped countries are often outshone. In that respect, the contributions could lead to an incomplete global climate framework. The preeminence of this group increases global inequality and weakens the potential impacts of climate policies by leaving the most vulnerable nations without support along the road to a climate-neutral economy.

4. **Long-Term Implications of Current Policies**

The long-term implications of the direction to climate neutrality is very ominous for the developing and underdeveloped world. Without proper financial and technical support, such countries will not be able to keep pace with the climate goals, which would mean a further gap between rich and poor. It would probably overemphasize rapid cuts in emissions without taking into consideration the necessity of developing sustainably, which may dampen the growth of a country's economy. It would in turn make the underdeveloped country less capable of dealing with the impact caused by climate change. Furthermore, extensive use of offsets by developed countries means keeping high levels of emissions, thus damaging the fight of the global community against climate change and placing an additional burden on poorer countries. These outcomes, listed above, must be prevented by following an approach that would show more equity in climate action-that which takes into consideration the differential conditions every country finds itself in and ensures that the pursuit of climate neutrality isn't to be pursued at the cost of global justice.

12.9 Conclusion

Reaching climate neutrality is a vital yet complex challenge, especially given the differing development levels of nations. Developed countries are more capable of achieving ambitious climate targets, but applying the same targets to all countries disregards differences in economic capacity, technology, and historical emissions. This chapter highlights the ethical dilemmas and power imbalances in the current global climate framework. The principle of "common but differentiated responsibilities" must be central to climate policy implementation. Developing and underdeveloped countries, with their minimal emissions and urgent development needs, cannot be held to the same standards as wealthier nations. Climate neutrality must be achieved with equity and justice, with developed countries leading in emission reductions and providing financial and technological support to less affluent nations. A cooperative, differentiated approach will help build a more equitable and sustainable world. Success depends on balancing the urgency of climate action with fairness. Global frameworks must allow flexibility for country-specific contexts, enabling tailored climate strategies. Innovation and international cooperation will be key to overcoming challenges posed by differing resources and capacities. Sharing knowledge, technology, and best practices will ensure no nation is left behind in the transition to sustainability.

References

Agarwal S, Jain H (2024) Investigating the potential of floral waste as a vermicompost and dual-functional biosorbent for sustainable environmental management. Water Air Soil Pollut 235:322. https://doi.org/10.1007/s11270-024-07144-y

Alola AA, Adebayo TS (2023) The potency of resource efficiency and environmental technologies in carbon neutrality target for Finland. J Clean Prod 389:136127. https://doi.org/10.1016/j.jclepro.2023.136127
Ardaillon L, Roebel A (2020) GCI detection from raw speech using a fully-convolutional network. In: ICASSP 2020–2020 IEEE International conference on acoustics, speech and signal processing (ICASSP). IEEE, pp 6739–6743. https://doi.org/10.1109/ICASSP40776.2020.9053032
Armin Razmjoo A, Sumper A, Davarpanah A (2020) Energy sustainability analysis based on SDGs for developing countries. Energy Sourc Part A Recov Utiliz Environ Effects 42(9):1041–1056. https://doi.org/10.1080/15567036.2019.1607947
Arora NK, Mishra I (2023) Sustainable development goal 13: recent progress and challenges to climate action. Environ Sustain 6(3):297–301. https://doi.org/10.1007/s42398-023-00206-w
Bhardwaj LK, et al (2025) Resilience in the digital age: emerging technologies for climate adaptation. In: Choudhury M, Dixit G, Majumdar S (eds) Sustainable synergy: harnessing ecosystems for climate resilience. Climate change management. Springer, Cham. https://doi.org/10.1007/978-3-031-77957-2_11
Bieszk-Stolorz B, Dmytrów K (2023) Decent work and economic growth in EU countries—static and dynamic analyses of sustainable development goal 8. Sustainability 15(18):13327. https://doi.org/10.3390/su151813327
Caetano T, Winkler H, Depledge J (2020) Towards zero carbon and zero poverty: integrating national climate change mitigation and sustainable development goals. Clim Policy 20(7):773–778. https://doi.org/10.1080/14693062.2020.1763900
Carpejani P, de Jesus ÉT, Bonfim Catapan BLS, Gouvea da Costa SE, Pinheiro de Lima E, Tortato U, Richter BK (2020) Affordable and clean energy: a study on the advantages and disadvantages of the main modalities. Int Bus Trade Instit Sustain 615–627. https://doi.org/10.1007/978-3-030-26759-9_35
Chen L, Msigwa G, Yang M, Osman AI, Fawzy S, Rooney DW, Yap PS (2022) Strategies to achieve a carbon neutral society: a review. Environ Chem Lett 20(4):2277–2310. https://doi.org/10.1007/s10311-022-01444-9
Chen F, Zhang W, Chen R, Jiang F, Ma J, Zhu X (2024) Adapting carbon neutrality: tailoring advanced emission strategies for developing countries. Appl Energy 361:122845. https://doi.org/10.1016/j.apenergy.2023.122845
Chovancová J, Vavrek R (2022) On the road to affordable and clean energy: assessing the progress of European countries toward meeting SDG 7. Polish J Environ Stud 31(2):1587–1600. https://doi.org/10.15244/pjes.2022.02743
Ciambra A, Siragusa A, Proietti P, Stamos I (2023) Monitoring SDG localisation: an evidence-based approach to standardised monitoring frameworks. J Urban Ecol 9(1):juad013. https://doi.org/10.1093/jue/juad013
Cracknell R, Bastaert D, Houille S, Châtelain J, Larguier O, Beaugé Y, Hamje H (2020) Assessing the efficiency of a new gasoline compression ignition (GCI) concept (No. 2020–01–2068). In: SAE technical paper. https://doi.org/10.4271/2020-01-2068
Da Rosa AV, Ordonez JC (2021) Fundamentals of renewable energy processes. Academic Press. https://doi.org/10.1016/b978-0-12-397219-4.00015-1
Destek MA, Sarkodie SA, Asamoah EF (2021) Does biomass energy drive environmental sustainability? An SDG perspective for top five biomass consuming countries. Biomass Bioenerg 149:106076. https://doi.org/10.1016/j.biombioe.2021.106076
Dong F, Li Y, Gao Y, Zhu J, Qin C, Zhang X (2022a) Energy transition and carbon neutrality: exploring the non-linear impact of renewable energy development on carbon emission efficiency in developed countries. Resour Conserv Recycl 177:106002. https://doi.org/10.1016/j.resconrec.2022.106002
Dong F, Zhu J, Li Y, Chen Y, Gao Y, Hu M, Sun J (2022b) How green technology innovation affects carbon emission efficiency: evidence from developed countries proposing carbon neutrality

targets. Environ Sci Pollut Res 29(24):35780–35799. https://doi.org/10.1007/s11356-022-20573-w

Emmerling J, Andreoni P, Tavoni M (2024) Global inequality consequences of climate policies when accounting for avoided climate impacts. Cell Rep Sustain 1(1). https://doi.org/10.1016/j.xcrs.2024.100003

Fitawek W, Hendriks SL (2024) Sustainable development goal 15: life on land. In: Handbook on public policy and food security. Edward Elgar Publishing, pp 354–364. https://doi.org/10.4337/9781839105449.00039

Foláyan MO, Schroth RJ, Duangthip D, Al-Batayneh OB, Virtanen JI, Sun IG, El Tantawi M (2024) A scoping review on the association between early childhood caries and life on land: the sustainable development goal 15. PLoS ONE 19(7):e0304523. https://doi.org/10.1371/journal.pone.0304523

Gajdosova K (2023) Role of GDP in the sustainable growth era. SocioEcon Chall 7(3):94–112. https://doi.org/10.21272/sec.7(3).94-112.2023

Giannetti BF, Agostinho F, Almeida CM, Liu G, Contreras LE, Vandecasteele C, Poveda C (2020) Insights on the United Nations sustainable development goals scope: are they aligned with a 'strong' sustainable development? J Clean Prod 252:119574. https://doi.org/10.1016/j.jclepro.2019.119574

Gjorgievski VZ, Mihajloska E, Abazi A, Markovska N (2022) Sustainable development goals—climate action nexus: quantification of synergies and trade-offs. Clean Technol Environ Policy 24(1):303–313. https://doi.org/10.1007/s10098-021-02158-2

Gulseven O (2020) Measuring achievements towards SDG 14, life below water, in the United Arab Emirates. Mar Policy 117:103972. https://doi.org/10.1016/j.marpol.2020.103972

Hailemariam A, Dzhumashev R, Shahbaz M (2020) Carbon emissions, income inequality, and economic development. Empir Econ 59(3):1139–1159. https://doi.org/10.1007/s00181-019-01785-x

He X, Khan S, Ozturk I, Murshed M (2023) The role of renewable energy investment in tackling climate change concerns: environmental policies for achieving SDG-13. Sustain Dev 31(3):1888–1901. https://doi.org/10.1002/sd.2472

Hu K, Raghutla C, Chittedi KR, Zhang R, Koondhar MA (2021) The effect of energy resources on economic growth and carbon emissions: a way forward to carbon neutrality in an emerging economy. J Environ Manage 298:113448. https://doi.org/10.1016/j.jenvman.2021.113448

Huan Y, Zhu X (2023) Interactions among sustainable development goal 15 (life on land) and other sustainable development goals: knowledge for identifying global conservation actions. Sustain Dev 31(1):321–333. https://doi.org/10.1002/sd.2341

Huyer S, Acosta M, Gumucio T, Ilham JIJ (2020) Can we turn the tide? Confronting gender inequality in climate policy. Gend Dev 28(3):571–591. https://doi.org/10.1080/13552074.2020.1817134

Jain H (2025) Synergizing carbon reduction technologies and nature-based solutions for a climate-neutral future. Mitig Adapt Strateg Glob Change 30:42. https://doi.org/10.1007/s11027-025-10237-6

Jakovljevic M, Timofeyev Y, Ranabhat CL, Fernandes PO, Teixeira JP, Rancic N, Reshetnikov V (2020) Real GDP growth rates and healthcare spending—comparison between the G7 and the EM7 countries. Glob Health 16:1–13. https://doi.org/10.1186/s12992-020-00572-2

Khalid AM, Sharma S, Dubey AK (2021) Concerns of developing countries and the sustainable development goals: case for India. Int J Sust Dev World 28(4):303–315. https://doi.org/10.1080/13504509.2021.1889072

Khan I, Muhammad I, Sharif A, Khan I, Ji X (2024) Unlocking the potential of renewable energy and natural resources for sustainable economic growth and carbon neutrality: a novel panel quantile regression approach. Renew Energy 221:119779. https://doi.org/10.1016/j.renene.2023.11.026

Kılkış Ş, Krajačić G, Duić N, Rosen MA (2020) Advances in integration of energy, water and environment systems towards climate neutrality for sustainable development. Energy Convers Manage 225:113410. https://doi.org/10.1016/j.enconman.2020.113410

Kreinin H, Aigner E (2022) From "Decent work and economic growth" to "Sustainable work and economic degrowth": a new framework for SDG 8. Empirica 49(2):281–311. https://doi.org/10.1007/s10663-021-09460-9

Küfeoğlu S (2022) SDG-14: life below water. In: Emerging technologies: value creation for sustainable development. Springer International Publishing, pp 453–468. https://doi.org/10.1007/978-3-030-92647-3_22

Lakshani S, Sandaruwan C, Fernando C, Vidyapathirana G, Jayathilaka R, Munasinghe S (2023) From short to long term: Dynamic analysis of FDI and net export in global regions. PLoS ONE 18(9):e0291301. https://doi.org/10.1371/journal.pone.0291301

Leal Filho W, Azul AM, Brandli L, Lange Salvia A, Wall T (eds) (2022) Life below water. Springer International Publishing. https://doi.org/10.1007/978-3-030-84251-3

Mantero G, Morresi D, Marzano R, Motta R, Mladenoff DJ, Garbarino M (2020) The influence of land abandonment on forest disturbance regimes: a global review. Landscape Ecol 35(12):2723–2744. https://doi.org/10.1007/s10980-020-01139-6

Meng F, Ding Z, Meng X, Ai X, Zhang Z, Ma W, Wang K (2020) Research on different cooling methods in the machining of CGI and GCI. Appl Nanosci 10:2177–2188. https://doi.org/10.1007/s13204-020-01369-7

Miao J, Song X, Zhong F, Huang C (2023) Sustainable development goal 6 assessment and attribution analysis of underdeveloped small regions using integrated multisource data. Remote Sens 15(15):3885. https://doi.org/10.3390/rs15153885

Organisation for Economic Co-operation and Development (OECD) (2020) Beyond GDP: measuring what counts for economic and social performance. OECD Publishing. https://doi.org/10.1787/9789264307292-en

Ortiz-Ospina E, Roser M (2023) Government spending. In: Our world in data. https://ourworldindata.org/government-spending

Paprotny D (2021) Convergence between developed and developing countries: a centennial perspective. Soc Indic Res 153(1):193–225. https://doi.org/10.1007/s11205-020-02514-5

Penglong WA, Bao WA, Bingxin XU, Yueyue HU, Feng GA, Xueyan ZH, Chunlin HU, Xiaoyu SO, Liang ZH (2022) Progress and prospect of sustainable development goals in underdeveloped mountainous areas. Adv Earth Sci 37(9):937. https://doi.org/10.11867/j.issn.1001-8166.2022.09.0937

Saini J, Gupta AK, Dhupper R, Shrivastava A (2025) Spatio-temporal study of urban dynamics with implications on land surface temperature of Gurugram City, India. Environ Monit Assess 197(8):1–25

Saini J, Gupta AK, Shrivastava A, Dhupper R (2024) Study of urban warming in Chandigarh union territory, India: geospatial approach: J. Saini et al. Theoret Appl Climatol 155(8):8245–8261

Salvia M, Reckien D, Pietrapertosa F, Eckersley P, Spyridaki NA, Krook-Riekkola A, Heidrich O (2021) Will climate mitigation ambitions lead to carbon neutrality? An analysis of the local-level plans of 327 cities in the EU. Renew Sustain Energy Rev 135:110253. https://doi.org/10.1016/j.rser.2020.110253

Saxena A, Ramaswamy M, Beale J, Marciniuk D, Smith P (2021) Striving for the United Nations (UN) sustainable development goals (SDGs): what will it take? Discov Sustain 2:1–14. https://doi.org/10.1007/s43953-021-00002-4

Sinha A, Sengupta T, Saha T (2020) Technology policy and environmental quality at crossroads: designing SDG policies for select Asia Pacific countries. Technol Forecast Soc Chang 161:120317. https://doi.org/10.1016/j.techfore.2020.120317

Sobkowiak M, Cuckston T, Thomson I (2020) Framing sustainable development challenges: accounting for SDG-15 in the UK. Account Audit Account J 33(7):1671–1703. https://doi.org/10.1108/AAAJ-10-2019-4091

Soergel B, Kriegler E, Bodirsky BL, Bauer N, Leimbach M, Popp A (2021a) Combining ambitious climate policies with efforts to eradicate poverty. Nat Commun 12(1):2342. https://doi.org/10.1038/s41467-021-22551-6

Soergel B, Kriegler E, Weindl I, Rauner S, Dirnaichner A, Ruhe C, Popp A (2021b) A sustainable development pathway for climate action within the UN 2030 Agenda. Nat Clim Chang 11(8):656–664. https://doi.org/10.1038/s41558-021-01058-x

Tan XC, Wang Y, Gu BH, Kong LS, Zeng A (2022) Research on the national climate governance system toward carbon neutrality—a critical literature review. Fundament Res 2(3):384–391. https://doi.org/10.1016/j.fmre.2022.04.007

Tan J, Chew C (2020) Gross Domestic Product (GDP). https://doi.org/10.1007/springerreference_1614

Tiwari A (2021) Advancing materials towards climate neutrality by 2050. Adv Mater Lett 12(8):1–3. https://doi.org/10.5185/amlett.2021.082701

United Nations (2015) Sustainable development goals. Department of Economic and Social Affairs, United Nations. https://sdgs.un.org/goals

Versey HS (2021) Missing pieces in the discussion on climate change and risk: intersectionality and compounded vulnerability. Policy Insights Behav Brain Sci 8(1):67–75. https://doi.org/10.1177/2372732221991328

Wang R, Hsu SC, Zheng S, Chen JH, Li XI (2020) Renewable energy microgrids: economic evaluation and decision making for government policies to contribute to affordable and clean energy. Appl Energy 274:115287. https://doi.org/10.1016/j.apenergy.2020.115287

Wang F, Harindintwali JD, Yuan Z, Wang M, Wang F, Li S, Chen JM (2021) Technologies and perspectives for achieving carbon neutrality. Innovation 2(4):100180. https://doi.org/10.1016/j.xinn.2021.100180

Wang Q, Zhang C, Li R (2022) Towards carbon neutrality by improving carbon efficiency—a system-GMM dynamic panel analysis for 131 countries' carbon efficiency. Energy 258:124880. https://doi.org/10.1016/j.energy.2022.124880

Wannisinghe P, Jayakody S, Rathnayake S, Wijayasinghe D, Jayathilaka R, Madhavika N (2023) Determining the influence of LPI, GCI and IR on FDI: a study on the Asia and Pacific Region. PLoS ONE 18(2):e0281246. https://doi.org/10.1371/journal.pone.0281246

Wiedenhofer D, Virág D, Kalt G, Plank B, Streeck J, Pichler M, Haberl H (2020) A systematic review of the evidence on decoupling of GDP, resource use and GHG emissions, part I: bibliometric and conceptual mapping. Environ Res Lett 15(6):063002. https://doi.org/10.1088/1748-9326/ab8429

Wolf S, Teitge J, Mielke J, Schütze F, Jaeger C (2021) The European Green Deal—more than climate neutrality. Intereconomics 56:99–107. https://doi.org/10.1007/s10272-021-0963-z

World Health Organization (2021) Tracking SDG 7: The energy progress report 2021. International Energy Agency, International Renewable Energy Agency, United Nations Statistics Division, World Bank, & WHO. https://trackingsdg7.esmap.org/data/files/download-documents/2021-Tracking%20SDG7-Full%20Report.pdf

Zhao X, Ma X, Chen B, Shang Y, Song M (2022) Challenges toward carbon neutrality in China: strategies and countermeasures. Resour Conserv Recycl 176:105959. https://doi.org/10.1016/j.resconrec.2021.105959

Zhao J, Dong K, Dong X (2024) How does energy poverty eradication affect global carbon neutrality? Renew Sustain Energy Rev 191:114104. https://doi.org/10.1016/j.rser.2024.114104

Chapter 13
Vision for a Circular and Sustainable Future: Challenges and Opportunities

Harshita Jain and Anamika Shrivastava

13.1 Introduction

The urgent environmental, economic, and social challenges of the twenty-first century have heightened the worldwide conversation on sustainability and resource management (Correia 2018; Dantas et al. 2021). At the heart of this discussion is the idea of a circular economy (CE), which fundamentally differs from the conventional linear economic model of "take-make-dispose" (Elisha 2020). Rather, the circular economy focuses on closed-loop systems where resources are kept at their optimal utility for extended periods using approaches like reuse, refurbishment, remanufacturing, and recycling. This comprehensive method to economic growth is inherently regenerative, seeking to separate expansion from the use of limited resources and harm to the environment (Wani and Hameed 2024). Sustainability, in its most comprehensive definition, includes fulfilling current requirements without jeopardizing future generations' capacity to fulfil theirs. It is supported by three interconnected pillars: ecological integrity, economic sustainability, and social fairness (Correia 2018). A circular economy puts sustainability into action by offering a pragmatic framework that enables industries, governments, and societies to uphold ecological balance while fostering economic resilience and social inclusion. Despite their different theoretical backgrounds, sustainability and circular economy are inherently linked (Gil-Lamata and Latorre-Martínez 2022). The circular economy provides tangible approaches for attaining sustainable development by focusing on resource efficiency, reducing waste production, and encouraging systemic innovation. The shift to these models, however, requires a change in production and consumption patterns, business practices, policy decisions, and societal values (Schroeder et al. 2019).

H. Jain (✉) · A. Shrivastava
Amity Institute of Environmental Sciences, Amity University Uttar Pradesh, Noida, India
e-mail: hjain@amity.edu

A. Shrivastava
e-mail: ashrivastava1@amity.edu

H. Jain et al. (eds.), *The Circular Fallacy: Closing the Loop in a Consumerist World*,
Circular Economy and Sustainability, https://doi.org/10.1007/978-3-032-22687-7_13

The existing global economic framework, mainly focused on extraction and wastefulness, is progressively seen as unsustainable within long-term planetary limits (Ritter et al. 2024). Linear models not only exhaust natural resources but also play a major role in climate change, loss of biodiversity, and environmental contamination (Ghisellini et al. 2016). Additionally, socio-economic disparities worsen as at-risk groups disproportionately endure the consequences of environmental harm and resource shortages. In response, the circular economy arises not just as an alternative economic model but as an essential evolution—one that requires systemic transformation throughout all levels of society (Kristensen and Mosgaard 2020). Systemic change denotes a core overhaul in the frameworks, procedures, and models that support existing socio-economic systems. This transformation requires reconsidering material movements, reengineering products for longevity and repairability, and creating systems that facilitate circular efforts, including resource recovery and industrial symbiosis (Bocken et al. 2016). The intricate nature and interrelations of global systems indicate that solitary actions are inadequate. Rather, an integrative, systems-oriented perspective is crucial. This involves harmonizing financial systems, legal structures, social customs, and technological advancements in a cohesive way. Shifting to a circular economy necessitates not just innovation and investment but also cultural changes and inclusive policy adjustments (Korhonen et al. 2018). Importantly, the transition needs to be fair and equitable, making certain that advantages are distributed among various socio-economic groups and that marginalized communities are not left behind during the change. This involves collaborative governance, inclusive innovation methods, and specific capacity development to promote community support and prevent the emergence of new inequalities (Kirchherr et al. 2018). Figure 13.1 illustrates how circular approaches (such as resource efficiency, waste reduction, and product lifespan extension) function as methods to reach global sustainability objectives, supported by advancements in technology, changes in policy, and shifts in societal behavior.

The circular economy's contribution to sustainable development is becoming more widely recognized in the global policy environment. Notably, the Sustainable Development Goals (SDGs) of the United Nations, a global call to action to eradicate poverty, safeguard the environment, and guarantee prosperity, provide a thorough framework for assessing how well circular policies connect with global priorities (Schroeder et al. 2019).

A number of SDGs are closely related to the concepts of the circular economy, including:

- SDG 6 (Clean Water and Sanitation), by means of effective treatment technology and water reuse;
- SDG 7 (Affordable and Clean Energy), through energy recovery and integration of renewable energy;
- SDG 9 (Industry, Innovation, and Infrastructure), via the promotion of sustainable industrialization and robust infrastructure;

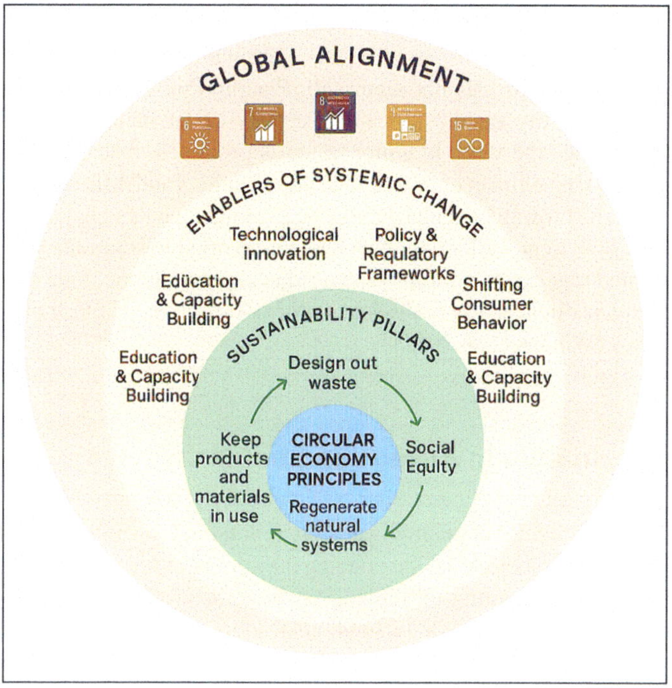

Fig. 13.1 Connections among systemic change, sustainability, and the circular economy

- SDG 13 (Climate Action) by lowering greenhouse gas emissions related to waste and production; SDG 12 (Responsible Consumption and Production) by preventing waste and using lifecycle thinking.

Furthermore, by improving waste management and urban resilience, circular approaches can support SDG 11 (Sustainable Cities and Communities) and SDG 8 (Decent Work and Economic Growth) by generating green jobs and new market opportunities (Schroeder et al. 2019). The increasing institutional support for circularity is demonstrated by policy tools like the European Green Deal, the Circular Economy Action Plan, and national policies in nations like the Netherlands, Finland, and Japan. In an effort to incorporate circularity into governmental and commercial frameworks, these frameworks support eco-innovation, extended producer responsibility, circular public procurement, and sustainable product design (Kirchherr et al. 2018). Global acceptance is still unequal, though. Operating circular models presents unique challenges for developing nations, which are sometimes hampered by restricted access to technology and infrastructural limitations. This emphasizes the significance of context-specific paths that take institutional capabilities, socioeconomic circumstances, and regional development priorities into account (Korhonen et al. 2018).

The circular economy presents a convincing and workable plan for tackling the interconnected problems of social injustice, climate change, and environmental degradation, even though it is not a cure-all. Placing this vision within the larger context of sustainability emphasizes the necessity of an inclusive, revolutionary strategy that rethinks the ways in which value is produced, distributed, and maintained in society (Ghisellini et al. 2016). The opportunities and difficulties that define the continuous shift toward circularity will be examined as this chapter progresses, along with the new trends affecting its course and workable solutions for achieving a sustainable and regenerative future. The need is clear: systemic, long-term change based on equity, resilience, and ecological integrity is required in place of disjointed, temporary fixes (Winans et al. 2017).

13.2 Opportunities in Advancing Circularity

The shift toward a circular economy opens great potential to solve the most serious global sustainability issues alongside opening new doors of innovation, competition, and social welfare. According to Table 13.1, there are four general categories of opportunities areas which include technological innovations, policy and regulatory enablers, market and consumer behaviour shifts, and the economic and environmental appeal of certain dual benefits involving circularity (Geissdoerfer et al. 2017).

Table 13.1 Opportunities in circular economy: drivers, examples, and impact

Opportunity domain	Key drivers	Examples	Potential impact
Technological innovations	Industry 4.0, green chemistry, AI, blockchain	Smart waste tracking, 3D printing, bio-based materials	Reduced waste, efficient resource use, product longevity
Policy and regulatory incentives	CE action plans, tax incentives, EPR, eco-design laws	EU CEAP, Japan's sound material cycle society act	Policy-driven market shifts, eco-innovation
Market and consumer shifts	Sustainability trends, awareness, ESG pressures	Clothing resale apps, refill systems, PaaS (Product-as-Service)	Demand for durable and ethical goods, behavior change
Economic and environmental benefits	Resource productivity, carbon mitigation, job growth	Circular SMEs, local remanufacturing, green job training	Climate mitigation, local employment, reduced pollution

13.2.1 Technological Innovations Enabling Circular Models

With technological transformations particularly in the sphere of IT, this aspect has reached the level of central enabler of the process of taking the circular economy strategies further into implementation into various fields (Yadav and Jain 2024; Jain 2025a; b). Digital solution, process optimization, and innovations in material science become the new rules for the extraction of the resources and their processing, utilization, and recovery (Rejeb et al. 2025). As an example, complex waste streams previously not recyclable, including plastic and textiles, are becoming recyclable through the use of advanced recycling-based technologies e.g., chemical recycling and the use of enzymes in depolymerization. With the emergence of Industry 4.0 technologies, such as the Internet of Things (IoT), artificial intelligence (AI), blockchain and data analytics, new efficiencies and transparency in the product life cycles have been established (Fernandes et al. 2023). Product-as-a-service and asset longevity promoting predictive maintenance through the use of IoT sensors and AI are taking off and these technologies will eventually drive product longevity upwards (Davila et al. 2023). The blockchain is also being used more widely to support traceability of materials, open up the possibility of closed loop supply chains and minimise the danger of resource leakage. Furthermore, additive manufacturing (3D printing) enables customisation of products at minimal strain to create unnecessary waste materials (Rejeb et al. 2025). Biomaterials and green chemistry advances are also enabling alternatives to the resource intensive or hazardous ingredients to be substituted by greener equivalents. These technologies, in telegram, are changing the paradigm of production-consumption and making the circular design and resource recovery concepts feasible (Fernandes et al. 2023).

13.2.2 Policy and Regulatory Incentives

Policy frameworks are important in determining the course of adoption of circular economies. The concept of circularity has become highly sought after by governments around the world in their ambitions towards the long-term sustainability and economic robustness (Siedschlag et al. 2022). An increasing number of countries and regional blocs have started their own circular economy action plans indicating a move away from voluntary action towards planned interventions. The Circular Economy Action Plan (CEAP) of the European Union is an example, and it provides a complete legislative program on sustainable design of products, consumer rights, waste mitigation, and business models that can be circular (Kirchherr et al. 2018). In Japan, the Netherlands, Finland and India, the same initiatives are being put in place which is an indicator of the emerging policy platform on the necessity of transitions. Such measures include incentive structures (e.g. an extended producer responsibility (EPR) scheme, green public procurement (GPP) policy and tax reform incentives (e.g. lower VAT rate on repair services) (She and Wee 2025). Furthermore, research and experimental development as well as infrastructure construction aid and subsidies have helped speed up deployment of circular solutions, at least in areas related to waste management and renewable energy (Agarwal and Jain 2024; Agarwal et al. 2025). Laws which mandate eco-design standards, limit single-use plastics, promote use of secondary raw materials are also playing part in establishing circular markets (Aristi et al. 2023). Notably, properly constructed policies enhance not only innovation but also facilitate the establishment of level playing field and long-running security of investments.

13.2.3 Market and Consumer Behaviour Shifts

Growing environmental consciousness and the obvious effects of climate change are causing a significant shift in consumer preferences and awareness (Saini et al. 2024). Nowadays, an increasing number of customers place a high value on sustainable consumption, looking for goods that are resource-efficient, long-lasting, repairable, and manufactured ethically (Vidal-Ayuso et al. 2023). Alternative consumption patterns including product leasing, second-hand markets, and sharing platforms are expanding as a result of this behavioral shift (Polyportis et al. 2022). In response to this demand, manufacturers and retailers are reconsidering their value propositions and integrating social and environmental considerations into the branding, packaging, and design of their products (Gomes et al. 2022). In an effort to preserve consumer loyalty and lessen their environmental impact, businesses are progressively implementing take-back initiatives, providing repair services, and creating closed-loop product lines. Furthermore, broader supply chain dynamics are being impacted by company sustainability pledges (Vidal-Ayuso et al. 2023). These days, a lot of global firms incorporate circularity into their environmental, social, and governance (ESG) plans, establishing goals for material circularity, carbon neutrality, and zero waste. Innovation throughout value chains is being stimulated by this top-down effect, which is creating market pull for circular goods and services (Pandey and Yadav 2023). Digital platforms that link consumers to peer-to-peer sharing networks, repair services, and marketplaces for reconditioned items can facilitate behavioral change. These platforms support the feasibility of circular consumption by democratizing access to sustainable solutions and lowering transaction barriers (Corsini et al. 2023).

13.2.4 Economic and Environmental Benefits

By providing both environmental stewardship and economic competitiveness at the same time, the circular economy presents an alluring value proposition. Circular models have the potential to lower material prices, increase resource productivity, and foster innovation from an economic standpoint (Roci and Rashid 2023). Businesses can reduce input costs and protect themselves from raw material price volatility by minimizing waste and maintaining the value of commodities over several usage cycles. Another important economic potential is the development of jobs (Hysa et al. 2020). The circular economy is labor-intensive by nature, especially in industries like remanufacturing, recycling, repair, and refurbishing. In emerging economies, where labor markets can be used to promote inclusive development, this potential for green employment creation is particularly pertinent. In terms of the environment, circular strategies help to mitigate pollution, conserve biodiversity, and lower greenhouse gas emissions (Gerhards and Greenwood 2021). By extending the life of materials and products, less virgin resource extraction is required, which minimizes water contamination and habitat devastation (Hysa et al. 2020). Circular systems also reduce emissions and enhance air quality by decreasing reliance on landfills and incineration. Furthermore, by facilitating the switch to renewable energy sources, lowering the amount of carbon embedded in materials, and promoting low-carbon lifestyles, circularity helps mitigate the effects of climate change (Nikolaou and Tsagarakis 2021; Khanna et al. 2022). All of these advantages put the circular economy in a strong position to meet international commitments like the SDGs and the Paris Agreement as well as national climate targets (Abdirahman et al. 2025).

13.3 Challenges to Achieving Circularity

The potential of the circular economy to address global sustainability issues is becoming more widely acknowledged, but there are still major obstacles in the way of the general acceptance and application of circular models (Fig. 13.2). These issues span the technological, economical, sociocultural, and institutional spheres and are multifaceted and systemic (Tirado et al. 2022). The absence of supporting infrastructure for the fundamental operations of a circular economy, including material recovery, remanufacturing, reuse, repair, and recycling, is a major obstacle to the advancement of circularity (Uloma et al. 2024). Waste management systems are still crude, dispersed, and underfunded in many places, particularly in the Global South. Closing resource loops is severely hampered by the lack of sophisticated recycling facilities, effective collection logistics, and consistent segregation at the source (Wani and Hameed 2024). Furthermore, stakeholders in the public and private sectors are unable to invest in circular economy infrastructure due to financial limitations. Reverse logistics networks, facility retrofits, and large upfront capital investments in new technologies are frequently necessary for circular solutions (Corsini et al. 2023). In contrast, linear models are more economically appealing in the short run since they are well-established and still enjoy the advantages of financial incentives and established infrastructure.

Small and medium-sized businesses (SMEs), who frequently play a major role in regional circular initiatives, find it difficult to obtain reasonably priced financing or de-risking tools for circular technology investments (Roci and Rashid 2023). Innovation and adoption are stifled by the absence of appropriate financing sources and ambiguous ROI schedules. Furthermore, the long-term social and environmental

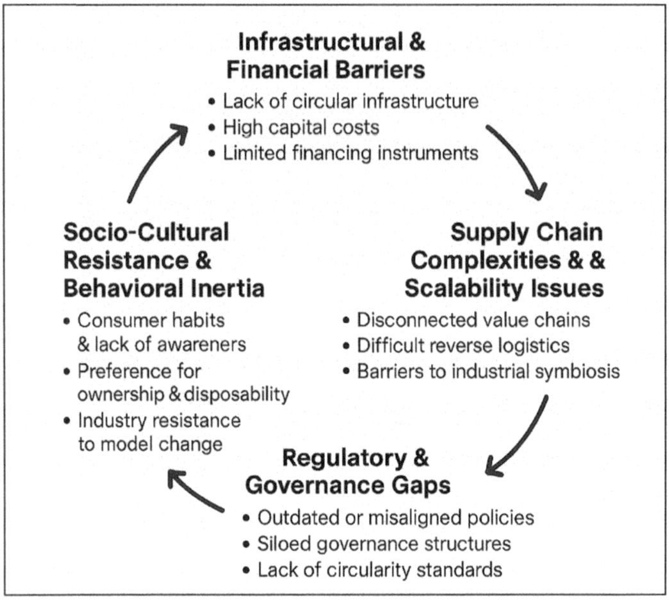

Fig. 13.2 Major challenges to circular economy implementation

advantages of circularity are frequently overlooked by conventional accounting and valuation frameworks, leading to misaligned economic signals (Hysa et al. 2020). Adoption of the circular economy is also significantly hampered by societal conventions and human behavior. A change in both production and consumption patterns is implied by circularity. Nonetheless, consumer behavior is still dominated by ingrained demands for ownership, disposability, and convenience (Vidal-Ayuso et al. 2023*)*. Efforts to market durable and repairable items are hampered by the pervasive belief that new is preferable than used or that repairs are expensive and inconvenient. Reluctance to embrace sharing economy ideas can also be a sign of cultural resistance, particularly in cultures where private ownership is strongly ingrained in status and identity (Gomes et al. 2022). The "value-action gap" is the term used to describe the discrepancy that frequently occurs between attitudes and actual purchase behavior, even in cases where customers show concern for sustainability.

Producers are also subject to behavioral inertia. Many businesses are resistant to changing profitable business models. The transition to circular practices frequently necessitates reevaluating supplier networks, modifying production procedures, and rethinking product design—steps that carry risk and uncertainty (Polyportis et al. 2022; Pandey and Yadav 2023). Opposition to change continues to be a strong obstacle in the absence of strong incentives or peer pressure from industry ecosystems. These difficulties are further exacerbated by deficiencies in awareness and education. Adoption is hampered by a lack of knowledge of circular principles and advantages among firms, customers, and legislators (Pandey and Yadav 2023). In order to create a welcoming and encouraging cultural climate for circularity, it is imperative that capacity building, behavior change communication, and stakeholder involvement be prioritized. Closed-loop or semi-closed-loop supply chains, which are by nature more complicated than linear supply chains, are commonly the foundation of circular business models (Hysa et al. 2020). Product tracking, reverse logistics, refurbishment operations, and secondary material markets are just a few of the services that must be integrated in order to implement circularity at scale. These tasks need cooperation between several parties, frequently from various industries and regions (Geissdoerfer et al. 2017).

Just-in-time manufacturing and outsourcing are hallmarks of globalized supply chains, which make putting circular ideas into practice more difficult. It is challenging for many producers to guarantee circularity throughout a product's life cycle because they lack awareness or control over upstream and downstream flows (Franco 2017). Furthermore, there are practical restrictions on material recovery because many supply chains are not made for simple disassembly, reuse, or recycling. Scalability is still an issue, especially for pilot programs or regional trials of the circular economy (Bressanelli et al. 2019). Community-level initiatives frequently face challenges in scaling up because of a lack of defined methodology, data interoperability, and market access. Because dominant industrial ecosystems are designed for linear throughput and economies of scale, circular enterprises may likewise have trouble integrating (Gomes et al. 2022). Circular loops cannot be implemented smoothly due to value chain fragmentation, a lack of data sharing, and a lack of responsibility. Digital infrastructure, supply chain transparency technologies, and cross-sectoral platforms that facilitate collaboration and traceability are therefore desperately needed (Pehlken et al. 2024).

Lastly, fragmented policy environments and poor governance systems also constitute one of the great challenges to the implementation of circular economies. Old laws in most jurisdictions were geared towards facilitating linear versions of industries and might be inadvertently penalizing circular innovations (She and Wee 2025). An example can be found in the waste classification laws where secondary materials are more problematically being seen as liabilities and not resources thus there have been legal and logistical barriers to recyclers and remanufacturers (Polyportis et al. 2022). Lack of consistency in quality, labeling and certifications of material further exacerbates the circular trade and market development of the secondary materials. Lack of coordinated regulatory regimes means that companies do not know what to expect particularly where they operate across boundaries (Khanna et al. 2022). Inability to have a long-term policy vision, to connect the policies seamlessly across various sectors (e.g., waste, energy, and transportation, and urban planning) hinders a systemic circular transition. Policymaking is reactive and it is not concerted, the responsibility lying within more than one department or one agency (She and Wee 2025). The siloed approach restricts the capabilities of developing flexible circular strategies and enforcement which gets diluted. Moreover, policy makers and regulators fail to achieve effective and acceptable regulatory frameworks due to lack of adequate participation of stakeholders in policy formulation and implementation procedures (Vidal-Ayuso et al. 2023). It requires the design of practice that is inclusive and incorporates the views of businesses, civil society, academia, and local communities when it comes to shaping pathways of the circular economy.

13.4 Emerging Trends Shaping Circular Futures

A number of new themes that represent the changing dynamics of technology, design, behavior, and governance are driving the shift to a circular economy. These developments are changing how systems and products are conceived, created, and used in addition to hastening the adoption of circular strategies (Kalmykova et al. 2018) (Table 13.2).

13.4.1 Role of Digital Technologies and Industry 4.0

The digital technologies related to Industry 4.0, such as the Internet of Things (IoT), artificial intelligence (AI), machine learning, big data analytics, and blockchain, are transforming ways the circular economy principles can be implemented in the practice (Dantas et al. 2021). They increase transparency, traceability, and optimization

Table 13.2 Key emerging trends supporting the circular economy transition

Trend	Description	Examples	Contribution to circularity
Digital technologies and industry 4.0	Use of smart technologies to optimize product life cycles and resource flows	IoT sensors, AI for waste sorting, blockchain for traceability	Enhances efficiency, transparency, and product longevity
Biomimicry and nature-inspired design	Applying principles from nature to design circular and regenerative systems	Self-cleaning surfaces, biodegradable packaging, closed-loop product design	Promotes regenerative design and reduces environmental impact
Collaborative consumption and sharing economy	Models that emphasize access over ownership and maximize asset use	Car sharing, tool libraries, clothing rental platforms	Reduces resource demand and minimizes underutilized assets
Cross-sectoral partnerships and stakeholder engagement	Collaboration across industries, governments, and communities to co-develop solutions	Public–private partnerships, industry-academia R&D hubs, citizen co-creation	Scales innovation, ensures inclusivity, and bridges silos

in the product life cycle, thus allowing using the resources more efficiently and optimally making decisions. Connected sensors and IoT technology support real-time tracking of the products and systems to enable predictive maintenance, condition-based monitoring and optimised assets utilisation (Abadian et al. 2024). It has led to new business models like product-as-a-service in which donors retain ownership of products and are rewarded to design products to last long and be reused (Bonilla et al. 2018). Blockchain use is fundamental to the authentication of traceability of materials and products, so a simple knowledge of the origin and sustainability claims of a product is more readily established. This is critical in the development of trust in the secondary markets and the promotion of extended producer responsibility (EPR) measures (Bahrin et al. 2016; Laskurain-Iturbe et al. 2021). In addition, data analytics and AI are deployed to improve production process, minimize waste, and plans material flows. As an illustration, AI may be used to detect manufacturing systems inefficiency, propose alternative products, or segregate in recycling centers automation to enhance the recovery of materials (Bahrin et al. 2016). These digital enablers are making the linear take-make-and-dispose system more intelligent, responsive, and regenerative network.

13.4.2 Biomimicry and Nature-Inspired Design

For a long time, nature has been used as an example of effectiveness, resiliency, and circularity. Biomimicry is a new design concept that informs sustainable design solutions by taking inspiration from natural systems, structures, and processes. This strategy encourages systems that optimize material and energy flows, regenerate, and adapt, which is intrinsically consistent with the circular economy (Ilieva et al. 2022). Modularity, multifunctionality, and biodegradability are characteristics that biomimetic designs frequently include because they mirror how biological organisms reduce waste and optimize efficiency (Andersen 2007). For instance, materials modeled after lotus leaves or spider silk are being created to improve product performance without compromising the environment and lessen need on hazardous synthetic inputs. Green roofs, biophilic structures, and water-sensitive urban design are examples of nature-based solutions in architecture and urban planning that show how biomimicry may be incorporated into larger systems to support resource efficiency and resilience (Ilieva et al. 2022). The development of self-healing materials, biodegradable packaging, and closed-loop production techniques based on ecological cycles all contribute to increased circularity in design (Bressanelli et al. 2019). By adopting biomimicry, engineers and designers can go beyond sustainability as a means of minimizing harm and instead pursue regenerative design, which actively improves and restores ecosystems by human action.

13.4.3 Collaborative Consumption and Sharing Economies

The emergence of sharing economy models and collaborative consumption, which go against conventional ideas of ownership and consumption, is another revolutionary movement influencing circular futures (Geissdoerfer et al. 2017). These models make it possible to pool unused resources, which lowers the need for new goods and lessens their negative effects on the environment. Peer-to-peer (P2P) sharing services, which allow users to access goods and services without owning them, have grown in popularity thanks to digital platforms (Aristi et al. 2023). Examples of these services include tool libraries, clothes rental services, ride-sharing, and co-working spaces. This access-over-ownership paradigm lowers material throughput in the economy and encourages more effective use of resources (Polyportis et al. 2022). Sharing models have social and economic benefits in addition to environmental ones, including equal access to resources, cost savings, and community involvement. Car-sharing programs, for example, can ease traffic congestion and cut emissions in urban areas by reducing the number of vehicles required (Hysa et al. 2020). However, user behavior, platform trust, and openness all affect how effective collaborative consumption models are. To make sure that these technologies promote circularity rather than consume more, it is essential to set up trustworthy digital platforms, equitable pricing schemes, and accountability procedures (Pehlken et al. 2024).

13.4.4 Cross-Sectoral Partnerships and Stakeholder Engagement

Transitioning to a circular economy necessitates the active participation and alignment of a variety of entities, including corporations, governments, academic institutions, civil society, and local communities (Kalmykova et al. 2018). Cross-sectoral cooperation is becoming a vital tool for systemic transformation and circular innovation. Scaling infrastructure development, funding circular businesses, and co-creating regulatory frameworks that encourage sustainable behaviors are all made possible by public–private partnerships, or PPPs (Ritter et al. 2024). New materials, technologies, and business models that are suited to circular goals are developed through partnerships between industry and research institutions. Platforms with several stakeholders, like the Platform for Accelerating the Circular Economy (PACE), provide venues for discussion, information sharing, and cooperative action (World Economic Forum 2017). In order to organize pilot projects at scale, reduce fragmentation, and align goals, these platforms bring together stakeholders from various value chains. At the neighbourhood level, citizen science projects and participatory governance enable people to support local circular economies, urban farming, trash reduction, and repair networks (Palea et al. 2024). This kind of bottom-up involvement is essential to integrating circular ideas into routine procedures and guaranteeing an equitable and inclusive transformation. Coherent policies and institutional congruence are also essential for successful circular transformations (Christensen 2021). The environment, trade, education, and urban development ministries must collaborate to make sure that policies in different areas support rather than contradict one another. This alignment is made possible by stakeholder engagement, which promotes communication and creates a common understanding across domains that have historically been divided (World Economic Forum 2017).

13.5 Pathways to Realizing the Vision

The attainment of circular and sustainable future is not only about the identification of the opportunities and knowledge of the challenges, much more a clear and definite pathways of actualizations. Such routes need to be based on system thinking and established on the foundations of inclusivity, innovation, and teamwork (Fig. 13.3) (Geissdoerfer et al. 2017). The design of an effective policy is crucial to the maximization of the pace of circular economy adoption. But inconsistent and piecemeal regulation frequently gets in the way. Hence, integrative policy frameworks that support the alignment of the environmental, industrial, trade, education, and urban development policies so that they revolve around the objectives of the circular economy are urgently required (Bocken et al. 2016). These paradigms need to show systemic coherence, which implies that practices carried out in one sector should not act against sustainability goals in another. As an example, bioresources circularity, and renewable energy need to be reconciled so that waste-to-energy policies do not lead to a decrease in the circularity of materials. In the same manner, products that fulfill requirements in terms of durability, reparability, and recyclability should be more appealing in regard to procurement policies (Ranta et al. 2018). Consistent implementation is crucial and cannot be avoided in multi-level governance, which also implies coordination between local, regional, national, and international governments (Bocken et al. 2016). Although the national policies are in a position to provide strategic guidance and regulation standards, the local governments are instrumental in implementation, mainly by using urban planning, waste infrastructure, and community participation, along with public procurement (Geissdoerfer et al. 2017). Eco-industrial parks, green building codes, and closed-loop urban metabolism systems The Circle (particularly in cities) is fertile ground, and circular experimentation in cities is evident in eco-industrial parks, green building codes, and systems of closed-loop urban metabolism (World Economic Forum 2017).

Furthermore, circular practices can be institutionalized within governance systems by including circularity into sustainability measures, climate plans, and long-term development planning (Saini et al. 2025). Cross-border policy learning and harmonization can be facilitated by international cooperation through forums like the Global Alliance on Circular Economy and Resource Efficiency (GACERE), UNEP's International Resource Panel, and the G20 Resource Efficiency Dialogue (UNEP 2020). Making the shift to a circular economy necessitates not only structural change but also a change in mindset and culture. Therefore, education and capacity building are

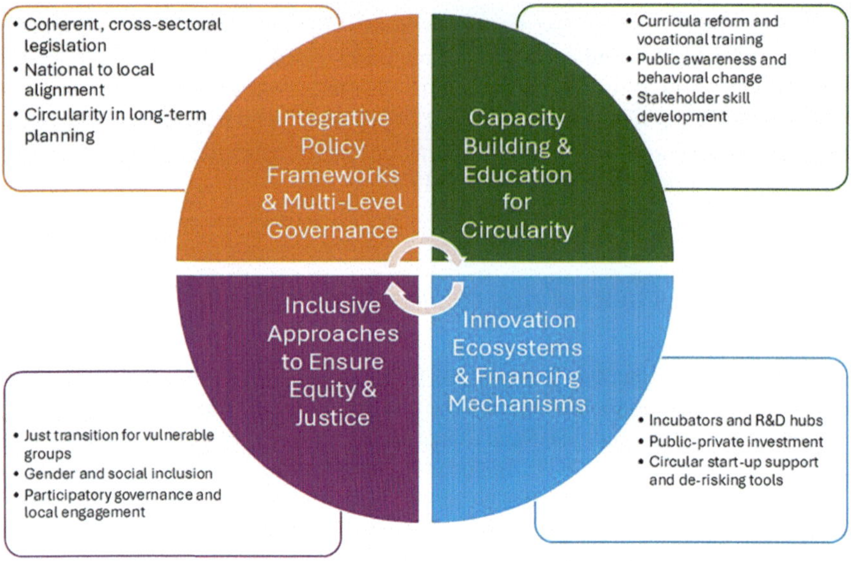

Fig. 13.3 Circular pathways to realizing a circular economy vision

essential to developing the abilities, attitudes, and institutional capacities required for change (Vidal-Ayuso et al. 2023). All levels of formal education curriculum, from elementary school to university, need to be updated to incorporate the concepts of lifecycle analysis, circular design, systems thinking, and sustainable ethics. In order to match worker competencies with new job markets, vocational and technical training programs should focus on skills linked to sustainable product development, recycling, remanufacturing, and repair (Gomes et al. 2022). Public servants, business executives, entrepreneurs, and civil society groups must all be included in capacity building. Decision-makers can benefit from workshops, online courses, policy toolkits, and knowledge-sharing platforms as they comprehend the usefulness of circular models and implement them in their industries (Khanna et al. 2022).

Furthermore, community-based projects and awareness-raising efforts are crucial for fostering a shift in consumer and citizen behavior. People are more inclined to support and take part in the shift when they are aware of the advantages of sharing, reusing, or selecting circular items (Gomes et al. 2022). The social underpinning of long-lasting and inclusive circular economies is made up of empowered people and organizations. The creation, expansion, and spread of circular economy solutions require a thriving innovation ecosystem. To jointly develop and test new technologies, business models, and value chains, these ecosystems need to include startups, government organizations, businesses, investors, and research institutes (Ranta et al. 2018). In addition to providing entrepreneurs with incubation support, circular innovation hubs and accelerators can promote cross-sectoral collaboration and information spillovers. Innovations in digital tools for circularity, industrial symbiosis, and material science are all produced in large part by universities and R&D centers (Christensen 2021). Encouraging collaborations between academics and business guarantees that inventions are both economically feasible and scientifically sound. However, innovation necessitates risk reduction and financial assistance, both of which are frequently absent, particularly in emerging nations. To promote circular enterprises, governments and financial institutions need to provide specialized financing tools including impact investing, blended finance plans, green bonds, and public–private co-investment funds (Andersen 2007).

Insurance, concessional loans, and guarantees are examples of de-risking tools that can further promote private sector involvement in high-impact circular initiatives. To refocus capital flows toward sustainability, mainstream financial institutions must also incorporate circularity into their credit evaluations, portfolio strategies, and ESG (Environmental, Social, and Governance) criteria (Palea et al. 2024). Transitions to a circular economy must be driven by the values of environmental justice and social fairness as they gain traction. Narrowly concentrating on economic and technical optimization runs the risk of excluding vulnerable groups or perpetuating already-existing disparities (Laskurain-Iturbe et al. 2021). It is imperative for inclusive circular transitions to guarantee that alterations in urban infrastructure, labor markets, or resource accessibility do not disproportionately impact underprivileged areas. Through education, legal protection, and respectable employment prospects, workers in the unorganized sector, especially in waste management should be incorporated into official circular systems (Dantas et al. 2021). Given that women frequently encounter structural obstacles while trying to obtain work, funding, and education in the green industries, gender equity should also be given top priority. In a same vein, indigenous people' demands and expertise must be honored, particularly when circular initiatives touch on land rights and natural resource management (UNEP 2020). Participatory techniques can help guarantee that transitions are responsive to real-world demands and sensitive to context by incorporating individuals, local organizations, and impacted groups in the planning and management of circular efforts (Ranta et al. 2018). Fairness and accountability in the transition to circularity can be operationalized with the aid of frameworks like social impact evaluations, inclusive stakeholder mapping, and Just Transition principles. In the end, the circular economy must foster social cohesiveness, economic democracy, and intergenerational justice in addition to regenerating ecosystems and reducing waste (UNEP 2020). For the transition to be legitimate and long-lasting, it must be socially and environmentally inclusive.

13.6 Conclusion

The path towards the circular and sustainable future is an insidious dilemma in addition to a revelation. As discussed in this chapter, the concept of circular economy has the potential to provide a feasible solution to the global issues including resource shortage, climate change and environmental deterioration and social unbalance.

The opportunities are immense—the potentials of the technology, practical policies, market changes, economical and environmental profit provide a solid ground to the circularity. Nonetheless, fulfilling this potential is not a walk in the park. There are infrastructural shortages, economic constraints, social-cultural stagnancy, complicated source chains, and disintegrated control frameworks that hinder advancement. It requires the joint effort and systematic change to overcome these challenges. It needs coherent policy systems, effective education programs, intersectoral innovation networks, and open governance designs where equity and justice are central to the transition plans. There is an urgent need to change the thinking and perception-changing linear and short-term to a regenerative and long-term visioning. Taking steps into the future, the vision of a circular future should be strong, equal and capable of restoration. In order to cope with environmental and economical shocks, resilience is required; equity guarantees that no community will be left behind and regeneration is concentrated on ecosystems restoration instead of sustaining them. Through the adoption of these values, the stakeholders in every sector will be in a position to establish a future in which the economic development is conducted in accordance with the planetary boundaries and the social welfare. The vision is not a dreamy future but a concrete reality–as long as we work together, take decisive action across boundaries, and start thinking inclusively with the idea of putting our systems, institutions, and behavior into the mode of circularity.

References

Abadian M et al (2024) Exploring backcasting as a tool to co-create a vision for a circular economy: A case study of the polyurethane foam industry. J Circ Econ [Preprint] (2). https://doi.org/10.55845/UZXQ5070

Abdirahman AA, Asif M, Mohsen O (2025) Circular economy in the renewable energy sector: A review of growth trends, gaps and future directions. Energy Nexus 17:100395. https://doi.org/10.1016/J.NEXUS.2025.100395

Agarwal S, Jain H (2024) Investigating the potential of floral waste as a vermicompost and Dual-Functional biosorbent for sustainable environmental management. Water Air Soil Pollut 235(5):322

Agarwal S, Jain H, Dhupper R, Mathur A (2025) Optimizing methyl orange dye removal from aqueous solutions using white chrysanthemum floral waste-derived bioadsorbent: a study of kinetics, thermodynamics, isotherms, and RSM optimization. Int J Environ Res 19(3):104

Andersen MS (2007) An introductory note on the environmental economics of the circular economy. Sustainability Science 2(1):133–140. https://doi.org/10.1007/S11625-006-0013-6

Aristi Capetillo A, Bauer F, Chaminade C (2023) Emerging technologies supporting the transition to a circular economy in the plastic materials value chain. Circ Econ Sustain. Springer Nature, pp 953–982. https://doi.org/10.1007/s43615-022-00209-2

Bahrin MAK et al (2016) Industry 4.0: A review on industrial automation and robotic. Jurnal Teknologi 78(6–13):137–143. https://doi.org/10.11113/JT.V78.9285

Bocken NMP et al (2016) Product design and business model strategies for a circular economy. J Ind Prod Eng 33(5):308–320. https://doi.org/10.1080/21681015.2016.1172124

Bonilla SH et al (2018) Industry 4.0 and sustainability implications: A scenario-based analysis of the impacts and challenges. Sustaina (Switzerland) 10(10). https://doi.org/10.3390/SU10103740

Bressanelli G, Perona M, Saccani N (2019) Challenges in supply chain redesign for the circular economy: a literature review and a multiple case study. Int J Prod Res 57(23):7395–7422. https://doi.org/10.1080/00207543.2018.1542176

Christensen TB (2021) Towards a circular economy in cities: Exploring local modes of governance in the transition towards a circular economy in construction and textile recycling. J Clean Prod 305:127058. https://doi.org/10.1016/J.JCLEPRO.2021.127058

Correia MS (2018) Sustainability. Int J Strat Eng 2(1): 29–38. https://doi.org/10.4018/ijose.2019010103

Corsini F, Gusmerotti NM, Frey M (2023) Fostering the circular economy with blockchain technology: insights from a bibliometric approach. Circ Econ Sustain 3(4):1819–1839. https://doi.org/10.1007/S43615-023-00250-9

Dantas TET et al (2021) How the combination of circular economy and industry 4.0 can contribute towards achieving the sustainable development goals. Sustain Prod Consum 26:213–227. https://doi.org/10.1016/j.spc.2020.10.005

Elisha OD (2020) Moving beyond take-make-dispose to take-make-use for sustainable economy. https://www.researchgate.net/publication/349964391

Davila MFR et al (2023) Sustainability digital twin: a tool for the manufacturing industry. Procedia CIRP 116:143–148. https://doi.org/10.1016/j.procir.2023.02.025

Fernandes L et al (2023) A review on adopting blockchain and IoT technologies for fostering the circular economy in the electrical and electronic equipment value chain. Sustain (Switzerland). MDPI. https://doi.org/10.3390/su15054574

Franco MA (2017) Circular economy at the micro level: A dynamic view of incumbents' struggles and challenges in the textile industry. J Clean Prod 168:833–845. https://doi.org/10.1016/J.JCLEPRO.2017.09.056

Geissdoerfer M et al (2017) The circular economy–a new sustainability paradigm? J Clean Prod 143:757–768. https://doi.org/10.1016/j.jclepro.2016.12.048

Gerhards J, Greenwood D (2021) One planet living and the legitimacy of sustainability governance: From standardised information to regenerative systems. J Clean Prod 313. https://doi.org/10.1016/j.jclepro.2021.127895

Ghisellini P, Cialani C, Ulgiati S (2016) A review on circular economy: The expected transition to a balanced interplay of environmental and economic systems. J Clean Prod 114:11–32. https://doi.org/10.1016/j.jclepro.2015.09.007

Gil-Lamata M, Latorre-Martínez MP (2022) The circular economy and sustainability: A systematic literature review. Cuad Gest 22(1):129–142. https://doi.org/10.5295/CDG.211492MG

Gomes GM, Moreira N, Ometto AR (2022) Role of consumer mindsets, behaviour, and influencing factors in circular consumption systems: A systematic review. Sustain Prod Consum 32:1–14. https://doi.org/10.1016/j.spc.2022.04.005

Hysa E et al (2020) Circular economy innovation and environmental sustainability impact on economic growth: An integrated model for sustainable development. Sustain (Switzerland) 12(12). https://doi.org/10.3390/SU12124831

Ilieva L et al (2022) Biomimicry as a sustainable design methodology—introducing the "biomimicry for sustainability" framework. Biomimetics 7(2). https://doi.org/10.3390/biomimetics7020037

Jain H (2025a) Data analytics enabled by the Internet of Things and artificial intelligence for the management of Earth's resources. In: Data analytics and artificial intelligence for earth resource management. Elsevier, pp 19–36

Jain H (2025b) Synergizing carbon reduction technologies and nature-based solutions for a climate-neutral future. Mitig Adapt Strat Glob Change 30(6):42

Kalmykova Y, Sadagopan M, Rosado L (2018) Circular economy—from review of theories and practices to development of implementation tools. Resour, Conserv Recycl 135:190–201. https://doi.org/10.1016/J.RESCONREC.2017.10.034

Khanna M, Gusmerotti NM, Frey M (2022) The relevance of the circular economy for climate change: An exploration through the theory of change approach. Sustain (Switzerland) 14(7). https://doi.org/10.3390/SU14073991

Kirchherr J et al (2018) Barriers to the circular economy: Evidence from the European Union (EU). Ecol Econ 150:264–272. https://doi.org/10.1016/j.ecolecon.2018.04.028

Korhonen J et al (2018) Circular economy as an essentially contested concept. J Clean Prod 175:544–552. https://doi.org/10.1016/j.jclepro.2017.12.111

Kristensen HS, Mosgaard MA (2020) A review of micro level indicators for a circular economy—moving away from the three dimensions of sustainability? J Clean Prod 243. https://doi.org/10.1016/j.jclepro.2019.118531

Laskurain-Iturbe I et al (2021) Exploring the influence of industry 4.0 technologies on the circular economy. J Clean Prod 321:128944. https://doi.org/10.1016/J.JCLEPRO.2021.128944

Nikolaou IE, Tsagarakis KP (2021) An introduction to circular economy and sustainability: Some existing lessons and future directions. Sustain Prod Consum 28:600–609. https://doi.org/10.1016/j.spc.2021.06.017

Palea V, Migliavacca A, Gordano S (2024) Scaling up the transition: The role of corporate governance mechanisms in promoting circular economy strategies. J Environ Manag 349:119544. https://doi.org/10.1016/J.JENVMAN.2023.119544

Pandey M, Yadav PS (2023) Understanding the role of individual concerns, attitude, and perceived value in green apparel purchase intention; the mediating effect of consumer involvement and moderating role of generation Z&Y. Clean Responsible Consum 9. https://doi.org/10.1016/j.clrc.2023.100120

Pehlken A et al (2024) Digital twins: Enhancing circular economy through digital tools. Procedia CIRP 122:563–568. https://doi.org/10.1016/J.PROCIR.2024.01.082

Polyportis A, Mugge R, Magnier L (2022) Consumer acceptance of products made from recycled materials: A scoping review. Resour, Conserv Recycl 186. https://doi.org/10.1016/j.resconrec.2022.106533

Ranta V et al (2018) Exploring institutional drivers and barriers of the circular economy: A cross-regional comparison of China, the US, and Europe. Resour, Conserv Recycl 135:70–82. https://doi.org/10.1016/J.RESCONREC.2017.08.017

Rejeb A et al (2025) When industry 5.0 meets the circular economy: A systematic literature review. Circ Econ Sustain [Preprint]. Springer Nature. https://doi.org/10.1007/s43615-025-00570-y

Ritter M et al (2024) Towards achieving the sustainable development goals: a collaborative action plan leveraging the circular economy potentials, Gruppe. Interaktion. Organisation. Z Fur Angew Organ 55(2):175–187. https://doi.org/10.1007/s11612-024-00733-9

Roci M, Rashid A (2023) Economic and environmental impact of circular business models: A case study of White Goods-as-a-Service using multi-method simulation modelling. J Clean Prod 407. https://doi.org/10.1016/j.jclepro.2023.137147

Saini J et al (2024) Study of urban warming in Chandigarh union territory, India: geospatial approach. Theor Appl Climatol 155(8):8245–8261. https://doi.org/10.1007/s00704-024-05128-w

Saini J et al (2025) Climate vulnerability and resilience in Peri-urban settings of Dehradun city in Uttarakhand, India: spatio-temporal and perception analysis. Environ Conserv J 26(2):501–515. https://doi.org/10.36953/ecj.31683032

Schroeder P, Anggraeni K, Weber U (2019) The relevance of circular economy practices to the sustainable development goals. J Ind Ecol 23(1):77–95. https://doi.org/10.1111/jiec.12732

She TW, Wee ST (2025) Advancing to a circular plastic economy: A systematic review of policy instruments. Int J Environ Sci 11(7s):214–238. https://doi.org/10.64252/dwpfsy14

Siedschlag I, Meneto S, Tong Koecklin M (2022) Enabling green innovations for the circular economy: What factors matter? Sustain (Switzerland) 14(19). https://doi.org/10.3390/su141912314

Tirado R et al (2022) Challenges and opportunities for circular economy promotion in the building sector. Sustain (Switzerland) 14(3). https://doi.org/10.3390/su14031569

Uloma SN et al (2024) Challenges and opportunities in implementing circular economy models in FMCG Industries. Int J Front Res Sci Technol 3(2):073–091. https://doi.org/10.56355/ijfrst.2024.3.2.0048

UNEP (2020) Global resources outlook 2019. United Nations

Vidal-Ayuso F, Akhmedova A, Jaca C (2023) The circular economy and consumer behaviour: Literature review and research directions. J Clean Prod. Elsevier Ltd. https://doi.org/10.1016/j.jclepro.2023.137824

Wani, Khalil S, Hameed H (2024) Future perspectives and challenges of circular economy and sustainable business performance management: A systematic literature review. In: Emerging technology, environment and social justice-a sustainable approach. QTanalytics India, pp 44–60. https://doi.org/10.48001/978-81-966500-3-2-4

Winans K, Kendall A, Deng H (2017) The history and current applications of the circular economy concept. Renew Sustain Energy Rev 68:825–833. https://doi.org/10.1016/j.rser.2016.09.123

World Economic Forum (2017) Platform for accelerating the circular economy. https://www3.weforum.org/docs/WEF_PACE_Platform_for_Accelerating_the_Circular_Economy.pdf (Accessed 17 July 2025)

Yadav K, Jain H (2024) integrating urban master planning with environmental impact assessment for sustainable and resilient cities. In: Cities of tomorrow: Urban resilience and climate change preparedness. Springer Nature Singapore, Singapore, pp 3–23

The manufacturer's authorised representative in the EU is Springer
Nature Customer Service Centre GmbH, Europaplatz 3, 69115 Heidelberg,
Germany. If you have any concerns regarding our products, please
contact ProductSafety@springernature.com

Printed and bound by CPI Group (UK) Ltd, Croydon, CR0 4YY
21/09/2026
02220013-0005